Emerging Materials for Photodegradation and Environmental Remediation of Micro- and Nano-Plastics

Series Editor
Gilles Pijaudier-Cabot

Emerging Materials for Photodegradation and Environmental Remediation of Micro- and Nano-Plastics

Recent Developments and Future Prospects

Edited by

Laxman Singh
Sunil Kumar

iSTE

WILEY

First published 2025 in Great Britain and the United States by ISTE Ltd and John Wiley & Sons, Inc.

ISTE Ltd
27-37 St George's Road
London SW19 4EU
UK

www.iste.co.uk

John Wiley & Sons, Inc.
111 River Street
Hoboken, NJ 07030
USA

www.wiley.com

Library of Congress Control Number: 2024950207

British Library Cataloguing-in-Publication Data
A CIP record for this book is available from the British Library
ISBN 978-1-83669-009-2

Contents

Foreword

Functional materials are the building blocks for empowering today's modern applied sciences, engineering and technologies. Functional materials are classified in different categories: inorganic, carbonaceous, organic and composite materials/ nanomaterials. Their unique contribution is from the size-dependent variation in optical, electronic, electrical and chemical properties. These materials play a vital role, are superior to bulk materials and cater to the needs of environmental remediation, solar energy industry, hydrogen evolution, photocatalytic conversion, green energy and many others, covering almost the entire range of significant global crises.

This book, entitled *Emerging Materials for Photodegradation and Environmental Remediation of Micro- and Nano-Plastics: Recent Developments and Future Prospects* (published by ISTE Ltd, London and John Wiley & Sons, New York), edited by Laxman Singh and Sunil Kumar, strives to address the current aspects of functional materials and their applications and to provide comprehensive literature on ongoing research, write-up and review to express the current scenario in the synthesis approaches, which constitute up-to-date reports and findings of their fabrication and applications of various reactive materials, that is, metal oxides, MXenes, metal nanoparticles, green materials, carbon nanotube and composite materials produced using different synthetic approaches of either physical or chemical routes. The academic and research credentials of the experts known in the diverse field of nanostructured materials for energy, biomedical and environmental applications have been employed in preparing, compiling and editing the chapters of this book. It has the potential to serve the community interested in materials engineering and the advancement of nanomaterials toward applications in modern technologies.

I wish the publisher, editors and contributors great success and the book readers, policymakers and other stakeholders will find this book valuable for further research in the materials synthesis, processing, designing and its systematic study in energy and environment for the mitigation of micro- and nano-plastic pollution.

Youngil LEE
University of Ulsan
South Korea

Preface

The purpose of this book is to provide comprehensive knowledge of ongoing research, write-up and review with respect to the current scenario in a wide range of materials design, diversity and environmental application to establish a better quality ecosystem. The plastic industry is one of the fastest developing industries worldwide. The large amount of plastic material waste in nature undergoes degradation into smaller pieces, which ultimately becomes micro-plastics (MPs) and nano-plastics (NPs). MPs are distinctively light, unbreakable and have the capacity to float and are able to travel far throughout the globe. On the other hand, there is very little research into the amounts, kinds and toxic effects of NPs and their impacts on human health.

Current studies on the size distribution properties of the plastic remains have suggested that the probability of continuous breakdown of MPs into nanosized particles is very high. Most of these plastic particles reach the environment via effluent discharge from wastewater treatment plants, owing to the lack of processes for its removal. Once released into the environment, plastic materials are susceptible to structural and chemical degradation from their exposure to biotic and abiotic factors such as friction and UV light. It has been found that MPs and NPs are becoming a global threat due to their intrinsic physicochemical characteristics and the potential effect on ecosystems (including humans). Literature says that MPs and NPs are mostly present in soil and, due to natural erosion, these particles get into water resources, which are mainly rivers and oceans. Photodegradation does not require additional chemicals and does not generate significant by-products or waste.

Photocatalysis occurs when a semiconductor material is irradiated with solar energy to promote the electron's (e^-) migration from the valence band to the conduction band, forming a hole (h^+) in the latter. These photogenerated species react with the H_2O and O_2 adsorbed on the semiconductor's surface, giving rise to the formation of reactive hydroxyl and superoxide radicals. Here, the photodegradation/degradation of the NPs and MPs will help to clean the environment and are also useful in decreasing the CO_2 emission and environmental pollution. For this purpose, functional materials/nanomaterials can remarkably change their catalytic, adsorption, electrical, thermal, optical and magnetic properties under the influence of parameters, that is, biochemical, pH, humidity content, atmospheric pressure, temperature, stress, electromagnetic waves, electrical or magnetic field, light and moisture. Functional materials also have a wide range of applications in medicine, industry, energy and engineering sectors, and the rise in demand for these materials facilitates the belief that there is a great scope for advance materials in the future. The effectiveness of functional materials and their magnificent properties such as high surface area-to-volume ratio, more reactive sites, which often results in higher reactivity, makes them particularly suitable for photodegradations, adsorptions and catalytic processes. Environmental remediation of pollutants depends mainly on utilizing the different technologies such as chemical reactions, absorption, adsorption, filtration and photocatalysis for the ousting of contaminants from different environmental media such as soil, land, surface, surface strata, surface waters, ponds, streams, groundwater, bedrock, drinking water supply, stream sediments, atmosphere, air, and vegetation and natural resource. Engineered materials are applied to remediate different hazardous contaminants such as heavy metals, pesticides, herbicides, fertilizers, oil spills, toxic gases, industrial effluents, sewage, and organic compounds, organic dyes, antibiotics, chlorinated organic compounds, organophosphorus compounds, volatile organic compounds, and halogenated herbicides.

This book provides an overview on a diverse class of functional materials and/or nanomaterials: metal oxide, metal nanoparticle, MXenes, carbonaceous (carbon nanotube and graphene), composites, hybrid, metal-organic frameworks (MOFs) and green materials, which are better photocatalysts for mitigation of MP and NP pollution. This book also discusses the progress and future perspective of these materials in photocatalytic degradation of MP and NP, exploring the innovative approaches in depth.

This book includes fifteen chapters mainly focusing on functional materials that describe the evolution of design, fabrication and application in photocatalytic degradation of MP and NP to remediate an environmental ecosystem. In this way, this book focuses on materials design, environmental impact, issues and solution at the basic level and is therefore a suitable problem, challenges and opportunity solving resource for a next-generation research environment.

Laxman SINGH
Department of Chemistry
Siddhartha University
Sidharthnagar
India

Sunil KUMAR
Department of Chemistry
L.N.T. College
B.R.A. Bihar University
Muzaffarpur
India

January 2025

Acknowledgments

I would like to take this opportunity to thank those who made a significant contribution in shaping this book. I would like to thank the entire team at Cambridge Scholars Publishing Press for collaborating with this book and providing all the necessary assistance. My special thanks go to the publisher ISTE Ltd., who were there to promptly respond to and address all my queries and provide the necessary assistance from the commencement of this book until its completion. I acknowledge and thank all of the authors who contributed their works in the book. A very special thanks to Professor K.D. Mandal and Professor U.S. Rai for encouraging me to work on the contemporary issues related to the environment and energy. I would also like to express my thanks to Professor Prakriti Rai, Dean of Science and my colleagues at the Department of Chemistry, Siddharth University, Kapilvastu, for supporting me and always being there to encourage me in my achievements and accomplishments. I would like to acknowledge the support and love of my parents (Late Shri Indrapal Singh and Smt. Shanti Devi), my elder brother (Late Shri Veerendra Singh), who always supported and encouraged me in my academic career, my loving and beautiful wife Radha and other members of my family. I acknowledge the love of my daughter, Mansi Singh, and my son, Pulkit Singh, who bring joy and happiness into my life, which rejuvenates me in work.

Laxman SINGH

I am highly grateful to all authors for their scientific contributions to make this book energetic and inspiring. The authors were very kind in addressing the comments that enhanced the quality of the chapters. I would like to acknowledge my parents (Sri Jayhind Prasad and Smt. Sunaina Devi) for their deep love and emotional support. A special acknowledgment goes to family members, friends, editors, proofreaders, academic colleagues, research assistants, and content contributors for shaping and managing the structure of this book. Further, I would like to thank my wife (Rangoli Jaiswal) and my son (Rishwik Jaiswal) for their love, affection and cooperation in helping me complete this scientific book. Great thanks goes to my PhD supervisor Dr. Pralay Maiti, Professor, School of Materials Science and Technology, IIT (BHU), Varanasi (U.P.) India. I would also like to acknowledge the Head of the University Department of Chemistry, B.R.A. Bihar University, Muzaffarpur, for supporting in research and development at the academic grounds. Special thanks goes to Dr. Avay Kumar Singh, Principal, L.N.T. College, Muzaffarpur, for his catalytic support and to Professor Dinesh Chandra Rai, Vice-Chancellor and Dr. Aprajita Krishna, Registrar, B.R.A. Bihar University, Muzaffarpur, for emotional support in helping me complete this technical task under the academic ecosystem.

Sunil KUMAR

1

Micro- and Nano-Plastic Pollution: Present Status on Environmental Issues and Photocatalytic Degradation

Plastic is an artificially produced synthetic polymer synthesized through a process referred to as cracking. Upon infiltrating the environment, plastic waste is gradually split down into smaller particles known as micro-plastics (MPs, with size ranging from 0.1 to 5 mm) and nano-plastics (NPs, which range in size from 1 to 100 nm), together we call these micro-nano-plastics (MNPs). However, MNP particles are highly complex and detailed in their shape, size, density, polymer structure, surface properties, etc. While particle concentrations across various media can differ by as many as ten orders of magnitude, examining such intricate samples can be comparable to searching for a needle in a haystack. MNPs have recently been identified as a significant global environmental pollutant. Studies indicate that as particles travel through the environment, their functional groups bind to organic pollutants such as heavy metals and persistent poisonous chemicals. Advancing eco-friendly plastic conversion technologies is vital for transitioning to a sustainable, less plastic-dependent future.

As a result, in recent times, efforts have been undertaken to incorporate nanomaterials and nanostructures into photocatalytic plastic degradation on the basis of developments in smart material technology using simple photocatalysts. Compared to conventional methods, photodegradation of MNPs can offer a more sustainable alternative for waste plastic reprocessing, as it utilizes solar energy as an energy source and operates at room temperature and pressure. Current efforts focus on the strategic design and surface modification to accomplish smart materials able to capture, transport and disperse MPs with varying shapes and chemical compositions. Catalytic materials used in photocatalysis show significant potential for degrading common plastics. As a result, recent advancements in these small, self-moving equipment are anticipated to drive a major breakthrough in environmental rehabilitation.

Chapter written by Monika VERMA, Yashaswini and Sujata KUNDAN.

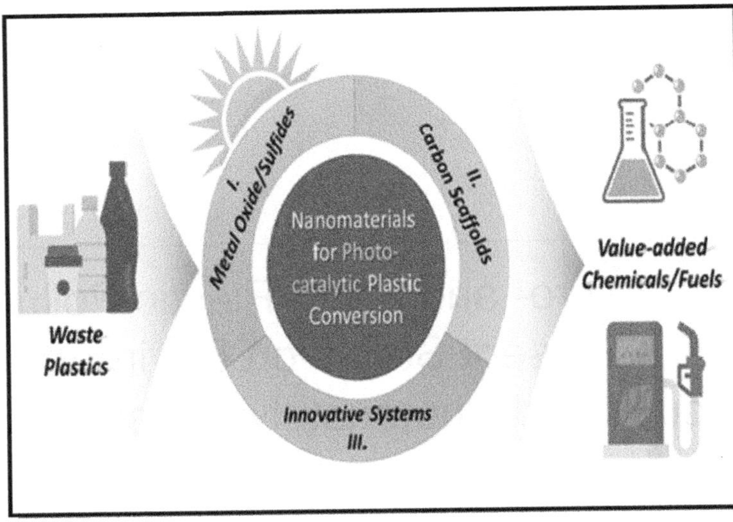

Figure 1.1. *Nanomaterials as advanced photocatalysts for plastic conversion. For a color version of this figure, see www.iste.co.uk/singh/photodegradation.zip*

1.1. Introduction

Plastic products have undeniably transformed our daily lives, providing unparalleled convenience owing to their remarkable characteristics such as lightweight quality of plastic makes it an ideal choice for packaging, contributing to reduced transportation costs and energy consumption (Du et al. 2021). Its impressive chemical stability ensures the preservation of goods, preventing decay and prolonging the longevity of numerous products. Moreover, the high durability of plastic means that products made from this material enjoy a prolonged lifespan, reducing the frequency of replacements in comparison to alternative materials. Additionally, the cost-effectiveness of plastic production translates to affordable products for consumers. This has led to a substantial surge in plastic production over the years (Llorente-García et al. 2020). In 2018 alone, an astonishing 360 million tons of plastic products were manufactured globally, highlighting the pervasive use of this versatile material (Long et al. 2019). Projections for 2025 estimate an even more substantial production figure of 500 million tons, underscoring the integral role that plastic plays in various industries (Miao et al. 2020). However, the widespread use of plastic has led to environmental concerns due to the durability that makes plastic products so desirable, which results in their persistence in the environment, contributing significantly to pollution. Insufficient recycling infrastructure and inappropriate waste-disposal activities have been directed to the accretion of plastic waste, threatening to ecosystems and wildlife (Duet al. 2021). While recycling efforts are increasingly prevalent, a noteworthy fraction of plastics, approximately one-third, remains either of reduced size or exhibits complex structures that present challenges for an economically feasible recovery (Garcia and Robertson 2017). Initially, there were concerns about plastic contamination, with the alarming statistic that 79% of all plastics were either ending up in landfills or being illegally dumped (Geyer et al. 2017). However, in the present day, we recognize that the problem of pollution does not end there but the issue extends beyond landfills. Plastics have the capacity to migrate within the environment, traversing rivers to reach freshwater bodies like lakes and lagoons, and ultimately making their way to seas and oceans. Using these marine currents, they can travel vast distances, reaching remote parts of the world (Golden et al. 2016). Owing to their natural water-repellent properties and strong resistance to both physical and chemical breakdown, plastics have the ability to migrate from land-based environments to aquatic ecosystems. The ubiquity of plastics has been identified in diverse environmental systems, including groundwater, soil, and even the air. This broad distribution underscores the extensive reach and impact of plastic pollution across various ecological domains (Alimi et al. 2018; Ng et al. 2018; Astner et al. 2019). New studies indicate that there could be around 5.3 trillion plastic particles presently suspended in the sea, equal to 268,940 tons (Eriksen et al. 2014). These particles consist of various plastics prevalent in our communities, such as polyester, polyethylene (PE), polypropylene (PP), polyvinyl chloride (PVC) and nylon (Zhang et al. 2016; Sruthy and Ramasamy 2017). Of these, the specific concern is the notable occurrence of micro-sized PE debris (less than 5 mm), which has attracted attention due to the physical harm it imposes on marine life and the growing risks it poses to public health (Weithmann et al. 2018) (Figures 1.2 and 1.3).

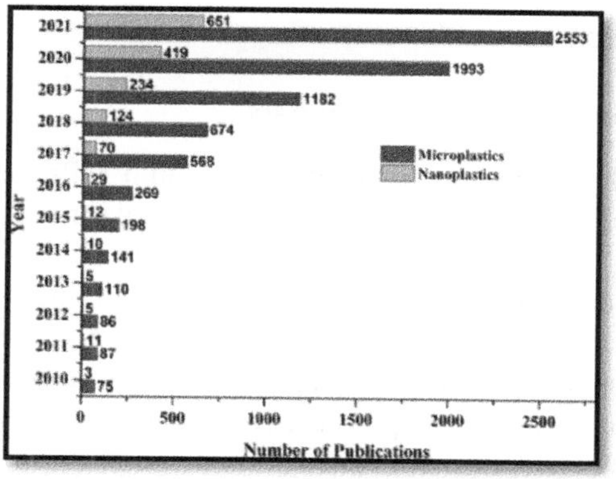

Figure 1.2. *Total annual research report on micro-plastics and nano-plastics (source: https://media.springernature.com/full/springerstatic/image/art%3A10.1007%2Fs11157-021-09609-6/MediaObjects/11157_2021_9609_Fig1_HTML.png?as=webp). For a color version of this figure, see www.iste.co.uk/singh/photodegradation.zip*

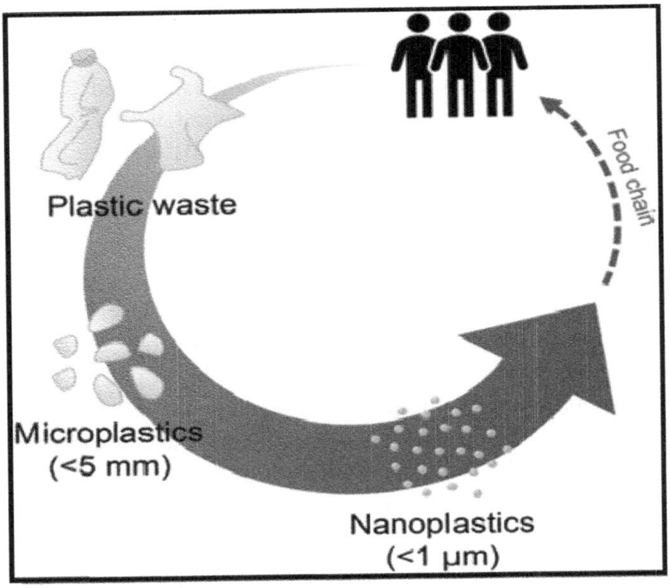

Figure 1.3. *Fragmentation of plastic waste into micro-plastics and nano-plastics (source: https://onlinelibrary.wiley.com/cms/asset/ae81c136-86ad-453e-b3f6ac94812dd059/adfm202112120-fig-0001-m.jpg). For a color version of this figure, see www.iste.co.uk/singh/photodegradation.zip*

1.2. MPs and NPs: Sources, impact and health hazards

1.2.1. *Micro-plastics*

1.2.1.1. *Sources and environmental impact*

Plastics released into the environment degrade into MPs (less than 5 μm) and NPs (less than 1 μm) through the combined effects of solar radiation, physical forces and biodegradation (Koelmans et al. 2015; van Weert et al. 2019) (Figure 1.4). Primary MPs are purposely developed in small sizes for specific uses, embracing microbeads in personal care products, microfibers in textiles and plastic pellets in manufacturing. Despite their small size, these particles can accumulate in both aquatic and terrestrial ecosystems, posing environmental risks (Cheng et al. 2021). In response to their environmental impact, several countries, involving Canada and the United States, have implemented bans on MPs in cosmetic products (Ballent et al. 2016). Secondary MPs result from the deprivation of larger plastic debris under conditions, for instance heat, sunlight and aeration (Auta et al. 2017). These elements have an extensive surface area relative to their volume, allowing them to absorb and transport harmful persistent bio-accumulative toxins (PBTs) that can adversely affect environmental and biological systems (Chen et al. 2019; Enfrin et al. 2020; Tang et al. 2020). Conversely, plastic degradation in cold and anoxic aquatic environments is extremely slow, taking centuries to progress (Zhang 2017). MPs manifest in various forms, including pellets, fibers, and fragments, as observed in environmental samples (Klein et al. 2015). This diversity in shape and size, coupled with their persistence and potential for harm, underscores significant challenges in managing and mitigating MP pollution.

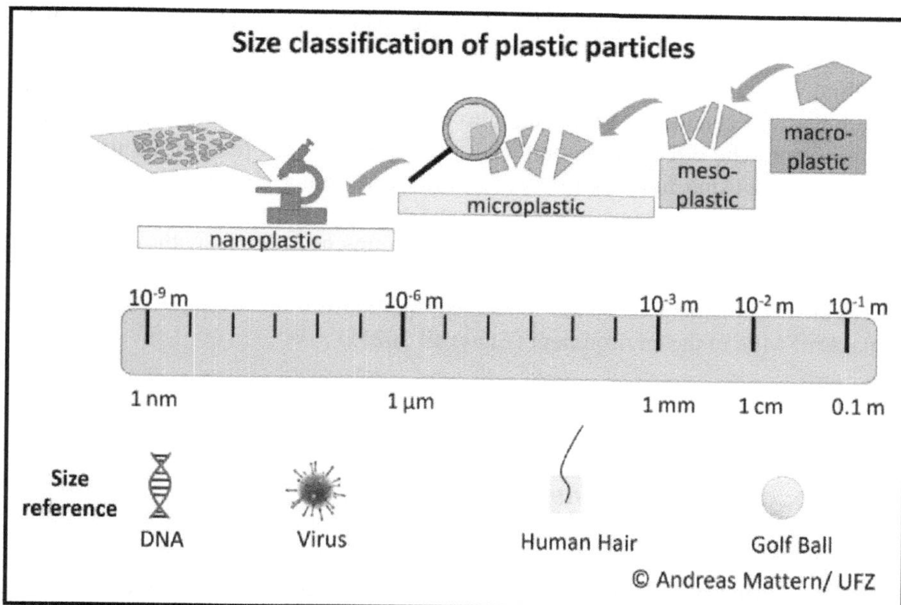

Figure 1.4. *Classification of plastic particles based on particle dimensions (source: https://nanopartikel.info/wp-content/uploads/2020/11/Seize_nanoplastic_final.png.jpg). For a color version of this figure, see www.iste.co.uk/singh/photodegradation.zip*

1.2.1.2. *MPs in aquatic systems and human exposure*

MPs have been sensed in numerous water bodies, with one prevalent source being the sewage discharged from wastewater treatment plants (WWTPs). Studies indicate that the treatment processes employed are not entirely effective in retaining MPs, allowing wastewater run-offs contaminated with MPs to enter municipal waters and rivers (Michielssen et al. 2016), eventually finding their way into the ocean. Once within aquatic ecosystems, MPs are ingested by a various range of marine organisms spanning different trophic levels facilitating their entrance into the food chain (Enfrin et al. 2020). These minuscule particles can pass in the human body through breath and absorption, resulting in cellular damage, inflammation and immune responses. Recent studies found MPs in human blood, with 17 out of 22 samples contaminated. Half had PET from drink bottles, one third had PS from food packaging and one quarter had PE from plastic bags (Leslie et al. 2022). MPs are also found in human faeces (Yan et al. 2022). This can lead to serious health issues such as cancer, chronic problems and harm to development. The consequences of plastic pollution extend beyond environmental damage, posing significant risks to human health (Koelmans et al. 2019; Yuan et al. 2022).

Both primary and secondary MPs pose ecological threats as they can be consumed by marine creatures, being incorporated into the food chain and potentially causing harm to aquatic life. Additionally, these particles may contain or absorb harmful chemicals, contributing to the transfer of contaminating agents in the environment. The effect on human health arises when MPs pass in the food chain, potentially exposing humans to MPs through consuming contaminated seafood or other food sources. In conclusion, the coexistence of primary and secondary MPs highlights the pervasive nature of this environmental issue, with wide-ranging implications for ecosystems and human health. To tackle the issue effectively, it is essential to implement thorough strategies that address the sources, pathways and impacts of MPs in the environment (Liu et al. 2021).

1.3. Nano-plastics

1.3.1. *Sources and environmental risks*

Similar to MPs, NPs are also classified by their origin into primary or secondary categories. Primary NPs are purposefully manufactured at the nanoscale, whereas minor NPs form from the breakdown and degradation of larger plastic pieces because of their biological or environmental factors (González-Pleiter et al. 2019). The occurrence of principal NPs in the environment has seen a significant rise, primarily attributed to the development and production of engineered nanomaterials intended for application in cosmetics, personal care items, research, various technologies and medical applications (Mattsson et al. 2018).

The small size of nano-sized plastics enhanced their capacity to penetrate living organisms, which anticipates posing a greater threat than MPs (Strungaru et al. 2019). Earlier studies have indicated that NPs serve as carriers for substantial metals and persistent toxic pollutants (POPs) found in the nearby environment (Bradney et al. 2019). Additionally, outcomes from various toxicological investigations have shown that the consumption of NPs and the associated contaminants adversely impact the immune system (Bergami et al. 2019), disrupt metabolism (Brandts et al. 2018), induce neurotoxicity (Sökmen et al. 2020) and interfere with the growth and reproduction processes in aquatic organisms. Nevertheless, there are emerging hypotheses regarding the capability of NPs to act as transporters of pathogens and viruses (Wang et al. 2021). Hence, the dimensions of NPs remain a crucial factor in ecological toxicity, and its impact on this event should be further investigated. While MPs have been identified and measured in diverse environmental samples, information regarding NPs (< 1,000 nm) is notably lacking. The investigation of NPs in actual field samples is in its early stages, with protocols still in the early phases of development and maturation (Mintenig et al. 2018). Urgent efforts are needed to deepen our comprehension of the prevalence of NPs in diverse environments and their potential toxicological impacts on various organisms (Fadare et al. 2020; Li et al. 2020; Materić et al. 2020; Sun et al. 2020).

1.4. Impact of Covid-19 on plastic pollution

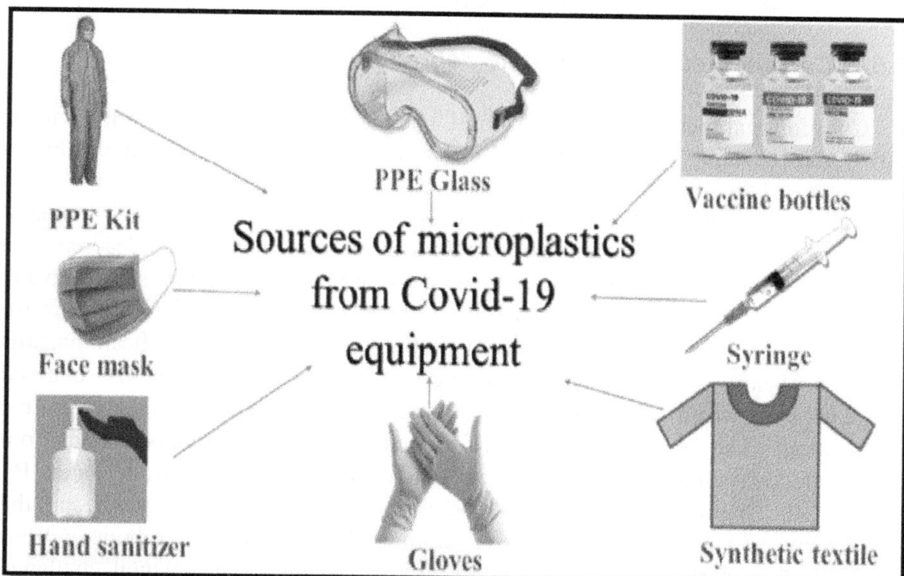

Figure 1.5. *Covid-19 materials as a source of micro-plastics in water environments. For a color version of this figure, see www.iste.co.uk/singh/photodegradation.zip*

The Covid-19 epidemic has significantly exacerbated plastic pollution (Figure 1.5) due to the widespread usage of PP-based personal protective equipment (PPE). This PPE, when exposed to environmental conditions, degrades into MPs and NPs. Current waste management practices, such as incineration and landfilling, are not sustainable and fail to effectively address the growing plastic waste issue (Fadare and Okoffo 2020). In contrast, pyrolyzing PP-based plastic waste presents a promising alternative. The pyrolysis process can convert over 75% of plastic waste into renewable crude oil, while slow pyrolysis can produce up to 33.50% solid char. This method also generates various volatile by-products, offering potential for diverse applications. To further address the problem, enhancing biodegradable PPE with graphene oxide, carbon nanotubes or Ag-ZnO nanocomposites can improve their self-sanitizing capabilities. Additionally, incorporating recycled PPE into carbon aggregates provides an effective and good approach for rapidly reducing plastic waste (Dey et al. 2023). This combination of innovative waste management techniques and material enhancements represents a comprehensive strategy for tackling the environmental impact of plastic pollution. The sources of MPs are also employed in Covid-19 equipment are shown in Figure 1.5 (Iheanacho et al. 2023).

1.5. Methods for plastic degradation

1.5.1. *Current methods for plastic degradation*

Conventional methods such as disposal, chemical treatment (Moharir and Kumar 2019) and recycling, typically applied to larger plastic items, are not applicable to MPs due to their diminutive size. Instead, current research is exploring biological degradation and photocatalysis as potential solutions to address this environmental challenge. Microbes can biodegrade MPs through enzymatic processes, decomposing the molecules into smaller parts that can undergo further mineralization steps (Silva et al. 2018). Toxic organic pollutants absorbed by plastic particles, their support for bacterial biofilm growth and their spread via the food chain pose significant hazards to human health. Consequently, nano-/micro-sized plastics pollution has evolved into a pressing worldwide challenge, necessitating the imperative for their definitive elimination. The emergence of self-propelled nano/microrobots stands out as an effective solution for the efficient eradication of NPs/MPs from water. By utilizing the improved physical and chemical characteristics of materials at the nano/microscale and leveraging active motion, these innovative robotic entities demonstrate remarkable effectiveness in addressing this environmental challenge (Urso and Pumera 2022).

1.5.2. *Emerging solutions for plastic degradation*

1.5.2.1. *Photocatalytic degradation of micro- and nano-plastics*

In this context, we suggest employing photocatalysis as a promising solution for degrading PE MPs, aiming to reduce their influx into the seas from land-based sources. This approach uses an eco-friendly method where PE is degraded with the help of an N-TiO_2 photocatalyst under visible light exposure. Photocatalytic degradation takes place when a semiconductor, illuminated by light with photon energy matching or exceeding its band gap, generates holes (h+) and excited electrons (e–). These holes react with water (H_2O) or hydroxyl groups (OH–) to create hydroxyl radicals (OH•), which are extremely effective at breaking down a range of organic pollutants (Cedillo-González et al. 2018) (Figure 1.6). To enhance sustainability, we can employ N-TiO_2 as a photocatalyst, which is synthesized using mussel proteins. This process involves using the proteins as a pore-forming agent and a nitrogen basis for doping the photocatalyst, as described by Zeng et al. (2015). This method not only advances the efficacy of the photocatalyst but also aligns with eco-friendly practices. We plan to apply this method to the degradation of PE MPs derived from facial scrubs, which are significant suppliers to marine MP pollution (Napper et al. 2015). By evaluating the effectiveness of this method on real-world samples, we aim to address a critical environmental issue and advance sustainable waste management solutions.

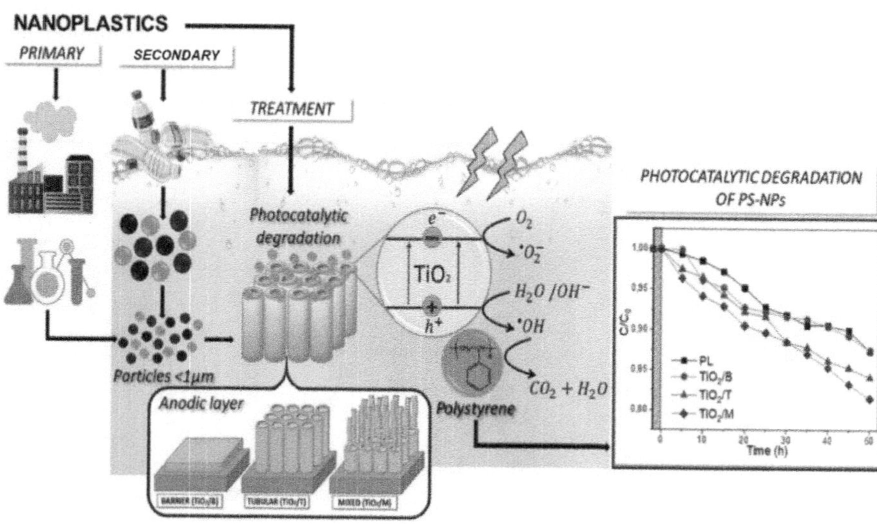

Figure 1.6. *Photocatalytic breakdown of primary nano-plastics using different anodized TiO_2 structures. For a color version of this figure, see www.iste.co.uk/singh/photodegradation.zip*

Recent research on the photocatalytic breakdown of plastics has largely concentrated on producing lower molecular weight intermediates, which have potential applications in organic synthesis. While limited, the studies indicate advantages in employing photocatalytic methods for plastic degradation. Through the use of solar radiation and photocatalysts, these methods transform waste into valuable intermediate products, offering potential for the synthesis of new materials. The development of photocatalytic processes utilizing renewable energy sources, particularly solar energy, is increasingly attractive economically, energetically and environmentally (Bratovcic 2019). A noteworthy breakthrough has occurred in the field of materials chemistry, featuring the creation of a singular TiO_2 superstructure with rod-like characteristics and inherent self-propulsion capabilities (Figure 1.7). In a singular synthetic process, an array of rod-like microrobots, distinguished by their diverse shapes and featuring multiple trapping sites, has been intricately designed to exploit the inherent asymmetry within the micro-robotic system, thereby optimizing their propulsion dynamics. These autonomous microrobots undergo in situ surface morphology transformation, forming intricate flower-like structures with the integration of a nano-seed array. Under light irradiation and photocatalysis, their surfaces transform into porous flower-like networks, enhancing functionality and enabling efficient entrapment of MPs. This reconfigurable micro-robotic technology marks a significant advance in addressing MP pollution, offering autonomous propulsion and adaptive surface transformations for environmentally conscious degradation (Ullattil and Pumera 2023). Figure 1.7 illustrates the excitation and subsequent transfer of TiO_2 to the conduction band (CB). This process underlies the fundamental mechanism for plastics degradation (Nabi et al. 2021).

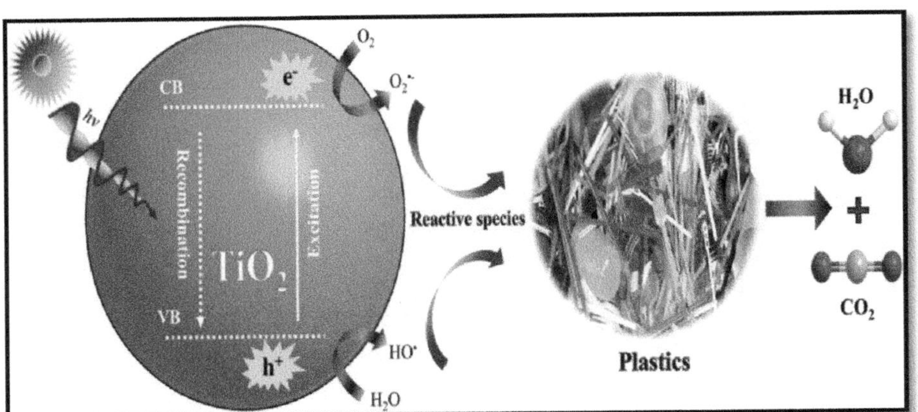

Figure 1.7. *Schematic illustration of general photocatalytic mechanism of TiO_2 under light exposure. For a color version of this figure, see www.iste.co.uk/singh/photodegradation.zip*

1.5.2.2. *Nanorobots in plastic degradation*

A new method for removing NPs from water employs self-propelled, light-activated FeHCF nanobots constructed from a metal-organic framework. When these nanobots are exposed to visible light, they transition to a metastable electronic state, creating localized electron density imbalances that drive their self-propulsion.

This process facilitates the aggregation of FeHCF@NP complexes and enhances the electrostatic attraction of negatively charged nanoparticles to the positively charged surfaces of nanobots. The light-induced charge transfer generates a bipolar effect on the nanobot surface, which significantly improves the binding efficiency of nanoparticles. With an impressive removal capacity of 3,060 $mg{\cdot}g^{-1}$ and a high-rate constant of 0.69 min^{-1}, this advanced approach demonstrates a substantial improvement in performance compared to existing materials, offering a highly effective solution for enhanced separation and purification applications. Beyond its effectiveness in removing plastic pollution, this clean and efficient light-driven method has promising applications in various fields, including nanopatterning, drug and gene delivery, cell manipulation and nano-surgery. Its potential for broad applications highlights its importance in advancing technologies for environmental and biomedical purposes (Jung et al. 2023). A TiO_2/graphite (TiO_2/C) cathode-based electro-Fenton-like (EF-like) technology was suggested for the efficient breakdown of PVC, a prevalent MP in water. This technology demonstrated significant efficacy in simultaneously achieving cathodic reduction dechlorination and hydroxyl radical (OH) oxidation during PVC degradation. The effect of reaction temperature and initial PVC concentration was examined, with the optimal conditions achieving an impressive 75% dechlorination efficiency following potentiostatic electrolysis at −0.7 V versus Ag/AgCl for 6 h. The research investigated the by-products generated during PVC MP breakdown, highlighting alterations in surface structures and molecular weight. The results suggest a potential degradation pathway for PVC, demonstrating the effectiveness of this TiO_2/C cathode-based heterogeneous EF-like method as an environmentally friendly solution for treating MP-contaminated wastewater (Miao et al. 2020).

The use of photocatalysis in water to address MP pollution by focusing on the degradation of primary high-density polyethylene (HDPE) MPs sourced from a commercial facial scrub. The research demonstrates that low pH and low temperature conditions significantly enhance the degradation process. Low pH introduces hydrogen ions (H+), which facilitate the breakdown of the plastic and improve the interrelation between colloidal nanoparticles and HDPE MPs. Simultaneously, reduced temperatures accelerate the breakdown of MPs, expanding their surface area and improving their interaction with C,N-TiO_2 photocatalysts. The study finds that varying operational parameters, such as pH and temperature, during photocatalysis can substantially accelerate the degradation of HDPE MPs. These conditions create a synergistic effect that improves the efficiency of the degradation process, creating it a promising method for mitigating MP pollution. This approach offers a potential solution to environmental issues caused by MPs and highlights the importance of optimizing photocatalytic conditions to enhance the effectiveness of plastic-waste treatment (Ariza-Tarazona et al. 2020).

1.6. Conclusion

Plastic, despite its widespread utility, has become one of the most significant environmental challenges of our time due to its persistence in various ecosystems and its ability to degrade into harmful MPs and NPs. While plastic's lightweight and durable properties offer economic and practical benefits, the lack of efficient disposal and recycling mechanisms has resulted in severe ecological impacts, with plastic waste polluting oceans, rivers, soils and even the air. MPs, originating from both primary sources like cosmetics and the breakdown of larger plastic items, are now found across ecosystems, including human food and water supplies. The potential health risks posed by these particles, from immune system disruption to cancer, demand urgent attention to curb plastic pollution, enhance waste management and explore alternatives to plastic use.

1.7. Future directions for plastic pollution control

Studies have already shown the presence of trillions of MPs in oceans, freshwater systems and even human blood, suggesting that plastic particles will continue to penetrate deep into the biosphere and food chain. In the coming decades, the issue of MP and NP will likely escalate as the production of plastic continues to grow. However, advancements in nanotechnology offer a promising solution, particularly through the development of photocatalytic nanomaterials such as titanium dioxide (TiO_2) that can degrade plastics using sunlight. Innovations like self-propelled nanorobots and sustainable nanomaterials could be deployed in water systems to break down plastics in real time, while their integration into filtration systems and waste management processes will help prevent further environmental contamination. Although challenges remain in certain ecosystems, future strategies involving nanomaterial-enhanced biodegradation, coupled with advancements in recycling and waste management technologies, will be pivotal in addressing the escalating global crisis of MP and NP pollution.

1.8. References

Alimi, O.S., Farner Budarz, J., Hernandez, L.M., Tufenkji, N. (2018). Microplastics and nanoplastics in aquatic environments: Aggregation, deposition, and enhanced contaminant transport. *Environmental Science and Technology*, 52, 1704–1724.

Ariza-Tarazona, M.C., Villarreal-Chiu, J.F., Hernández-López, J.M., De la Rosa, J.R., Barbieri, V., Siligardi, C., Cedillo-González, E.I. (2020). Microplastic pollution reduction by a carbon and nitrogen-doped TiO_2: Effect of pH and temperature in the photocatalytic degradation process. *Journal of Hazardous Materials*, 395, 122632.

Astner, A.F., Hayes, D.G., O'Neill, H., Evans, B.R., Pingali, S.V., Urban, V.S., Young, T.M. (2019). Mechanical formation of micro- and nano-plastic materials for environmental studies in agricultural ecosystems. *Science of the Total Environment*, 685, 1097–1106.

Auta, H.S., Emenike, C.U., Fauziah, S.H. (2017). Distribution and importance of microplastics in the marine environment: A review of the sources, fate, effects, and potential solutions. *Environment International*, 102, 165–176.

Ballent, A., Corcoran, P.L., Madden, O., Helm, P.A., Longstaffe, F.J. (2016). Sources and sinks of microplastics in Canadian Lake Ontario nearshore, tributary and beach sediments. *Marine Pollution Bulletin*, 110, 383–395.

Bergami, E., Krupinski Emerenciano, A., González-Aravena, M., Cárdenas, C.A., Hernández, P., Silva, J.R.M.C., Corsi, I. (2019). Polystyrene nanoparticles affect the innate immune system of the Antarctic sea urchin *Sterechinus neumayeri*. *Polar Biology*, 42, 743–757.

Bradney, L., Wijesekara, H., Palansooriya, K.N., Obadamudalige, N., Bolan, N.S., Ok, Y.S., Rinklebe, J., Kim, K.H., Kirkham, M.B. (2019). Particulate plastics as a vector for toxic trace-element uptake by aquatic and terrestrial organisms and human health risk. *Environment International*, 131, 104937.

Brandts, I., Teles, M., Tvarijonaviciute, A., Pereira, M.L., Martins, M.A., Tort, L., Oliveira, M. (2018). Effects of polymethylmethacrylate nanoplastics on *Dicentrarchus labrax*. *Genomics*, 110(6), 435–441.

Bratovcic, A. (2019). Degradation of micro- and nano-plastics by photocatalytic methods. *J. Nanosci. Nanotechnol. Appl.*, 3, 206.

Cedillo-González, E.I., Riccò, R., Costacurta, S., Siligardi, C., Falcaro, P. (2018). Below room temperature: How the photocatalytic activity of dense and mesoporous TiO_2 coatings is affected. *Applied Surface Science*, 435, 769–775.

Chen, Q., Zhang, H., Allgeier, A., Zhou, Q., Ouellet, J.D., Crawford, S.E., Luo, Y., Yang, Y., Shi, H. Hollert, H. (2019). Marine microplastics bound dioxin-like chemicals: Model explanation and risk assessment. *Journal of Hazardous Materials*, 364, 82–90.

Cheng, Y.L., Kim, J.G., Kim, H.B., Choi, J.H., Tsang, Y.F., Baek, K. (2021). Occurrence and removal of microplastics in wastewater treatment plants and drinking water purification facilities: A review. *Chemical Engineering Journal*, 410, 128381.

Dey, T.K., Rasel, M., Roy, T., Uddin, M.E., Pramanik, B.K., Jamal, M. (2023). Post-pandemic micro/nanoplastic pollution: Toward a sustainable management. *Science of the Total Environment*, 867, 161390.

Du, H., Xie, Y., Wang, J. (2021). Microplastic degradation methods and corresponding degradation mechanism: Research status and future perspectives. *Journal of Hazardous Materials*, 418, 126377.

Enfrin, M., Lee, J., Gibert, Y., Basheer, F., Kong, L., Dumée, L.F. (2020). Release of hazardous nanoplastic contaminants due to microplastics fragmentation under shear stress forces. *Journal of Hazardous Materials*, 384, 121393.

Eriksen, M., Lebreton, L.C., Carson, H.S., Thiel, M., Moore, C.J., Borerro, J.C., Galgani, F., Ryan, P.G., Reisser, J. (2014). Plastic pollution in the world's oceans: More than 5 trillion plastic pieces weighing over 250,000 tons afloat at sea. *PLOS ONE*, 9(12), e111913.

Fadare, O.O. and Okoffo, E.D. (2020). Covid-19 face masks: A potential source of microplastic fibers in the environment. *Science of the Total Environment*, 737, 140279.

Fadare, O.O., Wan, B., Liu, K., Yang, Y., Zhao, L., Guo, L.H. (2020). Eco-corona vs protein corona: Effects of humic substances on corona formation and nanoplastic particle toxicity in daphnia magna. *Environmental Science and Technology*, 54(13), 8001–8009.

Garcia, J.M. and Robertson, M.L. (2017). The future of plastics recycling. *Science*, 358(6365), 870–872.

Geyer, R., Jambeck, J.R., Law, K.L. (2017). Production, use, and fate of all plastics ever made. *Science Advances*, 3(7), e1700782.

Golden, N.H., Warner, S.E., Coffey, M.J. (2016). A review and assessment of spent lead ammunition and its exposure and effects to scavenging birds in the United States. *Reviews of Environmental Contamination and Toxicology*, 237, 123–191.

González-Pleiter, M., Tamayo-Belda, M., Pulido-Reyes, G., Amariei, G., Leganés, F., Rosal, R., Fernández-Piñas, F. (2019). Secondary nanoplastics released from a biodegradable microplastic severely impact freshwater environments. *Environmental Science: Nano*, 6(5), 1382–1392.

Iheanacho, S., Ogbu, M., Ekpeyoung, J., Tabi, E., Iheanacho, A., Bhuyan, M.S., Ogunji, J. (2023). COVID-19 waste as source of microplastics in the environment: Implication for aquatic species, human, and remediation measures – A review. *Pollution*, 9(2), 591–614.

Jung, Y., Yoon, S.J., Byun, J., Jung, K.W., Choi, J.W. (2023). Visible-light-induced self-propelled nanobots against nanoplastics. *Water Research*, 244, 120543.

Klein, S., Worch, E., Knepper, T.P. (2015). Occurrence and spatial distribution of microplastics in river shore sediments of the Rhine-Main area in Germany. *Environmental Science and Technology*, 49(10), 6070–6076.

Koelmans, A.A., Besseling, E., Shim, W.J. (2015). Nanoplastics in the aquatic environment. Critical review. *Marine Anthropogenic Litter*, XVIII, 325–340.

Koelmans, A.A., Mohamed Nor, N.H.M., Hermsen, E., Kooi, M., Mintenig, S.M., De France, J. (2019). Microplastics in freshwaters and drinking water: Critical review and assessment of data quality. *Water Research*, 155, 410–422.

Leslie, H.A., Van Velzen, M.J., Brandsma, S.H., Vethaak, A.D., Garcia-Vallejo, J.J., Lamoree, M.H. (2022). Discovery and quantification of plastic particle pollution in human blood. *Environment International*, 163, 107199.

Li, L., Luo, Y., Li, R., Zhou, Q., Peijnenburg, W.J., Yin, N., Yang, J., Tu, C., Zhang, Y. (2020). Effective uptake of submicrometre plastics by crop plants via a crack-entry mode. *Nature Sustainability*, 3(11), 929–937.

Liu, W., Zhang, J., Liu, H., Guo, X., Zhang, X., Yao, X., Cao, Z., Zhang, T. (2021). A review of the removal of microplastics in global wastewater treatment plants: Characteristics and mechanisms. *Environment International*, 146, 106277.

Llorente-García, B.E., Hernández-López, J.M., Zaldívar-Cadena, A.A., Siligardi, C., Cedillo-González, E.I. (2020). First insights into photocatalytic degradation of HDPE and LDPE microplastics by a mesoporous N–TiO2 coating: Effect of size and shape of microplastics. *Coatings*, 10(7), 658.

Long, Z., Pan, Z., Wang, W., Ren, J., Yu, X., Lin, L., Lin, H., Chen, H., Jin, X. (2019). Microplastic abundance, characteristics, and removal in wastewater treatment plants in a coastal city of China. *Water Research*, 155, 255–265.

Materić, D., Kasper-Giebl, A., Kau, D., Anten, M., Greilinger, M., Ludewig, E., van Sebille, E., Röckmann, T., Holzinger, R. (2020). Micro-and nanoplastics in alpine snow: A new method for chemical identification and (semi) quantification in the nanogram range. *Environmental Science and Technology*, 54, 2353–2359.

Mattsson, K., Jocic, S., Doverbratt, I., Hansson, L.A. (2018). Nanoplastics in the aquatic environment. In *Microplastic Contamination in Aquatic Environments: An Emerging Matter of Environmental Urgency*, Zeng, E.Y. (ed.). Elsevier, Amsterdam

Miao, F., Liu, Y., Gao, M., Yu, X., Xiao, P., Wang, M., Wang, S., Wang, X. (2020). Degradation of polyvinyl chloride microplastics via an electro-Fenton-like system with a TiO_2/graphite cathode. *Journal of Hazardous Materials*, 399, 123023.

Michielssen, M.R., Michielssen, E.R., Ni, J., Duhaime, M.B. (2016). Fate of microplastics and other small anthropogenic litter (SAL) in wastewater treatment plants depends on unit processes employed. *Environmental Science: Water Research and Technology*, 2(6), 1064–1073.

Mintenig, S.M., Bäuerlein, P.S., Koelmans, A.A., Dekker, S.C., Van Wezel, A.P. (2018). Closing the gap between small and smaller: Towards a framework to analyse nano- and microplastics in aqueous environmental samples. *Environmental Science: Nano*, 5(7), 1640–1649.

Moharir, R.V. and Kumar, S. (2019). Challenges associated with plastic waste disposal and allied microbial routes for its effective degradation: A comprehensive review. *Journal of Cleaner Production*, 208, 65–76.

Nabi, I., Ahmad, F., Zhang, L. (2021). Application of titanium dioxide for the photocatalytic degradation of macro-and micro-plastics: A review. *Journal of Environmental Chemical Engineering*, 9, 105964.

Napper, I.E., Bakir, A., Rowland, S.J., Thompson, R.C. (2015). Characterisation, quantity and sorptive properties of microplastics extracted from cosmetics. *Marine Pollution Bulletin*, 99(1–2), 178–185.

Ng, E.L., Lwanga, E.H., Eldridge, S.M., Johnston, P., Hu, H.W., Geissen, V., Chen, D. (2018). An overview of microplastic and nanoplastic pollution in agroecosystems. *Science of the Total Environment*, 627, 1377–1388.

Silva, A.B., Costa, M.F., Duarte, A.C. (2018). Biotechnology advances for dealing with environmental pollution by micro (nano) plastics: Lessons on theory and practices. *Current Opinion in Environmental Science & Health*, 1, 30–35.

Sökmen, T.Ö., Sulukan, E., Türkoğlu, M., Baran, A., Özkaraca, M., Ceyhun, S.B. (2020). Polystyrene nanoplastics (20 nm) are able to bioaccumulate and cause oxidative DNA damages in the brain tissue of zebrafish embryo (*Danio rerio*). *Neurotoxicology*, 77, 51–59.

Sruthy, S. and Ramasamy, E.V. (2017). Microplastic pollution in Vembanad Lake, Kerala, India: The first report of microplastics in lake and estuarine sediments in India. *Environmental Pollution*, 222, 315–322.

Strungaru, S.A., Jijie, R., Nicoara, M., Plavan, G., Faggio, C. (2019). Micro-(nano)plastics in freshwater ecosystems: Abundance, toxicological impact and quantification methodology. *TrAC Trends in Analytical Chemistry*, 110, 116–128.

Sun, X.D., Yuan, X.Z., Jia, Y., Feng, L.J., Zhu, F.P., Dong, S.S., Liu, J., Kong, X., Tian, H., Duan, J.L. et al. (2020). Differentially charged nanoplastics demonstrate distinct accumulation in Arabidopsis thaliana. *Nature Nanotechnology*, 15(9), 755–760.

Tang, S., Lin, L., Wang, X., Feng, A., Yu, A. (2020). Pb(II) uptake onto nylon microplastics: Interaction mechanism and adsorption performance. *Journal of Hazardous Materials*, 386, 121960.

Ullattil, S.G. and Pumera, M. (2023). Light-powered self-adaptive mesostructured microrobots for simultaneous microplastics trapping and fragmentation via in situ surface morphing. *Small*, 19, 2301467.

Urso, M. and Pumera, M. (2022). Nano/microplastics capture and degradation by autonomous nano/microrobots: A perspective. *Advanced Functional Materials*, 32(20), 2112120.

Wang, L., Wu, W.M., Bolan, N.S., Tsang, D.C., Li, Y., Qin, M., Hou, D. (2021). Environmental fate, toxicity and risk management strategies of nanoplastics in the environment: Current status and future perspectives. *Journal of Hazardous Materials*, 401, 123415.

van Weert, S., Redondo-Hasselerharm, P.E., Diepens, N.J., Koelmans, A.A. (2019). Effects of nanoplastics and microplastics on the growth of sediment-rooted macrophytes. *Science of the Total Environment*, 654, 1040–1047.

Weithmann, N., Möller, J.N., Löder, M.G., Piehl, S., Laforsch, C., Freitag, R. (2018). Organic fertilizer as a vehicle for the entry of microplastic into the environment. *Science Advances*, 4(4), eaap8060.

Yan, Z., Liu, Y., Zhang, T., Zhang, F., Ren, H., Zhang, Y. (2022). Analysis of microplastics in human feces reveals a correlation between fecal microplastics and inflammatory bowel disease status. *Environmental Science and Technology*, 56(1), 414–421.

Yuan, Z., Nag, R., Cummins, E. (2022). Human health concerns regarding microplastics in the aquatic environment – From marine to food systems. *Science of the Total Environment*, 823, 153730.

Zeng, H., Xie, J., Xie, H., Su, B.L., Wang, M., Ping, H., Wang, W., Wang, H., Fu, Z. (2015). Bioprocess-inspired synthesis of hierarchically porous nitrogen-doped TiO_2 with high visible-light photocatalytic activity. *Journal of Materials Chemistry A*, 3(38), 19588–19596.

Zhang, H. (2017). Transport of microplastics in coastal seas. *Estuarine, Coastal and Shelf Science*, 199, 74–86.

Zhang, K., Su, J., Xiong, X., Wu, X., Wu, C., Liu, J. (2016). Microplastic pollution of lakeshore sediments from remote lakes in Tibet plateau, China. *Environmental Pollution*, 219, 450–455.

2

Metal Oxide-based Smart Materials for Photocatalytic Degradation of Micro- and Nano-Plastics

Photocatalysis is an eco-friendly, significantly widespread and sustainable process that uses light to degrade micro-plastics (MPs) and nano-plastics (NPs) using various catalysts, as discussed in the previous chapters. This technique stands out because of its simplicity, low cost and environment-friendly nature. In addition to this, it provides higher degradation efficiency and completely non-toxic end products. This chapter discusses the remediation of environmental concerns involving MPs and NPs using metal oxide nanoparticles via photocatalytic degradation. Metal oxides are one of the most used catalysts in the photocatalytic degradation method. This chapter explains in detail the mechanism of photocatalytic degradation using metal oxides as catalysts and discusses the synthesis processes and characteristics of the metal oxide catalysts. The modifications made to enhance the photocatalytic efficiencies of the nano- and micro-structured metal oxide catalysts are also discussed thoroughly in this chapter.

2.1. Introduction

Plastics are undoubtedly an ideal choice in today's time because they are lightweight, cheap and extremely durable. However, the use of plastics poses some severe environmental hazards. During the degradation period of plastics, they lose their integrity and break down further into smaller particles. These smaller particles are called MPs and NPs, depending on their sizes. MPs are microscopic, typically in the range of 50 μm–5 mm (0.05–5 mm) (Barnes et al. 2009; Kang et al. 2019; Sun et al. 2019). Further fragmentation of MPs leads to the formation of NPs, which lie in the range of 1–100 nm (Murty et al. 2013; Lambert and Wagner 2016). Therefore, NPs are

Chapter written by Roopam GAUR and Satyendra SINGH.

extremely small particles, a thousand times smaller than an algal cell (da Costa et al. 2016).

These smaller particles now interact with the environment and living organisms, leading to another environmental hazard. Many organisms, ranging from phytoplanktons to mammals, birds and fishes, ingest these micro- and nano-plastics (MNPs) and worsen their effects (Wright et al. 2013). Both marine and terrestrial ecosystems are equally affected by these MNPs (Ghosh et al. 2023). With dimensions <1 mm, MPs and NPs are invading all terrestrial and marine environments (Rillig 2012; Duis and Coors 2016). In recent times, they have emerged as a global problem as scientific research shows that these fragments are found almost everywhere in the world. Initially, it was assumed that MNPs were primarily a marine pollution issue, but reports show that MNPs are even found in the remotest areas possible, even at places unspoiled before (Allouzi et al. 2021; Chen et al. 2021). During their degradation cycle, plastics undergo many physical, chemical and biological alterations that produce MPs and NPs. In other words, MNPs are mainly produced due to macro plastic's low ability to degrade with time and their depletion due to weather, climate and pathway conditions. MNPs can be classified into two categories: (a) primary MNPs and (b) secondary MNPs (Jambeck et al. 2015; Sun et al. 2019). The main source of primary MNPs are polymers of <5 mm in size, for example, components and precursors of plastic products, personal care products, small parts of toys, vehicle tire wear, etc. (Kang et al. 2019; Liu et al. 2022). Secondary MNPs are produced due to further fragmentation erosion or depletion of the bigger plastic wastes. When plastics are introduced into the environment, they are exposed to many physical abrasions that lead to the formation of MNPs. After facing weathering, exposure to UV radiations, wind, heat, salt, etc., large pieces of plastics turn into smaller particles, which then, according to their size, are classified as MNPs. Also, even when the macro-plastics are recycled, they unintentionally produce many secondary MNPs into the environment. MNPs, especially the secondary kind, are extremely harmful for the aquatic ecosystem. Researchers have proved that NPs can penetrate the blood–brain barrier of the fishes and affect the survival of zooplankton to a great extent (Besseling et al. 2013). This has a huge impact on the behavioral patterns of the fish, causing various disorders. However, MPs and NPs are also quite hazardous for all living organisms, including humans.

Hence, remediation of MNPs pollution from the environment is one of the most pressing problems of today's time. Researchers and scientists worldwide have employed various techniques to turn the tide against this environmental hazard.

One such way to achieve this is using photocatalysis (Ding et al. 2022; He et al. 2023). In the last decade, photocatalysis has been seen as the center of research for the remediation of many environment-related issues, after it gained popularity in the remediation of wastewater from various industries (Neelavannan et al. 2007; Gümüş and Akbal 2011; Kujur et al. 2022; Xie et al. 2023).

Degradation of MPs and NPs is one such application of photocatalysis (Chen et al. 2019; Liu et al. 2019a, 2019b; Mao et al. 2020; Ding et al. 2022; Xu et al. 2024). Photocatalysis is advantageous over other catalytic techniques as it uses all the readily available things like sunlight, room temperature ambiance, normal atmospheric pressure and very commonly available catalysts. The choice of catalysts is extremely flexible, making the photocatalytic degradation process even more interesting. Also, the degradation process is completely non-toxic and does not cause further pollution; even the end products are environmentally friendly. This book aims to cover the details of all types of materials that can be employed as catalysts in the photocatalytic degradation of MPs and NPs.

2.2. Metal oxide photocatalysts and their characteristics

For the past few decades, metal oxides have been at the forefront of research in the fields of physics, chemistry, and material science for addressing the remediation of environment-related issues (de Jesus et al. 2021, 2024; Hajra et al. 2024). These materials are preferred over other contemporary materials as they are low-cost, stable and efficiently respond to light-driven photocatalysis.

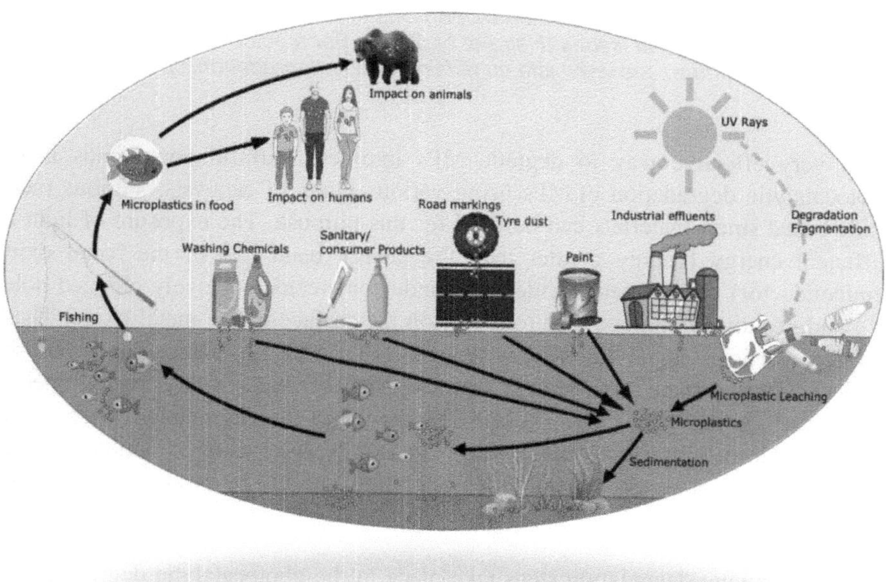

Figure 2.1. *Impact of micro- and nano-plastics on the ecosystem (Ghosh et al. 2023). For a color version of this figure, see www.iste.co.uk/singh/photodegradation.zip*

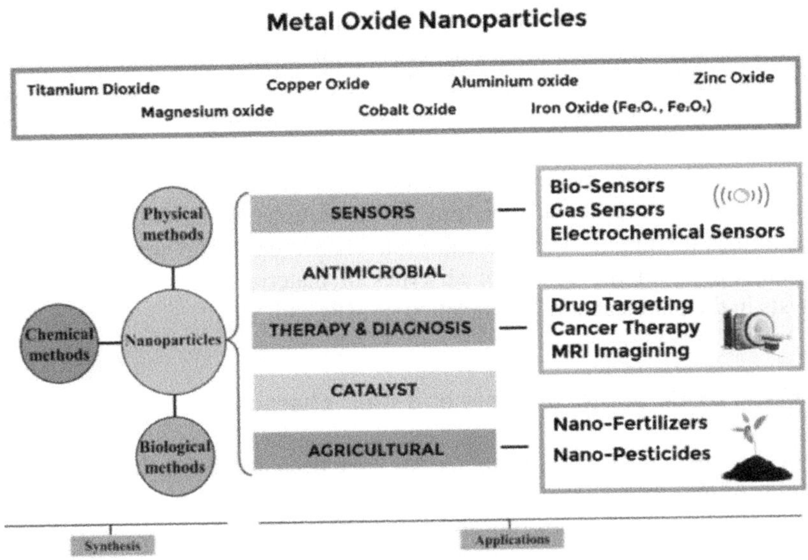

Figure 2.2. *Role of metal oxide nanoparticles in various fields for remediation of environmental issues (Hajra et al. 2024). For a color version of this figure, see www.iste.co.uk/singh/photodegradation.zip*

A very effective way to degrade MPs into less harmful compounds is the photocatalytic degradation of MPs using various kinds of catalysts. Various metal oxide-based smart materials can be used for this purpose. The exposure of light of sufficient energy (energy greater than the energy band gap of the metal oxide semiconductor) to any metal oxide semiconductor creates positively charged holes (h^+) and negatively charged electrons (e^+) on the surface of the metal oxide. These charged particles then interact with the H_2O and O_2 molecules near the semiconductor surface, resulting in various kinds of reactive oxygenated species (ROS) such as $OH^\bullet$ and $O_2^{\bullet-}$. These ROSs then start the degradation reactions by reacting with the polymeric chain (Ding et al. 2022). A series of oxidation steps are followed and as a result, MPs are degraded and various by-products are formed (Hamd et al. 2022).

The following degradation steps take place in the photocatalytic degradation of MPs:

$$M_xO_y\ h\nu \rightarrow M_xO_y\ (h^+{}_{VB}) + M_xO_y\ (e-{}_{CB}) \qquad [2.1]$$

$$h^+{}_{VB} + H_2O \rightarrow H^+ + OH^\bullet \qquad [2.2]$$

$$e^{-}_{CB} + O_2 \rightarrow O_2^{-\bullet} \quad [2.3]$$

$$O^{-\bullet}_2 + H_2O \rightarrow OH^{-} + OOH \quad [2.4]$$

$$2OOH^{\bullet} \rightarrow O_2 + H_2O_2 \quad [2.5]$$

$$2H_2O_2 \rightarrow 2OH^{\bullet} \quad [2.6]$$

$$OH^{\bullet} + \text{Micro-plastic} \rightarrow CO_2 + H_2O \quad [2.7]$$

$$e-_{CB} + \text{Micro-plastic} \rightarrow \text{Reduced products} \quad [2.8]$$

$$h^{+}_{VB} + \text{Micro-plastic} \rightarrow \text{Oxidized products} \quad [2.9]$$

As shown in equations [2.1]–[2.9], the ROS react with the immediately available polymer chains, and as a result of reactions, carbon-centred radicals are produced. These radicals then extend through the polymer chains. Oxygens enter the polymer due to these reactions, and the chains get disrupted; this forms compounds containing carboxyl and carbonyl groups. As a result, the oxygen content in the solution increases, which is denoted by carbonyl index (CI) (Liu et al. 2019a). The CI is one of the most common methods of indicating the chemical oxidation of plastics and is also used to analyze the loss of mechanical features of polymers.

Apart from degrading the MPs, the emphasis is also on converting them into useful industrial chemicals and products. However, for transforming MPs into useful products, other techniques, such as thermal, thermochemical, photothermal, etc., are better than photocatalysis. However, photocatalytic degradation of MPs using metal oxide semiconductors offers some other advantages over the above-mentioned methods, such as energy efficiency, moderate reaction conditions, environment friendliness, etc.

Some metal-oxide semiconductors used for photocatalytic degradation are ZnO, TiO_2, CuO, NiO, etc. (Tofa et al. 2019; Nabi et al. 2020; Acuña-Bedoya et al. 2021; He et al. 2023). Researchers have carried out many studies for modifications in surface morphology, crystallite sizes, phases, electron transfer, charge separation, etc., to obtain enhanced photocatalytic efficiencies. Doping metal oxide semiconductors with impurities (such as non-metallic elements C and N or transition metals such as Mn, Cr, Ni and Fe) can also modify their electronic structure, thereby increasing their light absorption range. Another important aspect is the size of the metal oxide nanoparticles as it plays a significant role in determining active sites at the surface. There are various methods to improve the degradation efficiencies of metal oxide semiconductors (Prasanna et al. 2019; Zarrintaj et al. 2019). However, in this chapter, we shall discuss in detail the efficiencies of the MP photodegradation of the metal oxide semiconductor systems commonly used for this purpose.

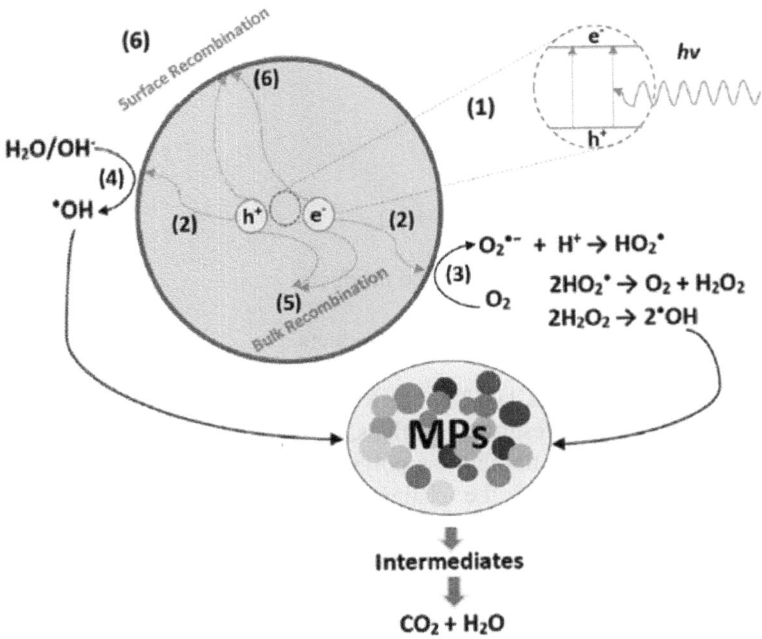

Figure 2.3. *Steps of degradation of microplastics at the surface via photocatalysis (Hamd et al. 2022). For a color version of this figure, see www.iste.co.uk/singh/photodegradation.zip*

2.2.1. TiO_2

TiO_2 has been a leading photocatalyst for more than four decades now, since the discovery of it degrading cyanide via photocatalysis in 1977 by Frank and Bard (1977). Since then, TiO_2 has been used to address many environment-related problems. Out of all, TiO_2 is the most popular among researchers because of its high oxidizing efficiency, low cost and high stability in all kinds of environments. Even for the degradation of MPs and NPs, TiO_2 has a special preference over its counterparts due to its high energy band gap value, which is around 3.2 eV (Dinoop lal et al. 2021). Even though TiO_2 has proved to be a reliable photocatalyst for quite some now, there are still some limitations attached to its usage. For example, the high recombination rate of electron and hole pairs, which are photogenerated, and its poor sensitivity, to name a few. This led to the search for a more efficient photocatalyst material by the researchers. In the last decade, plenty of photocatalysts have been explored for the remediation of environment-related issues using photocatalysis as the main technique. Researchers are also trying to improve the efficiency and overall catalytic activity of TiO_2 by making various modifications.

Very recently, in a report, the photocatalytic activity of TiO_2 was found to increase by decreasing the energy band gap of the TiO_2-based materials, which results in the reduction of recombination of electron and hole pair (Shang et al. 2003a; Thomas et al. 2013).

Many attempts have been made to improve upon the shortcomings of TiO_2.

TiO_2 is very popular worldwide and is considered a perfect photocatalyst for its high-performance ability of degrading organic pollutants. TiO_2 has performed exceptionally well in degrading organic pollutants using UV light, and it is completely logical to study its role in the degradation of MPs (Chattopadhyay et al. 2023). However, many researchers have attempted to fabricate photodegradable plastics using TiO_2/plastic composites. Many of the popular photodegradable plastics are polysterene (Shang et al. 2003b), polyethylene (Thomas et al. 2013), polypropylene (Kamrannejad et al. 2014), etc.

Copper phthalocyanine was used to enhance the photocatalytic activity of TiO_2 nanotubes in degrading low-density polyethelene under visible light (Ali et al. 2016).

Researchers have also tried to deposit noble metals such as Au, Ag, Pt and Pd on the TiO_2 surface to enhance its photocatalytic performance (Linsebigler et al. 1995; Liu et al. 2017; Plodinec et al. 2019; Malik et al. 2022; Li et al. 2023). This has successfully decreased the recombination rate as these materials act as electron acceptors. However, employing metal, especially noble metals, for deposition on TiO_2 to enhance photocatalytic performance is not an industrially viable option because it is expensive. Another issue with using metals is that they show photo corrosion during photocatalysis processes. Therefore, researchers have switched to depositing non-metals on the surface of TiO_2; and as a result, various catalysts with visible range efficiency are being developed. An efficient degradation in diclofenac of 60% was reported by Liu et al. (2017) using iodine-doped TiO_2 under visible light irradiation for 2 hours. One more method popular for enhancing the photocatalytic performance of TiO_2 is combining it with another semiconductor, thereby creating a heterojunction. This helps in slowing down the electron-hole recombination rate and modifies the energy band gap in such a way that more regions of the spectra can be used for the photocatalytic process.

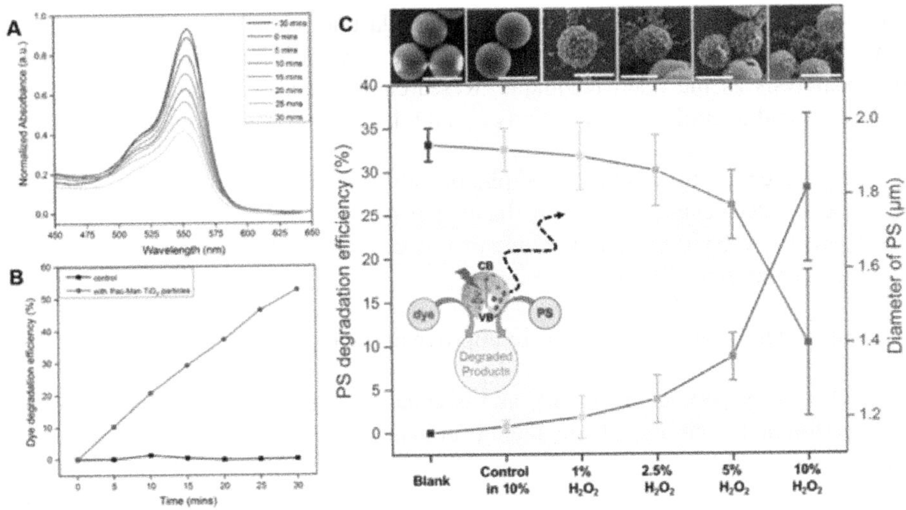

Figure 2.4. *(A) Degradation efficiency of micro Pac-Man particles using UV-Spectra, (B) analysis of catalytic efficiency with irradiation time, (C) dye-degradation efficiency and diameter change 2 μm PS particles post Pac-Man treatment in various H_2O_2 concentrations. (Chattopadhyay et al. 2023)*

2.2.1.1. *Doped TiO_2*

Doping TiO_2 with various elements of the periodic table has been another effective method for enhancing the photocatalytic degradation of the MNPs.

2.2.1.2. *TiO_2 heterojunctions*

Fabricating heterojunctions of TiO_2 with other compounds has been found to be one of the most powerful tools for preparing strong and effective photocatalysts. Out of many alternatives, Bi_2O_3 has been found to be one of the best choices for constructing heterostructures with TiO_2, because of its non-toxicity and easy availability.

2.2.1.3. *Photodegradation of polysterene using TiO_2 polyaniline*

The photodegradation of polysterene was done in UV irradiation using TiO_2-polyaniline (TiO_2-PANI) by Dinoop et al. (2021). It was seen that the photocatalytic activity showed enhancement when nano TiO_2 was coupled with PANI. It was derived that this enhancement is due to the charge separation caused by collective effect of both TiO_2 and PANI. Avoidance of recombination of holes and electrons caused by unification of TiO_2 and PANI acts as a key factor in improving the photocatalytic activity of TiO_2. It well known that the highest occupied energy level in PANI lies between the valence band and conduction band of TiO_2. After the electrons leave the TiO_2 on receiving phot energy, holes are left behind in the valence band which move to the intermediate energy level of PANI. Hence, after some time, when the electrons want to recombine with the holes, the holes are no longer available. This causes a delay in the recombination of holes and electrons of TiO_2, thereby enhancing their efficiency for photocatalytic degradation of polysterene (Zhang et al. 2008).

2.2.2. *ZnO*

ZnO is another metal oxide that is as popular as TiO_2. ZnO is an n-type semiconductor metal oxide that works with UV light as its band gap is ~3.37 eV (Koenig et al. 2017). Due to this, it can absorb light only in the UV region that comprises only 4–5% of the solar light. However, bandgaps can be modified via various techniques such as changing morphology, size, doping with appropriate dopants and synthesizing composites. Playing with the particle size is a very effective way to change the energy band gap.

ZnO has other advantages such as high redox potential, higher oxidation ability, excellent photocatalytic properties and electrochemical stability. It is also very cost-effective and non-toxic in nature, which makes it very suitable for environmental remediation applications. As far as the crystal structure of ZnO is concerned, it is found typically in three forms – rocksalt, zincblende and wurtzite with hexagonally shaped unit cells. However, ZnO found with wurtzite structure exhibits highest thermodynamic stability and is also considered “Safe” by FDA (Food and Drug Administration) (Mandal et al. 2014; Lee et al. 2016; Rajput et al. 2022).

Due to its defect chemistry, its band gap is tunable, and hence, ZnO has been extensively used in many photocatalytic degradations using both UV and visible light. Also, a wide range of morphologies have been reported for ZnO in the forms of nanoflowers, nanorods, nanoparticles, nanosheets, etc. Extensive research on various methods, such as physical, chemical and biological, have been carried out, putting ZnO at its center. However, it was observed that degradation of NPs by ZnO becomes weak in aqueous medium, as ZnO nanoparticles tend to agglomerate in it. Many things have been tried by researchers in order to control the surface characteristics of ZnO, as it is very crucial for the degradation of MPs using photocatalysis in aqueous media. Researchers, on the other hand, have been working on the alternative, which is growing ZnO nanostructures on solid substrates.

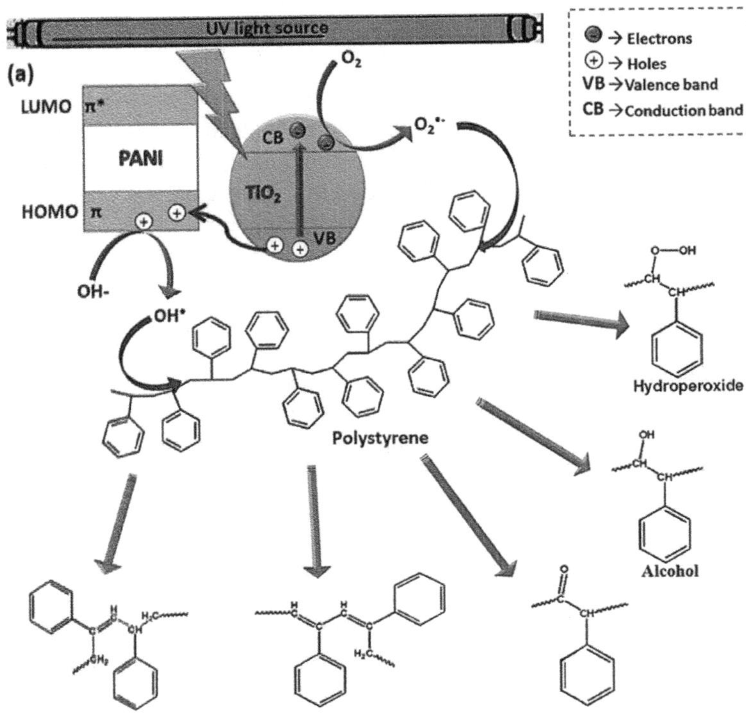

Figure 2.5. *Photodegradation mechanism of Polysterene (PS) using TiO_2-PANI composites (Dinoop et al. 2021). For a color version of this figure, see www.iste.co.uk/singh/photodegradation.zip*

2.2.2.1. *Degradation of MNPs using ZnO-based photocatalyst in solid phase*

The solid-state degradation process is preferred to establish any photodegradation system for MPs. This process was first demonstrated by Kamalian et al. (2019) where they used ZnO to degrade commercially available low-density polyethylene (commonly known as LDPE) using UV light irradiation. To determine the percentage of photodegradation of the LDPE, the weight loss in the samples was determined using the following formula (Lam et al. 2022):

$$\% \text{ Weight Loss} = (W_0 - W_d)/W_0 \times 100 \quad [2.10]$$

where W_0 and W_d are the weights of the LDPE films before and after the UV light irradiation.

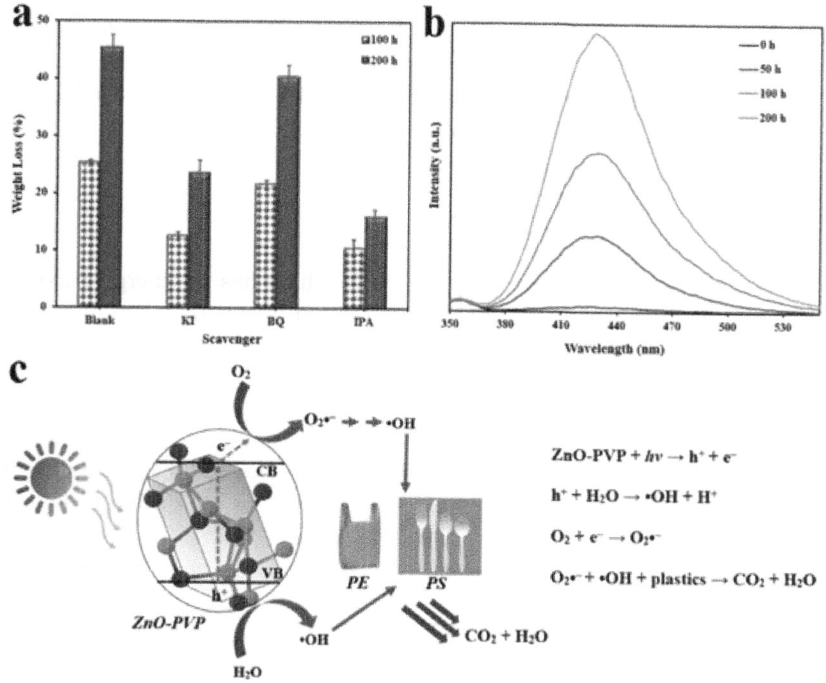

Figure 2.6. *(a) PE weight loss (%) of ZnO-PVP/PE using different capturing agents, (b) evolution of PL emission peak against sunlight irradiation time and (c) schematic (Lam et al. 2022). For a color version of this figure, see www.iste.co.uk/singh/photodegradation.zip*

2.2.3. *CuO*

There are two stable oxides of copper – cuprous oxide (Cu_2O) and cupric oxide (CuO) – with band gaps 1.2–1.8 eV and 2.0–2.2 eV, respectively (Balık et al. 2019). These oxides have been thoroughly studied for various types of applications such as sensors, batteries, transducers, actuators, supercapacitors and photo catalysts. These compounds are cheap, readily available, non-toxic and environment friendly, and because of their low band gap, they can even be used with visible light for the photocatalytic degradation. But since the energy band gap is too low, the recombination of electrons and holes occurs very easily (Acuña-Bedoya et al. 2021).

Acuña-Bedoya et al. (2021) reported that PS NPs of size ~350 nm got degraded in the presence of visible light with the help of Cu_2O/CuO photocatalysts. The concentration of PS was reduced by 23% as the photocatalysis in visible light promoted the polymer chain disintegration. The copper oxide semiconductors were grown in NH4F and NaOH mediums by anodizing technique and had varying morphologies with band gaps in the range of 1.6–2.0 eV. The CI for Cu_2O/CuO semiconductors grown in NH4F showed promising results as the long polymer chains got transformed into various chemical species such as ketones, aldehydes and carboxylic acid (Kumar 2023).

2.2.4. *NiO*

Another metal oxide in this series is NiO, which has an energy band gap of 3.4 eV. NiO helps in destructing various toxins and impurities when it is used as a catalyst in photocatalysis. As per a recent report, LDPE (low-density polyethylene) was degraded in solid phase by using NiO nanoparticles as catalysts in photocatalysis to break the polymer chain. It was reported that, in the presence of solar light for around ~240 h, LDPE showed a remarkable decrease in weight by almost up to 33%; and in the dark, the NiO nanoparticles could cause a weight loss of only ~1.90%. The CI of the degraded was found to be ~0.4 using FTIR (Olajire and Mohammed 2020).

2.3. Conclusion and future prospectives

The hazards caused by plastics, typically, MPs and NPs, is superficially known to all. However, the severity of it is something that everyone should be aware of. The disposal of plastics in the water bodies by individuals, factories, industries, etc., should be banned immediately. This chapter discusses in detail the consequences of intrusion of plastics into the water bodies and how MNPs are more dangerous than macroplastics. Many methodologies have been adopted to degrade MPs and NPs. Out of all the methods, photocatalytic degradation of MNPs has been proven to be the most popular. Different catalysts are used for the photocatalysis process; this chapter discusses the role of metal oxides as the catalysts in the photocatalytic degradation of MNPs. Various metal oxides such as TiO_2, ZnO, CuO/Cu_2O and NiO are also discussed in this chapter. Photocatalysis is a technique that does not require very high temperatures like thermal degradation. It is a simple process that can take in the presence of sunlight/visible light/UV light at room temperature. The by-products are usually non-toxic and environment friendly; however, sometimes, evolution of toxic gases take place during the degradation of MPs and NPs. Therefore, it is very important to conduct proper monitoring during the degradation of MNPs using photocatalysis. Researchers have also tried to combine photodegradation with other techniques such as thermal degradation. This technique has a lot of scope to degrade MPs in a larger range. Hence, future research should focus on optimizing photocatalysts and the photocatalysis conditions to attain better degradation results in terms of conversion rates and efficiencies. Considering all the pros and cons, photocatalytic degradation has great potential for addressing the hazards caused by plastics. However, any effort made would go waste unless humans are made aware of the severity of the toxicity caused by plastics.

2.4. Acknowledgments

Singh gratefully acknowledges the financial support received from SERB, Government of India (Project File No.: EEQ/2020/000373). Gaur acknowledges the Department of Science and Technology for DST – Women Scientist–A award (DST/WOS-A/PM-113/2021) for supporting the study.

2.5. References

Acuña-Bedoya, J.D., Luévano-Hipólito, E., Cedillo-González, E.I., Domínguez-Jaimes, L.P., Hurtado, A.M., Hernández-López, J.M. (2021). Boosting visible-light photocatalytic degradation of polystyrene nanoplastics with immobilized CuxO obtained by anodization. *Journal of Environmental Chemical Engineering*, 9, 106208 [Online]. Available at: https://api.semanticscholar.org/CorpusID:238687175.

Ali, S.S., Qazi, I.A., Arshad, M., Khan, Z., Voice, T.C., Mehmood, C.T. (2016). Photocatalytic degradation of low density polyethylene (LDPE) films using titania nanotubes. *Environmental Nanotechnology, Monitoring and Management*, 5, 44–53. doi: 10.1016/j.enmm.2016.01.001.

Allouzi, M.M.A., Tang, D.Y.Y., Chew, K.W., Rinklebe, J., Bolan, N., Allouzi, S.M.A., Show, P.L. (2021). Micro (nano) plastic pollution: The ecological influence on soil-plant system and human health. *Science of the Total Environment*, 788, 147815. doi: 10.1016/j.scitotenv.2021.147815.

Balık, M., Bulut, V., Erdogan, I.Y. (2019). Optical, structural and phase transition properties of Cu_2O, CuO and Cu_2O/CuO: Their photoelectrochemical sensor applications. *International Journal of Hydrogen Energy*, 44(34), 18744–18755. doi: 10.1016/j.ijhydene.2018.08.159.

Barnes, D.K.A., Galgani, F., Thompson, R.C., Barlaz, M. (2009). Accumulation and fragmentation of plastic debris in global environments. *Philosophical Transactions of the Royal Society of London. Series B, Biological Sciences*, 364(1526), 1985–1998. doi: 10.1098/rstb.2008.0205.

Besseling, E., Wegner, A., Foekema, E.M., van den Heuvel-Greve, M.J., Koelmans, A.A. (2013). Effects of microplastic on fitness and PCB bioaccumulation by the Lugworm Arenicola marina (L.). *Environmental Science and Technology*, 47(1), 593–600. doi: 10.1021/es302763x.

Chattopadhyay, P., Ariza-Tarazona, M.C., Cedillo-González, E.I., Siligardi, C., Simmchen, J. (2023). Combining photocatalytic collection and degradation of microplastics using selfasymmetric Pac-Man TiO2. *Nanoscale*, 15(36), 14774–14781. doi: 10.1039/D3NR01512B.

Chen, C., Chen, L., Yao, Y., Artigas, F., Huang, Q., Zhang, W. (2019). Organotin release from polyvinyl chloride microplastics and concurrent photodegradation in water: Impacts from salinity, dissolved organic matter, and light exposure. *Environmental Science and Technology*, 53(18), 10741–10752. doi: 10.1021/acs.est.9b03428.

Chen, Y., Awasthi, A.K., Wei, F., Tan, Q., Li, J. (2021). Single-use plastics: Production, usage, disposal, and adverse impacts. *Science of the Total Environment*, 752, 141772. doi: 10.1016/j.scitotenv.2020.141772.

da Costa, J.P., Santos, P.S.M., Duarte, A.C., Rocha-Santos, T. (2016). (Nano)plastics in the environment: Sources, fates and effects. *Science of the Total Environment*, 566–567, 15–26. doi: 10.1016/j.scitotenv.2016.05.041.

Ding, L., Yu, X., Guo, X., Zhang, Y., Ouyang, Z., Liu, P., Zhang, C., Wang, T., Jia, H., Zhu, L. (2022). The photodegradation processes and mechanisms of polyvinyl chloride and polyethylene terephthalate microplastic in aquatic environments: Important role of clay minerals. *Water Research*, 208, 117879. doi: 10.1016/j.watres.2021.117879.

Dinoop lal, S., Sunil Jose, T., Rajesh, C., Anju Rose Puthukkara, P., Savitha Unnikrishnan, K., Arun, K.J. (2021). Accelerated photodegradation of polystyrene by TiO_2-polyaniline photocatalyst under UV radiation. *European Polymer Journal*, 153, 110493. doi: 10.1016/j.eurpolymj.2021.110493.

Duis, K. and Coors, A. (2016). Microplastics in the aquatic and terrestrial environment: Sources (with a specific focus on personal care products), fate and effects. *Environmental Sciences Europe*, 28(1), 1–25. doi: 10.1186/s12302-015-0069-y.

Frank, S.N. and Bard, A.J. (1977). Heterogeneous photocatalytic oxidation of cyanide and sulfite in aqueous solutions at semiconductor powders. *Journal of Physical Chemistry*, 81(15), 1484–1488. doi: 10.1021/j100530a011.

Ghosh, S., Sinha, J.K., Ghosh, S., Vashisth, K., Han, S., Bhaskar, R. (2023). Microplastics as an emerging threat to the global environment and human health. *Sustainability*, 15(14). doi: 10.3390/su151410821.

Gümüş, D. and Akbal, F. (2011). Photocatalytic degradation of textile dye and wastewater. *Water, Air, and Soil Pollution*, 216(1–4), 117–124. doi: 10.1007/s11270-010-0520-z.

Hajra, K., Maity, D., Saha, S. (2024). Recent advancements of metal oxide nanoparticles and their potential applications: A review. *Advanced Materials Letters*, 15(1), 2401–1740. doi: 10.5185/amlett.2024.011740.

Hamd, W., Daher, E.A., Tofa, T.S., Dutta, J. (2022). Recent advances in photocatalytic removal of microplastics: Mechanisms, kinetic degradation, and reactor design. *Front. Mar. Sci.*, 9(7), 1–28. doi: 10.3389/fmars.2022.885614.

He, J., Han, L., Wang, F., Ma, C., Cai, Y., Ma, W., Xu, E.G., Xing, B., Yang, Z. (2023). Photocatalytic strategy to mitigate microplastic pollution in aquatic environments: Promising catalysts, efficiencies, mechanisms, and ecological risks. *Critical Reviews in Environmental Science and Technology*, 53(4), 504–526. doi: 10.1080/10643389.2022.2072658.

Jambeck, J.R., Geyer, R., Wilcox, C., Siegler, T.R., Perryman, M., Andrady, A., Narayan, R., Law, K.L. (2015) Marine pollution. Plastic waste inputs from land into the ocean. *Science*, 347(6223), 768–771. doi: 10.1126/science.1260352.

de Jesus, R.A., de Assis, G.C., de Oliveira, R.J., Costa, J.A.S., da Silva, C.M.P., Bilal, M., Iqbal, H.M.N., Ferreira, L.F.R., Figueiredo, R.T. (2021). Environmental remediation potentialities of metal and metal oxide nanoparticles: Mechanistic biosynthesis, influencing factors, and application standpoint. *Environmental Technology and Innovation*, 24, 101851. doi: 10.1016/j.eti.2021.101851.

de Jesus, R.A., de Assis, G.C., Oliveira, R.J., Costa, J.A.S., da Silva, C.M.P., Iqbal, H.M.N., Ferreira, L.F.R. (2024). Metal/metal oxide nanoparticles: A revolution in the biosynthesis and medical applications. *Nano-Structures and Nano-Objects*, 37, 101071. doi: 10.1016/j.nanoso.2023.101071.

Kamalian, P., Khorasani, S.N., Abdolmaleki, A., Neisiany, R.E. (2019). Grafted ZnO nanoparticles used for development in photocatalytic degradation performance of polyethylene. *Polymer Bulletin*, 76(7), 3593–3606. doi: 10.1007/s00289-018-2564-y.

Kamrannejad, M.M., Hasanzadeh, A., Nosoudi, N., Mai, L., Babaluo, A.A., Ave, H. (2014). Photocatalytic degradation of polypropylene/TiO_2 nano-composites. *Materials Research*, 17(4), 1039–1046.

Kang, J., Zhou, L., Duan, X., Sun, H., Ao, Z., Wang, S. (2019). Degradation of cosmetic microplastics via functionalized carbon nanosprings. *Matter*, 1(3), 745–758. doi: 10.1016/j.matt.2019.06.004.

Koenig, E., Jacobs, A., Lisensky, G. (2017) Properties of semiconductors: Synthesis of oriented ZnO for photoelectrochemistry and photoremediation. *Journal of Chemical Education*, 94(6), 738–742. doi: 10.1021/acs.jchemed.6b00887.

Kujur, V.S., Gaur, R., Gupta, V., Singh, S. (2022). Significantly enhanced UV-light-driven photocatalytic performance of ferroelectric (K0.5Na0.5)NbO_3: Effect of corona-poling and particle size. *Journal of Physics and Chemistry of Solids*, 167, 110751. doi: 10.1016/j.jpcs.2022.110751.

Kumar, R. (2023). Metal oxides-based nano/microstructures for photodegradation of microplastics. *Advanced Sustainable Systems*, 7(6), 1–24. doi: 10.1002/adsu.202300033.

Lam, S.M., Chew, K.C., Sin, J.C., Zeng, H., Lin, H., Li, H., Lim, J.W., Mohamed, A.R. (2022). Ameliorated photodegradation performance of polyethylene and polystyrene films incorporated with ZnO-PVP catalyst. *Journal of Environmental Chemical Engineering*, 10, 107594. doi: 10.1016/j.jece.2022.107594.

Lambert, S. and Wagner, M. (2016). Characterisation of nanoplastics during the degradation of polystyrene. *Chemosphere*, 145, 265–268. doi: 10.1016/j.chemosphere.2015.11.078.

Lee, K.M., Lai, C.W., Ngai, K.S., Juan, J.C. (2016). Recent developments of zinc oxide based photocatalyst in water treatment technology: A review. *Water Research*, 88, 428–448. doi: 10.1016/j.watres.2015.09.045.

Li, Y., Bu, D., Wang, J., Hou, X., Lin, C., Ma, X., Li, S., Zhang, Y., Ma, T. (2023). Study on method and mechanism of noble metals photoelectric deposition by directly utilizing solar energy. *Advanced Energy and Sustainability Research*, 4(10), 2300061. doi: 10.1002/aesr.202300061.

Linsebigler, A.L., Lu, G., Yates, J.T.J. (1995). Photocatalysis on TiO_2 surfaces: Principles, mechanisms, and selected results. *Chemical Reviews*, 95(3), 735–758. doi: 10.1021/cr00035a013.

Liu, H., Feng, J., Jie, W. (2017). A review of noble metal (Pd, Ag, Pt, Au)–zinc oxide nanocomposites: Synthesis, structures and applications. *Journal of Materials Science: Materials in Electronics*, 28(22), 16585–16597. doi: 10.1007/s10854-017-7612-0.

Liu, P., Qian, L., Wang, H., Zhan, X., Lu, K., Gu, C., Gao, S. (2019a). New insights into the aging behavior of microplastics accelerated by advanced oxidation processes. *Environmental Science and Technology*, 53(7), 3579–3588. doi: 10.1021/acs.est.9b00493.

Liu, Y., Hu, Y., Yang, C., Chen, C., Huang, W., Dang, Z. (2019b). Aggregation kinetics of UV irradiated nanoplastics in aquatic environments. *Water Research*, 163, 114870. doi: 10.1016/j.watres.2019.114870.

Liu, L., Xu, M., Ye, Y., Zhang, B. (2022). On the degradation of (micro)plastics: Degradation methods, influencing factors, environmental impacts. *Science of the Total Environment*, 806, 151312. doi: 10.1016/j.scitotenv.2021.151312.

Malik, A.S., Liu, T., Rittiruam, M., Saelee, T., Da Silva, J.L.F., Praserthdam, S., Praserthdam, P. (2022). On a high photocatalytic activity of high-noble alloys Au–Ag/TiO_2 catalysts during oxygen evolution reaction of water oxidation. *Scientific Reports*, 12(1), 2604. doi: 10.1038/s41598-022-06608-7.

Mandal, H., Shyamal, S., Hajra, P., Samanta, B., Fageria, P., Pande, S., Bhattacharya, C. (2014). Improved photoelectrochemical water oxidation using wurtzite ZnO semiconductors synthesized through simple chemical bath reaction. *Electrochimica Acta*, 141, 294–301. doi: 10.1016/j.electacta.2014.06.013.

Mao, R., Lang, M., Yu, X., Wu, R., Yang, X., Guo, X. (2020). Aging mechanism of microplastics with UV irradiation and its effects on the adsorption of heavy metals. *Journal of Hazardous Materials*, 393, 122515. doi: 10.1016/j.jhazmat.2020.122515.

Murty, B.S., Shankar, P., Raj, B., Rath, B.B., Murday, J. (2013). Unique properties of nanomaterials. In *Textbook of Nanoscience and Nanotechnology*, Murty, B.S., Shankar, P., Raj, B., Rath, B.B., Murday, J. (eds). Springer, Berlin, Heidelberg.

Nabi, I., Bacha, A.U.R., Li, K., Cheng, H., Wang, T., Liu, Y., Ajmal, S., Yang, Y., Feng, Y., Zhang, L. (2020). Complete photocatalytic mineralization of microplastic on TiO_2 nanoparticle film. *iScience*, 23, 101326 [Online]. Available at: https://api.semanticscholar.org/CorpusID:220521167.

Neelavannan, M.G., Revathi, M., Ahmed Basha, C.A. (2007). Photocatalytic and electrochemical combined treatment of textile wash water. *Journal of Hazardous Materials*, 149(2), 371–378. doi: 10.1016/j.jhazmat.2007.04.025.

Olajire, A.A. and Mohammed, A.A. (2020). Green synthesis of nickel oxide nanoparticles and studies of their photocatalytic activity in degradation of polyethylene films. *Advanced Powder Technology*, 31, 211–218 [Online]. Available at: https://api.semanticscholar.org/CorpusID:210806324.

Plodinec, M., Grčić, I., Willinger, M.G., Hammud, A., Huang, X., Panžić, I., Gajović, A. (2019). Black TiO_2 nanotube arrays decorated with Ag nanoparticles for enhanced visible-light photocatalytic oxidation of salicylic acid. *Journal of Alloys and Compounds*, 776, 883–896. doi: 10.1016/j.jallcom.2018.10.248.

Prasanna, S.R.V.S., Balaji, K., Pandey, S., Rana, S. (2019). Metal oxide based nanomaterials and their polymer nanocomposites. In *Nanomaterials and Polymer Nanocomposites*, Niranjak K. (ed.). Elsevier, Amsterdam, 123–144.

Rajput, R.B., Shaikh, R., Sawant, J., Kale, R.B. (2022). Recent developments in ZnO-based heterostructures as photoelectrocatalysts for wastewater treatment: A review. *Environmental Advances*, 9, 100264. doi: 10.1016/j.envadv.2022.100264.

Rillig, M.C. (2012). Microplastic in terrestrial ecosystems and the soil? *Environmental Science and Technology*, 46(12), 6453–6454. doi: 10.1021/es302011r.

Shang, J., Chai, M., Zhu, Y. (2003a). Solid-phase photocatalytic degradation of polystyrene plastic with TiO_2 as photocatalyst. *Journal of Solid State Chemistry*, 174(1), 104–110. doi: 10.1016/S0022-4596(03)00183-X.

Shang, J., Chai, M., Zhu, Y. (2003b). Photocatalytic degradation of polystyrene plastic under fluorescent light. *Environmental Science and Technology*, 37(19), 4494–4499. doi: 10.1021/es0209464.

Sun, J., Dai, X., Wang, Q., van Loosdrecht, M.C.M., Ni, B.J. (2019). Microplastics in wastewater treatment plants: Detection, occurrence and removal. *Water Research*, 152, 21–37. doi: 10.1016/j.watres.2018.12.050.

Thomas, R.T., Nair, V., Sandhyarani, N. (2013). TiO_2 nanoparticle assisted solid phase photocatalytic degradation of polythene film: A mechanistic investigation. *Colloids and Surfaces. Part A, Physicochemical and Engineering Aspects*, 422, 1–9. doi: 10.1016/j.colsurfa.2013.01.017.

Tofa, T.S., Ye, F., Kunjali, K.L., Dutta, J. (2019). Enhanced visible light photodegradation of microplastic fragments with plasmonic platinum/zinc oxide nanorod photocatalysts. *Catalysts*, 9 [Online]. Available at: https://api.semanticscholar.org/CorpusID:208713023.

Wright, S.L., Thompson, R.C., Galloway, T.S. (2013). The physical impacts of microplastics on marine organisms: A review. *Environmental Pollution*, 178, 483–492. doi: 10.1016/j.envpol.2013.02.031.

Xie, A., Jin, M., Zhu, J., Zhou, Q., Fu, L., Wu, W. (2023). Photocatalytic technologies for transformation and degradation of microplastics in the environment: Current achievements and future prospects. *Catalysts*, 13(5). doi: 10.3390/catal13050846.

Xu, Y., Ou, Q., van der Hoek, J.P., Liu, G., Lompe, K.M. (2024). Photo-oxidation of micro- and nanoplastics: Physical, chemical, and biological effects in environments. *Environmental Science and Technology*, 58(2), 991–1009. doi: 10.1021/acs.est.3c07035.

Zarrintaj, P., Khalili, R., Vahabi, H., Saeb, M.R., Ganjali, M.R., Mozafari, M. (2019). Polyaniline/metal oxides nanocomposites. In *Fundamentals and Emerging Applications of Polyaniline*, Mozafari, M. and Chauhan, P. (eds). Elsevier, Amsterdam.

Zhang, H., Zong, R., Zhao, J., Zhu, Y. (2008). Dramatic visible photocatalytic degradation performances due to synergetic effect of TiO_2 with PANI. *Environmental Science and Technology*, 42(10), 3803–3807. doi: 10.1021/es703037x.

3

WO_3-based Smart Material for Photocatalytic Degradation of Micro- and Nano-Plastic

Micro- and nano-plastics (MNPs) are a serious environmental hazard to ecosystems as well as human health resulting from extensive contamination. Nonetheless, successfully extracting and segregating MNPs from polluted water remains an uphill task given their minuteness and meager content (low amount). The zinc oxide (ZnO) and titanium dioxide (TiO_2) nanoparticles are an effective and popular choice for photocatalytic degradation due to their large bandgap and electron–hole pairs generation ability under UV light irradiation. However, tungsten trioxide (WO_3)-based smart materials could be a promising alternative. WO_3 nanoparticles can also be altered or functionalized with different materials called WO_3-based smart materials to improve their selectivity and photocatalytic efficacy for particular contaminants. In contrast to TiO_2, which is primarily responsive to UV light, WO_3 has photocatalytic activity under visible light, which is advantageous since it utilizes a greater percentage of the solar spectrum. WO_3 has emerged as a smart material exhibiting remarkable photocatalytic properties, making it an attractive candidate for the process of photocatalytic degradation of following micro-plastics (MPs) and nano-plastics (NPs). This chapter explores the most recent advancements in utilizing WO_3-based smart materials for enhanced photocatalytic degradation of MPs and NPs, highlighting their mechanisms, potential applications, challenges and its future aspects.

3.1. Overview of micro- and nano-plastics

Many researchers have arguably dubbed the late 20th century to the current period as the "plastic age" (Porta 2021). The first synthetic plastic was termed "Bakelite", which is a thermoset of phenol formaldehyde, and its commercial application was limited until the end of Second World War. Since plastics are

---

Chapter written by Rachana SAIN and Sudarshan SARKAR.

lightweight, chemical and corrosion resistant, durable and cheap, they replaced most traditional metal commodities and the aftermath of the war witnessed a boom in plastic manufacturing for most industrial and commercial products (Napper and Thompson 2020). The word "plastics" specifies a class of polymer and has its root in Latin, "plasticus", which itself comes from the Greek word "plastikós" or "plastos", which translates to "molded" or "formed". There has not been any satisfactory definition given for plastics, which falls under the class of polymers. Polymers are defined as long-chain molecules formed by joining smaller molecular moieties known as monomers through a process called polymerization. Though plastic and polymers are often used interchangeably, the stark distinction lies in the composition. When pure monomers join to form a long-chain molecular structure, it is termed polymers. However, when additives are added to pure polymers, the product formed is termed plastic. In this regard, we understand that: "all plastics belong to polymers but not all polymers belong to a class of plastic". The family of low- and high-density polyethylene, nylon, polypropylene, polyvinyl chloride (PVC), etc. are some of the most common plastics used in our daily lives (Mateos-Aparicio and Rodríguez-Moreno 2019; Crawford and Martin 2020).

Type of plastics	Chemical formula	Used in
Polylactic acid (PLA)	$(C_3H_4O_2)n$	Agricultural fabrics, packaging
Polyethylene (high-density PE)	$(–CH_2–CH_2-)n$	Rope, woven bags
Polyethylene (low-density PE)	$(–CH_2–CH_2-)n$	All types of packaging
Polyamide-66 (PA_66)	$(C_{12}H_{22}N_2O_2)n$	High-pressure sealing rings, chemical and electrical parts
Polyvinyl chloride (PVC)	$(–CH_2–CHCl-)n$	Containers
Polyethylene (PE)	$(–CH_2–CH_2-n$	Pipes, wires and cables
Polystyrene (PS)	$(CH_2CHC_6H_5)n$	Disposable foam packaging, optical chemical instrument parts, cosmetics
Polypropylene (PP)	$(–C_3H_6–)n$	Textiles, packaging, electrical appliances and architecture
Polyformaldehyde (POM)	$(CH_2O)n$	Gears, bearings and auto parts

Table 3.1. *Fewer nano- and micro-plastics products present in the environment (Deng et al. 2020; Wang et al. 2024)*

Though global-plastic utilization has improved technology and become an intricate part of our livelihoods, the threat from plastics and MPs has recently come under the spotlight. As of 2019, annual global production of total plastic was found to be approximately 368 million tons. The production of plastic-made materials such as medical personal protective equipment (PPEs), disposable masks, face shields, gloves, etc. increased and was utilized during and after the Covid-19 pandemic, as shown in Figure 3.1. These single-use disposable products are mostly made with plastic materials and fibers and due to their extensively long degradation period, they ended up as huge masses of landfill (Lee and Li 2021; Mehran et al. 2021). The statistical data from "United Nations Environment Programmer (UNEP)" reveal that the majority of products manufactured from plastic are disposable and have high molecular weight and significant hydrophobicity, which prevents chemical or microbial degradation of these materials (Wang et al. 2024).

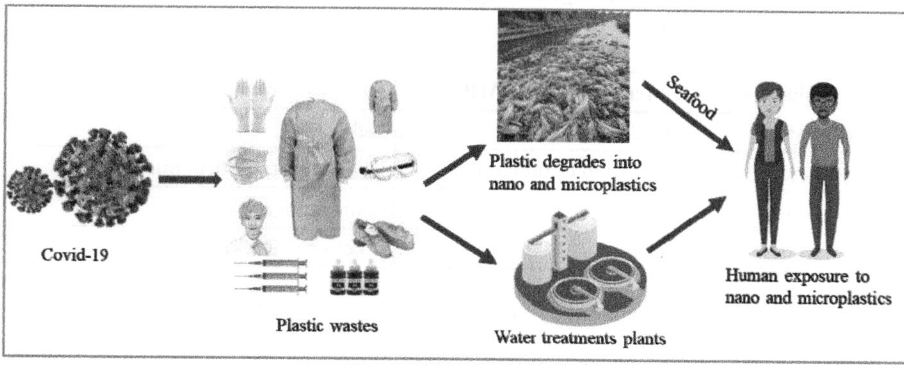

Figure 3.1. *Human exposure to nano- and micro-plastics produced during the Covid-19 pandemic. For a color version of this figure, see www.iste.co.uk/singh/photodegradation.zip*

In addition to these existing problems, MPs emerged as a critical threat. Thompson et al. (2004) were the first to introduce the concept of *micro-plastic* in 2004, when their study revealed the immediacy of plastic debris present in sea water as well as in the sediments. In 2008, MPs were defined as plastic debris, having a diameter less than 5 mm in the "first international research workshop on the occurrence, effects, and fate of micro-plastic marine debris". The categorization of plastic waste according to its size was determined from the Scientific Aspects of Marine Environmental Protection (GESAMP), as illustrated in Table 3.2 (Andrady 2017) and Figure 3.2. The threat of MPs comes from their ability to harbor waterborne pollutants and toxins. MPs themselves can be carcinogenic due to the presence of carcinogenic additives or chemicals (Toops 2020; Fadli et al. 2021; Wang et al. 2024). MPs may originate from either manufacturing processes or the fragmentation of bigger plastic materials due to abiotic or biotic influences. MP pollution is mainly due to plastic mismanagement and many of the plastic wastes generated are either dumped in landfill or in dumping yards and eventually finds its way into soil, water and air (Patil et al. 2022). Interestingly, among different types of MPs such as microbeads (spherical shape), microfibers (elongated, thread-like structures), micro-pellets (hard, globular shaped), microfilms (thin, transparent and flexible) and micro-fragments (rigid and irregularly shaped), microfibers are responsible for more than 90% of MP marine debris. Numerous studies have determined that the microfibers identified in marine water, sediments and various organisms are primarily composed of polyester. Globally, more than 9.5 million tons of synthetically manufactured fibers are put together annually and wastewater generated from the textile manufacturing and processing establishments are the most common origin of microfibers contaminations in the underwater (aquatic) environment (Carr 2017; Barrows et al. 2018; Deng et al. 2020; Prapanchan et al. 2023).

Class of plastics	Dimension (GESAMP)	Visualization technique
Macro-plastic	100–2.4 cm	Naked eye
Meso-plastic	2.4–0.1 cm (1,000 μm)	Naked eye or optical microscope
Micro-plastic	0.1 cm (1,000 μm)–1 μm	Optical microscope
Nano-plastic	< 1 μm	Electron microscope

Table 3.2. *Classification of plastic debris as per GESAMP 2015 (Van Cauwenberghe et al. 2015; Andrady 2017)*

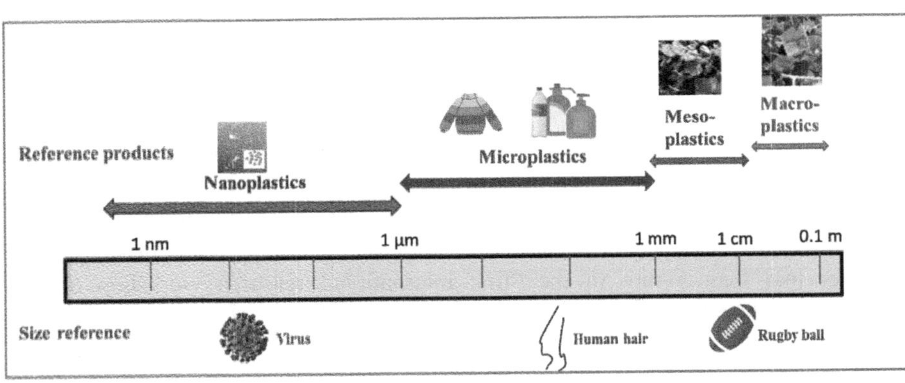

Figure 3.2. *Size classification of plastic particles concerning type of waste. For a color version of this figure, see www.iste.co.uk/singh/photodegradation.zip*

The adverse effects of the MPs in organisms due to bioaccumulation and disrupting biological cycles are well researched and documented. Kauts et al. (2023) have reported the toxicological profile of polyethylene terephthalate (PET) MPs on *Drosophila melanogaster* (Oregon R+), a species of fruit fly. Their research indicated that the concern lies in the concentration of MPs and is time dependent. If the concentration exceeds a threshold, the growth, movement, and (female) reproduction are adversely affected in fruit flies. The MPs' threat to marine life is not only through ingestion but also through intake by different orifices or through disposition. Many marine organisms like Bivalve molluscs are filter feeders meaning they filter food from seawater through their gills, and this also results in the uptake of MPs. Ikuta et al. (2022) studied the effect of phagocytosis of MP particles in the gill cells of coastal and deep-sea mussels, namely, *Bathymodiolus japonicus* and *Perna viridis*. The MPs histologically confirmed the presence of microbeads in an encapsulated form around the membrane-bound vacuoles of these species. Wenchao et al. (2023) investigated the impact of MPs containing dimethyl phthalate (DEP), a plasticizer (discussed in a later section), on the growth of *Chlorella pyrenoidosa*, a species of marine algae. Their findings revealed that elevated concentrations of DEP induce oxidative stress on marine algae, leading to a significant inhibition of growth. Therefore, it becomes a necessity for the degradation and removal of MPs/NPs and their additives. In the following chapter, we mainly elaborate the photocatalytic degradation of several forms of phthalates and MPs by employing WO_3-based smart materials as the heterocatalysts. However, MP degradation methods encompass a range of physical, chemical and biological processes aimed at breaking down or transforming NP and MP particles into smaller fragments or harmless by-products. Each degradation method has its advantages and limitations, and the method adopted depends on various factors such as the nature of the plastic, its environmental state and the desired outcomes for MP pollution mitigation. Ongoing research aims to develop more efficient and sustainable methods for MP degradation to address the growing environmental concern posed by MP pollution. The various methods commonly used or proposed for NP and MP degradation are as follows:

1) Mechanical fragmentation: mechanical forces such as abrasion, erosion and weathering can break down larger plastic items into smaller fragments over time. These fragments can further degrade into MPs through physical processes such as wave action, wind erosion and friction with surfaces (Andrady 2017).

2) Chemical degradation: this pertains to the involvement of chemical agents or reactions to break down plastic polymers. Common methods include hydrolysis, oxidation and pyrolysis. Hydrolysis breaks down polymers by reacting with water molecules, while oxidation involves reactions with oxygen or ozone. The pyrolysis is a thermochemical process where all types of plastic wastes are heated to high

temperatures in the non-oxygen environment to decompose them into smaller molecules (Rani and Shim 2017).

3) Biodegradation: this refers to the degradation of plastics with the influence of microorganisms like bacteria, archaea, fungi (yeasts and molds) and algae. Some microorganisms produce enzymes that can break down specific types of plastic polymers, converting them into simpler organic compounds. Biodegradation is influenced by numerous factors including temperature, moisture, oxygen availability and microbial activity (Geyer et al. 2017).

4) Advanced oxidation processes (AOPs): the following process requires the generation of highly reactive oxygen species (ROS) through chemical or physical processes, which can oxidize and degrade plastic polymers. Examples include photocatalysis using semiconductor materials like titanium dioxide (TiO_2) or zinc oxide (ZnO) to generate ROS under the irradiation of light (Wright and McManus 2019).

5) Enzymatic degradation: enzymes produced by certain microorganisms or engineered enzymes can catalyze the degradation of specific plastic polymers. For example, enzymes such as lipases, cutinases and esterase, have been identified for their ability to break down plastics like PET and polyethylene (PE) (Danso et al. 2019).

6) Sonolysis: this involves the application of ultrasonic waves to degrade plastics by generating cavitation bubbles, which implode and produce high temperatures and pressures capable of breaking chemical bonds in the plastic polymer chains (Imhof et al. 2013).

7) Combination methods: combining different degradation methods can enhance the efficiency and effectiveness of MP degradation. For example, combining biological degradation with chemical or physical processes can accelerate the breakdown of plastics and minimize environmental impacts (Chamas et al. 2020).

3.2. Photocatalytic degradation mechanism

The persistence of MPs in water poses a significant threat, primarily because of the limited availability of effective removal technologies capable of degrading these pollutants into environmentally sustainable products. Photocatalytic degradation process is regarded as an advanced-oxidation method that has garnered significant attention for its ability to demineralize pollutants without generating any detrimental residual by-products under visible or UV irradiation (Liu et al. 2020). There are numerous nanomaterials reported that have shown promising photocatalytic degradation under visible or UV irradiation. Tunguston dioxide (TiO_2) is the most admirable metal oxide-based inorganic semiconductors due to its affordability, stability and easy synthesis under UV irradiation (Wu et al. 2020). Graphitic carbon nitride is another notable photocatalyst that has shown promising applicability in water remediation. Its metal-free composition, cost-effectiveness, chemical stability, easy synthesis, visible-range bandgap and eco-friendly nature make it particularly interesting for such applications (Zhao et al. 2021).

Interestingly, WO_3 and its various polymorphs (cubic, monoclinic, trigonal, tetragonal, hexagonal and orthorhombic) have exhibited noteworthy photocatalytic activity. However, the monoclinic structure of WO_3 is the preferred choice, possibly due to the ease of synthesis. However, the narrow adsorption edge of pristine WO_3 results in the effective utilization of photons across the solar spectrum, thereby limiting the efficiency of the photocatalysis (Kumar and Rao 2017).

The introduction of hetero-catalysts for photocatalytic degradation has improved the efficiency of the process by mitigating electron–hole recombination process, a common bottleneck in the degradation through photocatalytic process. The application of hetero-catalysts for photocatalytic oxidation was initially proposed in 1967 (Chen et al. 2020). Photocatalytic efficiency greatly depends on the arrangement of the photocatalyst's heterojunction. The arrangements are classified into three categories based on the hetero-catalysts' bandgap energies:

– straddling-gap structure;

– staggered-gap structure;

– broken-gap structure.

To understand this, suppose there are two semiconductor materials (semiconductor A and semiconductor B) that form the heterojunction catalyst. Now, if the conduction band (CB) of semiconductor B exists as highly negative and the valance band (VB) is more positive compared to semiconductor A, then the arrangement is termed a straddling gap structure as shown in Figure 3.3(a).

This leads to the accumulation of charge carriers in semiconductor A which has a narrow bandgap. This results in higher recombination rates of charge carriers leading to poor photocatalytic efficiency. If the CB of semiconductor B is found to be more negative than the semiconductor A and the VB of semiconductor A is found to be more positive compared to semiconductor B, then the arrangement is termed a staggered gap structure as shown in Figure 3.3(b). This leads to the electron migration from semiconductor B to semiconductor A, while the holes effectively migrate from semiconductor A to semiconductor B. This type of hetero-catalyst improves charge separation thereby reducing the electron–hole recombination process. Most typical hetero-catalysts have staggered gap structures.

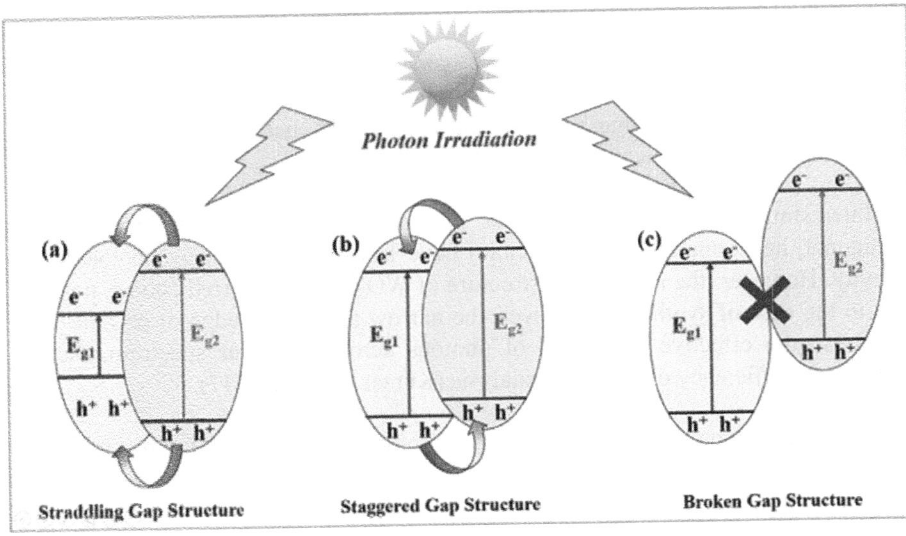

Figure 3.3. *Various structural configurations of semiconductor heterojunctions. For a color version of this figure, see www.iste.co.uk/singh/photodegradation.zip*

The third type is the broken gap structure in Figure 3.3(c), which arises when the charge transfer is impeded by a wide bandgap energy. This occurs because the CB and VB of semiconductor A are situated lower than those present in semiconductor B. As a result, the migration of charge carriers through the interface between the two semiconductors is restricted (Pirhashemi et al. 2018; Zhang and Jaroniec 2018; Koe et al. 2020).

The photocatalytic degradation process of MP can be elucidated by taking the example of titanium oxide, TiO_2 as an established photocatalyst. The degradation of MPs may involve several steps that highlight the complexity of MP degradation, thereby underscoring the importance of understanding each stage for effective mitigation strategies.

The photoaging process of MPs may be induced under solar (visible) light. The absorbed photons create an excited state, thereby breaking the complex bonds among the micro-molecules directly or through the oxidation process. The photocatalyst accelerates the process of bond rupturing in the MPs. The anatase phase of TiO_2 has a wide band gap of 3.2 eV, thus it absorbs photons in the UV spectrum. The absorption of photons leads to the excitation of the electrons (e^-)

from lower lying VB to migrate to the CB, leading to the generation of number of holes (h^+) in the low energy VB. These excited electrons migrate to the catalyst's surface and promptly react with the oxygen molecules adsorbed on the catalyst's surface, generating super-oxides ($O_2^{\bullet-}$) while the holes formed in the VB react with the adsorbed water to form hydroxyl radicals (•OH). These super-oxides and hydroxyl radicals are ROS that are essential for initiating the degradation process of MPs achieved through the cleaving of the adjacent polymer molecular chains. The photocatalytic degradation reaction proceeds when •OH targets the C–H bonds within the macromolecule, prompting the hydrogen atom to scavenge an e^- and then combine with the hydroxyl radical to generate water there. It results in the formation of free radicals centered around electron-rich carbon atoms. When introduced into the macromolecule, these carbon-centered radicals undergo continuous reactions with various ROS available, resulting in the breaking of polymeric molecular chains yielding CO_2, H_2O and intermediate derivatives such as hydroxyl, carboxyl and carbonyl compounds, among others. The figure illustrates the reaction pathway and production of ROS during the photocatalytic degradation process of MP utilizing titanium dioxide photocatalyst (Cho and Choi 2001; Pirhashemi et al. 2018; Chen et al. 2019; Li et al. 2019; Luo et al. 2021; Uheida et al. 2021; Ge et al. 2022).

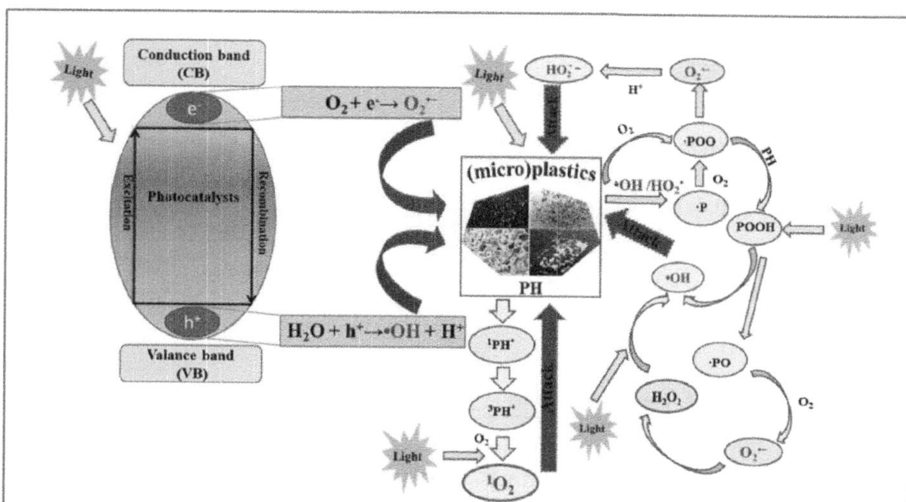

Figure 3.4. *The pathway of reactions and production of ROS triggered by the TiO_2 photocatalyst. Reprinted from Ge et al. (2022). Copyright 2022, with permission from Elsevier. For a color version of this figure, see www.iste.co.uk/singh/photodegradation.zip*

Here, photocatalytic degeneration of (MP) reactions induced by TiO_2 photocatalysts shown in Figure 3.4 can be summarized as follows (Ge et al. 2022):

$$TiO_2 \xrightarrow{h\upsilon} TiO_2^{*} \text{ (generation of } e^{-} \text{ and } h^{+}\text{)}$$

$$O_2 + e^{-} \longrightarrow O_2^{\bullet -}$$

$$TiO_2^{*} + O_2 \longrightarrow {}^{1}O_2$$

$$O_2^{\bullet -} + H_2O \longrightarrow HO_2^{\bullet} + OH^{-}$$

$$2HO_2^{\bullet} \longrightarrow H_2O_2 + O_2$$

$$H_2O_2 \xrightarrow{h\upsilon} 2 \bullet OH$$

$$OH^{-} + h^{+} \longrightarrow \bullet OH$$

$$h^{+} + H_2O \longrightarrow \bullet OH + H^{+}$$

$$P - H \text{ (polymer)} + \bullet OH \longrightarrow \bullet P + H_2O$$

$$\bullet P + O_2 \longrightarrow \bullet POO$$

$$P - H + HO_2^{\bullet} \longrightarrow \bullet P + H_2O_2$$

$$\bullet POO + P - H \longrightarrow POOH + \bullet P$$

$$POOH \xrightarrow{h\upsilon} \bullet PO + \bullet OH$$

$$\bullet PO + PH \longrightarrow POH + \bullet P$$

$$\bullet PO \xrightarrow{\text{breaking of polymeric chains}} \text{Intermediate products} + CO_2 + H_2O$$

Photocatalytic degradation offers a potential approach for the effective degradation of MPs in water remediation processes (Ouyang et al. 2021; Venkataramana et al. 2021). However, it is crucial to acknowledge that the threat extends not only from MPs themselves but also from the additives they release during natural weathering. These additives are termed plasticizers, which impart stability and flexibility to the plastics. Among such plasticizers, phthalate esters (PAEs) are one of the most predominantly used plastics for children's toys and water bottles. The US Environmental Protection Agency (EPA) has classified PAEs as priority pollutants. The toxicity and bioaccumulation of PAEs leached from MPs are identified as a critical environmental pollutant related to MPs (Keith and Telliard 1979; Cao et al. 2022). Phthalates are commonly found in PVC and children's toys and are known to cause health damage. Urbanization has led generations of MPs, subsequently leading to the release of phthalates and their esters through the leaching process in an alarming concentration (Henkel et al. 2022; Kumari and Pulimi 2023).

3.3. Tungsten trioxide (WO_3)

Tungsten trioxide is a versatile material known for its catalytic properties, particularly in photocatalytic applications. As a photocatalyst, WO_3 utilizes light energy to drive the chemical reactions, enabling the degeneration of organic compounds, like MPs and NPs. Tungsten trioxide is a semiconductor material with a wide bandgap, typically around 2.4–3.2 eV, which corresponds to the absorption in the near-ultraviolet (UV) and visible regime of the electromagnetic radiation spectrum (Rajeshkumar and Sudhakar 2019; Suriyakumar et al. 2019; Wu et al. 2019). When WO_3 absorbs photons having energies similar to its bandgap energy, then electrons get excited to high-energy CB from lower lying VB, leaving behind a hole and thus, creating an electron–hole (e^{-}-h^{+}) pair. These (e^{-}-h^{+}) pairs act as the highly reactive species that participate in the redox reactions with molecules adsorbed on the surface of the WO_3. MP and NP particles, along with other organic pollutants present in the environment, can adsorb onto the WO_3 catalyst surface due to physical interactions such as Van der Waals forces and electrostatic attraction. This brings the pollutants into contact with the reactive sites on the WO_3 surface, facilitating their degradation. Additionally, the photogenerated electrons in the CB of WO_3 reacts with adsorbed oxygen (O_2) to form superoxide radicals ($\bullet O^{-2}$). Meanwhile, the holes present in the VB of WO_3 catalyst can lead to the oxidation of water molecules (H_2O) or hydroxide ions (OH^{-}) to produce highly reactive hydroxyl radicals ($\bullet OH$). These ROS are highly oxidizing and can attack the chemical bonds in organic pollutants, including MPs and NPs, leading to their degradation into smaller, less toxic molecules. To maintain catalytic activity, it is essential to minimize the recombination rate of the photo-generated electron–hole pairs (e^{-}-h^{+}), as this can inevitably compromise the efficiency of this photocatalytic process. Strategies like doping with other transition metals or coupling with co-catalysts can enhance charge separation and reduce recombination rates, thereby enhancing the overall efficiency and stability of the WO_3-based photocatalytic performance. Regarding, MP and NP degradation, WO_3 photocatalysis offers a promising approach for breaking down plastic polymers into smaller molecular fragments or mineralizing into nontoxic by-products. By harnessing light energy, WO_3 can initiate oxidative reactions that lead to the cleavage of chemical bonds in plastic polymers, ultimately resulting in their degradation and removal from the environment. Overall, tungsten trioxide catalysis in photocatalytic degradation exploits its semiconductor properties to generate ROS that facilitate the oxidation of organic pollutants, including MPs and NPs, under illumination. This approach holds significant potential for addressing the challenge of plastic pollution and promoting environmental sustainability (Wang et al. 2018; Rajeshkumar and Sudhakar 2019; Wu et al. 2019; Zhu et al. 2020).

3.3.1. *(WO_3)-based smart materials*

Ttungsten trioxide has attracted significant attention as both a semiconductor and a promising catalyst for the photocatalytic degradation process because of its small bandgap (approximately 2.6 eV), enabling substantial near-UV and visible light absorption capacity along with chemical and thermal stability (Keith and Telliard 1979). The VB potential value of +3.1 V_{NHE} for WO_3 along with band gap of ~ 2.6 eV suggests its potential suitability for visible-light-induced photodegradation. Interestingly, the CB potential of +0.4 V_{NHE} is not sufficiently negative to utilize O_2 as the electron acceptor, resulting in insignificant photocatalytic oxidation by WO_3. The introduction of another suitable semiconducting material to form a hetero-junction with WO_3 becomes necessary to scavenge the electrons from CB, allowing the electrons to migrate to the hetero-catalyst's surface or water interface to facilitate visible-light-induced photocatalytic oxidation by WO_3 (Adhikari et al. 2018).

Numerous studies have reported the exceptional photocatalytic degradation of organic pollutants using transition metal-doped WO_3 (Kim et al. 2015), WO_3/Ag_2CO_3 heterocatalyst or $Ag/WO_3/g\text{-}C_3N_4$ (Zhou et al. 2021; Naeimi et al. 2022), compared to pure WO_3. Conversely, the formation and application of WO_3-based smart materials present an intriguing opportunity for the photocatalytic degradation of MPs. Remarkably, this aspect of the field remains relatively unexplored in the context of photocatalytic degradation of MPs for water remediation.

A graphitic carbon nitride ($g - C_3N_4$) is a 2D polymeric metal free and a n-type semiconductor compound that has garnered attention for being thermally and chemically stable and having a tunable bandgap suitable for the capacity for visible light absorption. The typical $g - C_3N_4$ structure is made up of tri-s-triazine moiety, making it an environmentally friendly material. It is identified as a smart material finding its application as supercapacitors (Kumaravel et al. 2023), photochromic materials (Devi et al. 2023) and visible-light-driven photocatalysts (Latiff et al. 2018; Fang et al. 2022; Abdul Rasheed et al. 2023; Sharath et al. 2023).

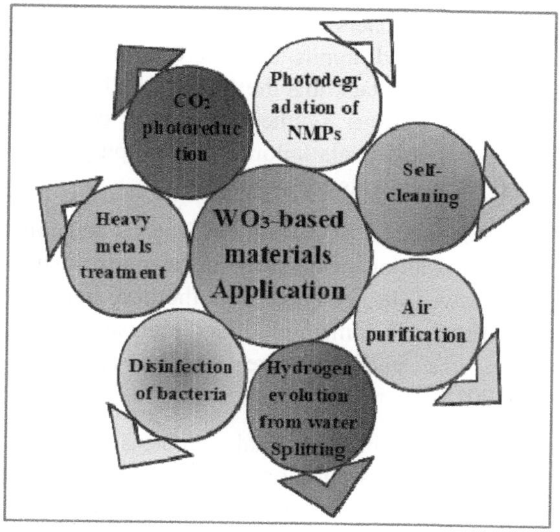

Figure 3.5. *Various applications of WO_3-based smart materials. For a color version of this figure, see www.iste.co.uk/singh/photodegradation.zip*

The bandgap range of WO_3 being 2.4–3.2 eV compliments the very bandgap of $g-C_3N_4$, thus improving charge carrier separation by lowering the recombination rate of the (e^--h^+) pairs and enhance the photocatalytic efficiency of the hetero-catalysts (Li et al. 2023). The preparation of $WO_3/g-C_3N_4$ smart material is easily achieved through the precipitation technique or by the self-assembly method. The precipitation technique involves first preparing graphitic carbon nitride separately using a straightforward polymerization method with appropriate precursors, including urea, thiourea or melamine. The resulting, as prepared $g-C_3N_4$, is dispersed into a solution, into which a sodium tungstate solution is gradually added while stirring constantly, followed by ultrasonication to ensure even distribution. Hydrochloric acid (HCl) is then slowly added to the mixture while stirring continuously. The product obtained after washing and drying is the $WO_3/g-C_3N_4$ heterocatalyst. In the self-assembly method, the $g-C_3N_4$ and WO_3 are individually synthesized. Subsequently, then both materials at varying loading percentages undergo ultrasonication-assisted mixing in a solvent, which can be either deionized water or an organic solvent, to form the heterojunction catalyst (Wang et al. 2023).

3.3.2. *Synthesis of WO_3-based smart material*

Synthesis of WO_3-based smart materials for the photocatalytic degradation of MPs and NPs requires precise control over the material's composition, morphology and surface properties. Researchers can choose the most appropriate synthesis method based on the intended application to optimize the performance and efficiency of WO_3 materials in tackling the global challenge of plastic pollution (Chong et al. 2010). The various synthesis methods commonly employed for producing WO_3-based smart materials tailored for this purpose as a photocatalytic degradation of MP and NP pollutants are as follows:

1) Sol-gel method: this method follows the hydrolysis and condensation reaction pathway for the precursor solutions containing tungsten precursors, such as tungsten alkoxides or salts, followed by gelation and thermal treatment to form WO_3 nanoparticles or thin films. This method offers precise commandment over the composition, particle size and structure of the resulting WO_3 materials (Cordero-García et al. 2017).

2) Hydrothermal synthesis: hydrothermal synthesis involves the reaction of tungsten precursors with aqueous solutions under high temperature and pressure conditions in a sealed reactor. This method enables the formation of crystalline WO_3 nanostructures with controlled size, shape and surface properties through nucleation and growth processes (Wu et al. 2020).

3) Template-assisted synthesis: template-assisted synthesis involves using preformed templates, such as porous materials or surfactant micelles, as scaffolds for the deposition or growth of WO_3 nanoparticles or nanowires. This method allows for the fabrication of ordered nanostructures with tailored morphologies and dimensions (Kumar and Rao 2017).

4) Chemical vapor deposition (CVD): CVD involves the deposition of tungsten oxide precursors onto substrates at high temperatures under controlled gas atmospheres. By adjusting the deposition parameters, including concentration of the precursor, temperature and gas flow rates, WO_3 thin films or nanostructures can be synthesized with precise control over film thickness and crystallinity (Uheida et al. 2021; Ge et al. 2022).

5) Sputtering and physical vapor deposition (PVD): these techniques participate the physical vapor deposition of tungsten targets onto substrates in a vacuum environment. These methods enable the fabrication of WO_3 thin films with uniform thickness and high purity, suitable for photocatalytic applications (Kumar and Rao 2017).

6) Electrochemical deposition: electrochemical deposition involves the electrodeposition of tungsten ions from aqueous solutions onto conductive substrates under applied electric fields. This method offers control over the deposition rate, film thickness and crystallinity of WO_3 films, making it suitable for large-area coatings and device fabrication (Chong et al. 2010).

7) Combustion synthesis: combustion synthesis involves the rapid combustion of precursor mixtures containing tungsten and oxygen sources to form WO_3 nanoparticles or powders. This method offers simplicity, cost-effectiveness for wide-range production of WO_3 materials but may require careful control of reaction parameters to achieve desired properties (Ouyang et al. 2021).

3.3.3. *A few WO_3-based smart materials*

1) WO_3/ultrathin carbon nitride $g - C_3N_4$ (UCN): this is a 2D photocatalyst, prepared via thermal polycondensation with ammonium chloride, demonstrated enhanced visible light-responsive photodegradation when combined with WO_3, forming WO_3/UCN (WU) via precipitation. The optimal synthesis conditions included a 1:1 WO_3 to UCN mass ratio, 9-h stirring with sodium tungstate and 6 mL of concentrated HCl. WUopt exhibited a 4–15 nm thickness, facilitating electron transmission and reducing electron–hole pair recombination. It possessed a wrinkled, porous structure with increased active sites, broadening infrared spectra and shifting the light absorption edge to 524 nm with a 2.13 eV band gap. WUopt demonstrated a 2.1-fold enhancement in RhB photocatalytic degradation efficiency (97.2%) compared to UCN (53.4%) within 50 min, maintaining a high efficiency (91.8%) after five cycles. The photodegradation mechanism involved the generation of active free radicals (O_2, OH) under visible light irradiation, where O_2 and OH played a significant role, while h+ was negligible (Li et al. 2023).

2) WO_3/TNAs photoelectrodes: these can be effectively fabricated via electrodeposition on TNAs substrate, which demonstrated an improved photocatalytic activity. For these optimal conditions included a deposition potential approximately 3 V, around 10- to 12-min deposition time, Na_2WO_4 concentration ~ 0.01 mol/dm^3 and 550°C annealing temperature. This yielded a DMP photodegradation rate of 77% after 60 min of illumination. Even after 10 cycles of reuse, the photodegradation rate remained above 76%. The incorporation of WO_3 onto TNAs enhanced photocatalytic performance, suggesting potential for water treatment applications (Xin et al. 2018; Henkel et al. 2022; Kumari and Pulimi 2023).

3) WO_3/g–C_3N_4 composite: this, as reported, was synthesized, characterized and utilized for the PET degeneration and H_2 generation under visible light irradiation. The WO_3 and $g - C_3N_4$ together produce a Z-type heterojunction that improved the electron–hole charge pair separation, thus enhancing overall performance of the photocatalytic reaction. The 30% $WO_3/g - C_3N_4$ composite exhibited remarkable stability for the evolution rate of H_2. The mineralized products of the PET were

investigated using ^1H NMR spectroscopy along with the use of EPR spectroscopy that identified $\cdot O_2$ as the active radicals taking part in the degradation reaction. The composite exhibited potential as a stable photocatalyst showcasing high efficiency for H_2 generation and the degeneration of PET through the photocatalytic degradation reaction pathway under the visible light irradiation. This solidifies its suitability for environmental remediation and sustainable energy production and storage applications (Wang et al. 2023).

4) WO_3/TiO_2-N catalysts: these, synthesized via sol-gel method, exhibit improved photocatalytic performance for the degeneration of the diclofenac under the simulated sun light irradiation. W atoms disperse as WO_3 on TiO_2 surface, with N partially substituting oxygen in TiO_2 lattice, modifying its electronic structure and narrowing the bandgap. This modification induces a strong response to solar light and reduces charge carrier recombination by oxidizing Ti^{3+} states. The synergistic effect of WO_3/N enhances TiO_2 photocatalytic activity, achieving complete diclofenac mineralization using TWN0.18 catalyst. Compared to bare TiO_2 and non-doped WO_3/TiO_2, TWN0.18 accelerates TOC removal, highlighting its potential for environmental remediation. This study underscores the significance of tailored catalyst design for efficient solar-driven photocatalysis, offering insights into addressing emerging contaminants in water treatment (Cordero-García et al. 2017).

5) ZnO@WO_3 nanocomposite: a novel green synthesis approach utilizing *M. koenigii* plant extract and water yielded coupled and crystalline ZnO@WO_3 nanocomposite, as evidenced by the shifted peaks in PXRD pattern indicating lattice distortion. The ZnO@WO_3 nanocomposite exhibited favorable properties for the efficient removal of AO and BPA under solar light irradiation, a large surface area, including a negative zeta potential of −14.5 mV and also small band gap energy (2.3 eV). Optimal removal efficiencies were achieved at low pollutant concentrations (5 mg/L) with a 15 mg catalyst dosage. First-order kinetics governed pollutant degradation, indicating radical species' attack on less hindered sites. Green synthesis avoided toxic chemicals, and the nanocomposite demonstrated excellent stability and reusability over eight cycles, suggesting its potential for real-world applications, thus reducing environmental impact (Rani et al. 2024).

3.4. Applications and future scope

Certainly, the future scope for WO_3-based smart materials in the photocatalytic degradation of NPs and MPs holds considerable promise. Future research could focus on further enhancing the catalytic activity of WO_3-based photocatalysts through innovative nano-structuring, surface modification, and doping strategies. By tailoring the properties of WO_3 nanoparticles or thin films, researchers can achieve higher degradation efficiencies and selectivity towards specific types of plastic polymers. There is also a need to develop WO_3-based photocatalysts with selectivity towards different types of plastic polymers, including commonly used plastics as well as emerging biodegradable polymers. By optimizing the photocatalytic properties and surface chemistry of WO_3 materials, it may be possible to target specific chemical bonds or structural features characteristic of different plastic types. Scaling up the production and utilization of the WO_3-based photocatalysts for industrial-scale photocatalytic degeneration of NPs and MPs is an important future direction. Research efforts should focus on developing cost-effective synthesis methods, scalable reactor designs and integrated water treatment systems that can efficiently process large volumes of plastic-contaminated wastewater or marine environments. Moreover, designing advanced photocatalytic reactor systems tailored for the degradation of NPs and MPs is crucial for practical implementation.

Future research could explore novel reactor configurations, such as immobilized WO_3 photocatalyst membranes, flow-through photocatalytic reactors and photocatalytic microfluidic devices, to optimize mass transfer, light utilization and reaction kinetics. WO_3-based smart materials could be integrated into multifunctional environmental remediation strategies targeting plastic pollution hotspots, such as urban wastewater treatment plants, coastal areas, and marine ecosystems. Combined approaches involving photocatalytic degradation, filtration and sorption techniques may offer synergistic benefits for comprehensive plastic waste management.

Exploring strategies for the recovery and recycling of WO_3-based photocatalysts after the degradation of NPs and MPs is essential for sustainability. Future research could investigate methods for regenerating or reactivating spent WO_3 materials, as well as exploring the potential valorization of degraded plastic waste as feedstock for other industrial processes.

It is very important to assess the environmental impacts and potential risks that can be associated from utilizing WO_3-based photocatalysts for plastic degradation. Future studies should investigate the fate and toxicity of degradation byproducts, as well as the long-term effects on aquatic ecosystems, microbial communities and human health.

Overall, the future scope for WO_3-based smart materials in the photocatalytic degradation of NPs and MPs is vast, offering opportunities for technological innovation, environmental stewardship and sustainable development. Continued interdisciplinary research and collaboration will be crucial in unlocking the full capabilities of these materials to tackle the global challenge of plastic pollution.

3.5. References

Abdul Rasheed, P., Thomas, R.T., Backer, S.N. (2023). Photocatalytic degradation and bacterial disinfection applications of graphitic carbon nitride. In *Synthesis, Characterization, and Applications of Graphitic Carbon Nitride*, Thomas, S., Anas, S., Joy, J. (eds). Elsevier, Amsterdam.

Adhikari, S., Sarath Chandra, K., Kim, D.H., Madras, G., Sarkar, D. (2018). Understanding the morphological effects of WO_3 photocatalysts for the degradation of organic pollutants. *Advanced Powder Technology*, 29, 1591–1600. doi: 10.1016/j.apt.2018.03.024.

Andrady, A.L. (2017). The plastic in microplastics: A review. *Marine Pollution Bulletin*, 119, 12–22. doi: 10.1016/j.marpolbul.2017.01.082.

Barrows, A.P.W., Cathey, S.E., Petersen, C.W. (2018). Marine environment microfiber contamination: Global patterns and the diversity of microparticle origins. *Environmental Pollution*, 237, 275–284. doi: 10.1016/j.envpol.2018.02.062.

Cao, Y., Lin, H., Zhang, K., Xu, S., Yan, M., Leung, K.M.Y., Lam, P.K.S. (2022). Microplastics: A major source of phthalate esters in aquatic environments. *Journal of Hazardous Materials*, 432, 128731. doi: 10.1016/j.jhazmat.2022.128731.

Carr, S.A. (2017). Sources and dispersive modes of micro-fibers in the environment. *Integrated Environmental Assessment and Management*, 13, 466–469. doi: 10.1002/ieam.1916.

Chamas, A., Moon, H., Zheng, J., Qiu, Y., Tabassum, T., Jang, J.H., Abu-Omar, M.M. (2020). Degradation rates of plastics in the environment. *ACS Sustainable Chemistry and Engineering*, 8, 3494–3511. doi: 10.1021/acssuschemeng.9b06540.

Chen, J., He, Z., Ji, Y., Li, G., An, T., Choi, W. (2019). OH radicals determined photocatalytic degradation mechanisms of gaseous styrene in TiO_2 system under 254 nm versus 185 nm irradiation: Combined experimental and theoretical studies. *Applied Catalysis B*, 257, 117912. doi: 10.1016/j.apcatb.2019.117912.

Chen, D., Cheng, Y., Zhou, N., Chen, P., Wang, Y., Li, K., Huo, S., Cheng, P., Peng, P., Zhang, R. et al. (2020). Photocatalytic degradation of organic pollutants using TiO_2-based photocatalysts: A review. *Journal of Cleaner Production*, 268, 121725. doi: 10.1016/j.jclepro.2020.121725.

Cho, S. and Choi, W. (2001). Solid-phase photocatalytic degradation of PVC-TiO_2 polymer composites. *Journal of Photochemistry and Photobiology A*, 143, 221–228. doi: 10.1016/S1010-6030(01)00499-3.

Chong, M.N., Jin, B., Chow, C.W.K., Saint, C. (2010). Recent developments in photocatalytic water treatment technology: A review. *Water Research*, 44, 2997–3027. doi: 10.1016/j.watres.2010.02.039.

Cordero-García, A., Turnes Palomino, G., Hinojosa-Reyes, L., Guzmán-Mar, J.L., Maya-Teviño, L., Hernández-Ramírez, A. (2017). Photocatalytic behavior of WO_3/TiO_2-N for diclofenac degradation using simulated solar radiation as an activation source. *Environmental Science and Pollution Research International*, 24, 4613–4624. doi: 10.1007/s11356-016-8157-0.

Crawford, R.J. and Martin, P.J. (2020). *Plastics Engineering*, 4th edition. Elsevier, Amsterdam. doi: 10.1016/C2015-0-00550-0.

Danso, D., Chow, J., Streit, W.R. (2019). Plastic waste and recycling: Environmental impact, sociocultural issues, prevention, and solutions. *Journal of Cleaner Production*, 214, 787–797. doi: 10.1016/j.jclepro.2018.12.219.

Deng, H., Wei, R., Luo, W., Hu, L., Li, B., Di, Y., Shi, H. (2020). Microplastic pollution in water and sediment in a textile industrial area. *Environmental Pollution*, 258, 113658. doi: 10.1016/j.envpol.2019.113658.

Devi, M., Upadhyay, S., Mir, R.A., Kumar, N., Sharma, S. (2023). Synthetic waste derived graphitic carbon nitride (g-CN) and g-CN/carbon hybrid for supercapacitors. *Journal of Energy Storage*, 73, 1–6. doi: 10.1016/j.est.2023.109067.

Dong, Q., Latiff, N.M., Mazánek, V., Rosli, N.F., Chia, H.L., Sofer, Z., Pumera, M. (2018) Triazine- and heptazine-based carbon nitrides: Toxicity. *ACS Applied Nano Materials*, 1, 4442–4449. doi: 10.1021/acsanm.8b00708.

Fadli, M.H., Ibadurrohman, M., Slamet, S. (2021). Microplastic pollutant degradation in water using modified TiO photocatalyst under UV-irradiation. *IOP Conference Series: Materials Science and Engineering*, 1011, 012055. doi: 10.1088/1757-899X/1011/1/012055.

Fang, X., Lu, Y., Chen, X., Cheng, H., Qiu, H., Zheng, Y., Zhu, J. (2022). Carbon nitride nanosheet-based photochromic physical unclonable functions for anticounterfeiting applications. *ACS Applied Nano Materials*, 5, 14722–14732. doi: 10.1021/acsanm.2c03055.

Ge, J., Zhang, Z., Ouyang, Z., Shang, M., Liu, P., Li, H., Guo, X. (2022). Photocatalytic degradation of (micro)plastics using TiO_2-based and other catalysts: Properties, influencing factor, and mechanism. *Environmental Research*, 209, 112729. doi: 10.1016/j.envres.2022.112729.

Geyer, R., Jambeck, J.R., Law, K.L. (2017). Production, use, and fate of all plastics ever made. *Science Advances*, 3, e1700782. doi: 10.1126/sciadv.1700782.

Henkel, C., Hüffer, T., Hofmann, T. (2022). Polyvinyl chloride microplastics leach phthalates into the aquatic environment over decades. *Environmental Science and Technology*, 56, 14507–14516. doi: 10.1021/acs.est.2c05108.

Ikuta, T., Tame, A., Takahashi, T., Nomaki, H., Nakajima, R. (2022). Microplastic particles are phagocytosed in gill cells of deep-sea and coastal mussels. *Frontiers in Marine Science*, 9, 1034950. doi: 10.3389/fmars.2022.1034950.

Imhof, H.K., Ivleva, N.P., Schmid, J., Niessner, R., Laforsch, C. (2013). Contamination of beach sediments of a subalpine lake with microplastic particles. *Current Biology*, 23, R867–R868. doi: 10.1016/j.cub.2013.09.001.

Kauts, S., Mishra, Y., Yousuf, S., Bhardwaj, R., Singh, S.K., Alshabrmi, F.M., Abdurahman, M., Vamanu, E., Singh, M.P. (2023). Toxicological profile of polyethylene terephthalate (PET) microplastic in ingested *Drosophila melanogaster* (Oregon R+) and its adverse effect on behavior and development. *Toxics*, 11(9), 782. doi: 10.3390/toxics11090782.

Keith, L.H. and Telliard, W.A. (1979). Priority pollutants. I. A perspective view. *Environmental Science and Technology*, 13, 416–423. doi: 10.1021/es60152a601.

Kim, H., Yoo, H.Y., Hong, S., Lee, S., Lee, S., Park, B.S., Park, H., Lee, C., Lee, J. (2015). Effects of inorganic oxidants on kinetics and mechanisms of WO_3-mediated photocatalytic degradation. *Applied Catalysis B*, 162, 515–523. doi: 10.1016/j.apcatb.2014.07.019.

Koe, W.S., Lee, J.W., Chong, W.C., Pang, Y.L., Sim, L.C. (2020). An overview of photocatalytic degradation: Photocatalysts, mechanisms, and development of photocatalytic membrane. *Environmental Science and Pollution Research International*, 27, 2522–2565. doi: 10.1007/s11356-019-07193-5.

Kumar, S.G. and Rao, K.S.R.K. (2017). Comparison of modification strategies towards enhanced charge carrier separation and photocatalytic degradation activity of metal oxide semiconductors (TiO_2, WO_3 and ZnO). *Applied Surface Science*, 391, 124–148. doi: 10.1016/j.apsusc.2016.07.081.

Kumaravel, S., Durai, M., Sepúlveda, R., Chicardi, E., Kumaravel, S., Kim, M.J., Balu, K., Hasan, I., Srinivasan, K., Ahn, Y.H. (2023). Fabrication of Ag/WO_3/g-C_3N_4 composites for the photocatalytic degradation of harmful dyes. *Optical Materials (Amst)*, 144, 2–7. doi: 10.1016/j.optmat.2023.114322.

Kumari, M. and Pulimi, M. (2023). Phthalate esters: Occurrence, toxicity, bioremediation, and advanced oxidation processes. *Water Science and Technology*, 87, 2090–2115. doi: 10.2166/wst.2023.119.

Lee, Q.Y. and Li, H. (2021). Photocatalytic degradation of plastic waste: A mini review. *Micromachines*, 12. doi: 10.3390/mi12080907.

Li, J., Chen, J., Ji, Y., Wang, J., Li, G., An, T. (2019). Solar light induced transformation mechanism of allyl alcohol to monocarbonyl and dicarbonyl compounds on different TiO_2: A combined experimental and theoretical investigation. *Chemosphere*, 232, 287–295. doi: 10.1016/j.chemosphere.2019.05.219.

Li, D., Zhang, W., Huang, Y., Feng, H., Wang, Z., Yang, Z., Chen, J., Zhang, X., Zhang, G., Chen, Y. (2023). Visible light-induced catalytic performance of composite photocatalyst synthesized with nanomaterials WO_3 and two-dimensional ultrathin g-C_3N_4. *Water Science and Technology*, 88, 1910–1925. doi: 10.2166/wst.2023.313.

Liu, H., Wang, C., Wang, G. (2020). Photocatalytic advanced oxidation processes for water treatment: Recent advances and perspective. *Chemistry, an Asian Journal*, 15, 3239–3253. doi: 10.1002/asia.202000895.

Luo, H., Xiang, Y., Tian, T., Pan, X. (2021). An AFM-IR study on surface properties of nano-TiO_2 coated polyethylene (PE) thin film as influenced by photocatalytic aging process. *Science of the Total Environment*, 757, 143900. doi: 10.1016/j.scitotenv.2020.143900.

Mateos-Aparicio, P. and Rodríguez-Moreno, A. (2019). The impact of studying brain plasticity. *Frontiers in Cellular Neuroscience*, 13, 66. doi: 10.3389/fncel.2019.00066.

Mehran, M.T., Raza Naqvi, S., Ali Haider, M., Saeed, M., Shahbaz, M., Al-Ansari, T. (2021). Global plastic waste management strategies (technical and behavioral) during and after COVID-19 pandemic for cleaner global urban life. *Energy Sources, Part A*, 1–10. doi: 10.1080/15567036.2020.1869869.

Naeimi, A., Sharifi, A., Montazerghaem, L., Abhari, A.R., Mahmoodi, Z., Bakr, Z.H., Soldatov, A.V., Ali, G.A.M. (2022). Transition metals doped WO_3 photocatalyst towards high efficiency decolourization of azo dye. *Journal of Molecular Structure*, 1250, 2–7. doi: 10.1016/j.molstruc.2021.131800.

Napper, I.E. and Thompson, R.C. (2020). Plastic debris in the marine environment: History and future challenges. *Global Challenges*, 4, 1900081. doi: 10.1002/gch2.201900081.

Ouyang, Z., Yang, Y., Zhang, C., Zhu, S., Qin, L., Wang, W., He, D., Zhou, Y., Luo, H., Qin, F. (2021). Recent advances in photocatalytic degradation of plastics and plastic-derived chemicals. *Journal of Materials Chemistry A*, 9, 13402–13441. doi: 10.1039/D0TA12465F.

Patil, P.B., Maity, S., Sarkar, A. (2022). Potential human health risk assessment of microplastic exposure: Current scenario and future perspectives. *Environmental Monitoring and Assessment*, 194, 898. doi: 10.1007/s10661-022-10539-1.

Pirhashemi, M., Habibi-Yangjeh, A., Rahim Pouran, S.R. (2018). Review on the criteria anticipated for the fabrication of highly efficient ZnO-based visible-light-driven photocatalysts. *Journal of Industrial and Engineering Chemistry*, 62, 1–25. doi: 10.1016/j.jiec.2018.01.012.

Porta, R. (2021). Anthropocene, the plastic age and future perspectives. *FEBS Open Bio*, 11, 948–953. doi: 10.1002/2211-5463.13122.

Prapanchan, V.N., Kumar, E., Subramani, T., Sathya, U., Li, P. (2023). A global perspective on microplastic occurrence in sediments and water with a special focus on sources, analytical techniques, health risks, and remediation technologies. *Water*, 15(11). doi: 10.3390/w15111987.

Rajeshkumar, S. and Sudhakar, Y.N. (2019). Photocatalytic degradation of organic pollutants using WO_3-based nanomaterials: A review. *Environmental Chemistry Letters*, 17, 1633–1651. doi: 10.1007/s10311-019-00900-y.

Rani, M. and Shim, W.J. (2017). Tackling marine plastic pollution: A review of current developments and future considerations. *Environmental Engineering Research*, 22, 133–146. doi: 10.4491/eer.2017.022.

Rani, M., Keshu, S., Pandey, S., Rishabh, U., Sharma, S., Shanker, U. (2024). Sunlight assisted highly efficient photocatalytic remediation of organic pollutants by green biosynthesized ZnO@ WO_3 nanocomposite. *Journal of Photochemistry and Photobiology A*, 446, 115160. doi: 10.1016/j.jphotochem.2023.115160.

Sharath, R.A., Fang, F., Futter, J., Trompetter, W.J., Singh, G., Vinu, A., Kennedy, J. (2023). Nitrogen defect engineering in porous g-C_3N_4 via one-step thermal approach. *Emergent Materials*, 6, 1117–1125.

Suriyakumar, S., Ramaraj, S.K., Perumal, K. (2019). Tungsten trioxide based nanomaterials for photocatalytic applications: A review. *Materials Today: Proceedings*, 18, 5472–5479. doi: 10.1016/j.matpr.2019.08.119.

Thompson, R.C., Olsen, Y., Mitchell, R.P., Davis, A., Rowland, S.J., John, A.W., McGonigle, D., Russell, A.E. (2004). Lost at sea: Where is all the plastic? *Science*, 304(5672), 838–838. doi: 10.1126/science.1094559.

Toops, T.J. (2020). Industrial scalable additives for enhanced decomposition of plastic waste through photocatalysis. *Academic Journal of Polymer Science*, 4, 1–20. doi: 10.19080/ajop.2020.04.555633.

Uheida, A., Mejía, H.G., Abdel-Rehim, M., Hamd, W., Dutta, J. (2021). Visible light photocatalytic degradation of polypropylene microplastics in a continuous water flow system. *Journal of Hazardous Materials*, 406, 124299. doi: 10.1016/j.jhazmat.2020.124299.

Van Cauwenberghe, L., Devriese, L., Galgani, F., Robbens, J., Janssen, C.R. (2015). Microplastics in sediments: A review of techniques, occurrence and effects. *Marine Environmental Research*, 111, 5–17. doi: 10.1016/j.marenvres.2015.06.007.

Venkataramana, C., Botsa, S.M., Shyamala, P., Muralikrishna, R. (2021). Photocatalytic degradation of polyethylene plastics by $NiAl_2O_4$ spinels-synthesis and characterization. *Chemosphere*, 265, 129021. doi: 10.1016/j.chemosphere.2020.129021.

Wang, Y., Cao, R., Xie, H., Chen, W., Jing, L. (2018). Recent progress in WO_3-based photocatalysts for organic pollutants degradation. *Journal of Materials Science: Materials in Electronics*, 29, 4427–4443. doi: 10.1007/s10854-017-8354-1.

Wang, X., Zhu, Z., Jiang, J., Li, R., Xiong, J. (2023). Preparation of heterojunction C_3N_4/WO_3 photocatalyst for degradation of microplastics in water. *Chemosphere*, 337, 139206. doi: 10.1016/j.chemosphere.2023.139206.

Wang, X., Dai, Y., Li, Y., Yin, L. (2024). Application of advanced oxidation processes for the removal of micro/nanoplastics from water: A review. *Chemosphere*, 346, 140636. doi: 10.1016/j.chemosphere.2023.140636.

Wenchao, W., Zhang, D., Sophocleous, M., Qu, Y., Jing, W., Chalermwisutkul, S., Russel, M. (2023). Measuring the effects of diethyl phthalate microplastics on marine algae growth using dielectric spectroscopy. *Science of the Total Environment*, 865, 161221. doi: 10.1016/j.scitotenv.2022.161221.

Wright, R.J. and McManus, G.B. (2019). Microplastic pollution in benthic environments: Implications for organisms. *Marine Pollution Bulletin*, 143, 92–98. doi: 10.1016/j.marpolbul.2019.03.05.

Wu, C.M., Naseem, S., Chou, M.H., Wang, J.H., Jian, Y.Q. (2019). Recent advances in tungsten-oxide-based materials and their applications. *Frontiers in Materials*, 6, 49. doi: 10.3389/fmats.2019.00049.

Wu, S., Hu, H., Lin, Y., Zhang, J., Hu, Y.H. (2020). Visible light photocatalytic degradation of tetracycline over TiO_2. *Chemical Engineering Journal*, 382, 122842. doi: 10.1016/j.cej.2019.122842.

Xin, Y., Wang, G., Liu, Y., Zang, Z., Gao, M. (2018). Effect of WO_3 on the morphology and photodegradation of dimethyl phthalate of TiO_2 nanotube array photoelectrode. *Environment Protection Engineering*, 44(3), 129–139. doi: 10.5277/epe180309.

Zhang, L. and Jaroniec, M. (2018). Toward designing semiconductor-semiconductor heterojunctions for photocatalytic applications. *Applied Surface Science*, 430, 2–17. doi: 10.1016/j.apsusc.2017.07.192.

Zhao, G.Q., Zou, J., Hu, J., Long, X., Jiao, F.P. (2021). A critical review on graphitic carbon nitride (g-C_3N_4)-based composites for environmental remediation. *Separation and Purification Technology*, 279, 1–7. doi: 10.1016/j.seppur.2021.119769.

Zhou, M., Tian, X., Yu, H., Wang, Z., Ren, C., Zhou, L., Lin, Y.W., Dou, L. (2021). WO_3/Ag_2CO_3 mixed photocatalyst with enhanced photocatalytic activity for organic dye degradation. *ACS Omega*, 6, 26439–26453. doi: 10.1021/acsomega.1c03694.

Zhu, B., Zhang, L., Shen, T., Wang, P. (2020). The catalytic properties and applications of WO_3-based materials in photocatalysis. *Materials Chemistry and Physics*, 250, 123028. doi: 10.1016/j.matchemphys.2020.123028.

4

The Chemistry of Carbon Nanotubes in Photocatalytic Degradation of Micro- and Nano-Plastic

The proliferation of micro- and nano-plastics in the surrounding shows a substantial hazard to environmental stability and human wellness. Conventional methods for degrading plastic are in effective and harmful to environment. Thus, the emergence of advanced resources for efficient plastic degradation is imperative. Carbon nanotubes (CNTs) have emerged as promising candidates because of their distinctive structural and electronic properties. This paper presents a comprehensive review of the recent advancement in utilizing CNTs as smart materials for photocatalytic degradation of micro- and nano-plastics. The synergetic effects between CNTs and photocatalytic material enhance the degradation efficiency by promoting adsorption, increasing surface area and facilitating charge separation. Various strategies for incorporating CNTs into photocatalytic systems, including hybrid composite and functionalization techniques, are discussed. Furthermore, the mechanisms underlying the photocatalytic degradation processes are elucidated, highlighting the role of CNTs in enhancing photocatalytic activity and selectivity. Challenges and future perspectives for the practical application of CNT based photocatalysts in plastic degradation are also addressed.

4.1. Introduction

CNTs stand out as remarkable materials in the realm of nanotechnology due to their unique properties. When it comes to addressing the issue of micro- and nano-plastic through photocatalytic degradation, CNTs exhibit promising characteristics. Their high surface area, remarkable mechanical strength and excellent electrical conductivity (including photocatalysis) make them perfect for various applications (Eder 2010). In the context of plastic degradation, CNTs can be utilized as catalysts

Chapter written by Manish KUMAR and Sunil KUMAR.

in the photocatalytic process. When combining suitable photocatalysts, such as titanium oxide, these nanostructures can significantly enhance the degradation of micro- and nano-plastics under light irradiation (Ouda et al. 2023). This process involves the release of reactive oxygen species (ROS) upon exposure to light, which aids in breaking down the plastic particles into smaller, less harmful components. Degradation can be done by different methods such as biological degradation, mechanical degradation, chemical degradation and photodegradation, which is shown in Figure 4.1.

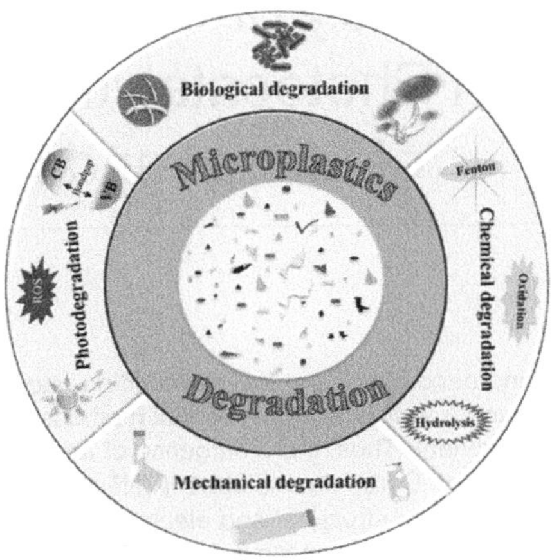

Figure 4.1. *Degradation of micro-plastic by different ways. For a color version of this figure, see www.iste.co.uk/singh/photodegradation.zip*

CNTs' ability to efficiently absorb light across a wide spectrum, including visible and UV light, makes them advantageous in promoting photocatalytic reactions for plastic degradation. Their structural versatility allows form functionalization and modification, further enhancing their catalytic performance and selectivity in breaking down plastic pollutants (Mennani et al. 2024).

This amalgamation of CNTs with photocatalytic materials showcase their potential as a smart material combating micro- and nano-plastic pollution, offering a promising avenue for mitigating the environmental impact of plastic waste.

Exposure to sun light, air, moisture and heat causes plastics to breakdown into micro-plastic in the environment. They contain harmful compounds like bisphenol which can seriosly affect the health of humans (Campanale et al. 2020). The prevalence of micro-plastic in wastewater is indeed widely documented, source from surface runoffs and personal care items. Fragments and fibers constitute the majority of their composition, which is noted in various studies. The array of methods developed for micro-plastic removal is impressive, with techniques such as adsorption, magnetic extraction, bio filtration, membrane filtration, chemical coagulation and photocatalysis being notable among them. Photocatalysis particularly holds promise in wastewater treatment due to its sustainable nature and minimal secondary pollution. Essentially, this method utilizes light energy and initiates a chemical reaction within a photocatalyst, creating reactive species that breakdown contaminants into harmless by-products (Bagheri et al. 2017; Khan 2021). It is a fascinating process with significant potential for addressing micro-plastic and nano-plastic pollution in wastewater. Nano-photocatalysts are designed to accelerate the breakdown of plastic by light energy. Through oxidation reaction, they facilitate the transformation of plastic into smaller, more environmentally friendly molecules such as carboxylic acid, ketones and aldehydes, which are either biodegradable or less harmful to the environment, compared to their original plastic forms (Radhika et al. 2019). Nano-photocatalysts possess high photocatalytic activity, leading to enhanced efficiency in breakdown plastics. Their advantage lies in the increased surface area relative to volume due to their very small size. This huge area of surface allows for more interactions between the catalyst and the plastic, boosting the degradation rate significantly. The reactive oxidizing species are generated in significant amounts based on the pH of the solution through a photocatalyst (Xie et al. 2023). It increases the electrical charge at the catalyst surface which affects the formation of electron–hole pairs critical for the catalytic process. Additionally, variation in pH modifies the surface charge of the photocatalyst particles, thereby modifying catalyst potential. This pH induces change, which affects the uptake of organic contaminants on the catalyst's surface and impacts the overall reaction rate (Xie et al. 2023).

4.2. Micro- and nano-plastic

The presence of micro- and nano-plastics (Figure 4.2) in the environment has indeed become a considerable concern. Their small size and widespread distribution pose potential risks to various ecosystems, wildlife and even human health. Their impact is actively being studied and ways being looked for to migrate their effects and prevent further pollution. It is fascinating and concerning how micro- and nano-plastics have become more prevalent in the environment over the past few decades. Understandings their origins and the processes involved in their degradation are crucial in addressing the complex global issue. This increased presence raises an important question about their impact on ecosystems and the environmental at large. Micro-plastic and nano-plastic are tiny plastic particles that measure less than 5 mm and are even smaller than that, respectively. They come from the breakdown of large plastic debris, as well as various products such as cosmetics, clothing fibers and industrial processes (Boyle and Örmeci 2020; Enfrin et al. 2021). These particles are a growing concern as they pervade the environment, potentially posing risks to ecosystems, wildlife and even human health due to their ability to accumulate and transport toxic substance. The impact of plastic pollution on our environment is alarming. Finding effective interventions and sustainable solutions is crucial to mitigating the issue and protecting our ecosystems. Awareness, recycling and innovative approaches to reduce plastic waste are key steps in addressing this growing problem. The current methods of handling plastic present their own set of challenges and environmental drawbacks. Landfill disposal, incineration and mechanical reprocessing have their limitations and can contribute to pollution or resource wastage (Lange 2021). Developing more sustainable, long-term solutions that minimize environmental impact is crucial to tackle this issue effectively. Innovation in recycling technologies and material design could help create a more circular and ecofriendly approach to managing plastic waste. The band gap of photocatalysts plays a pivotal role in photocatalytic processes (Guo et al. 2018).

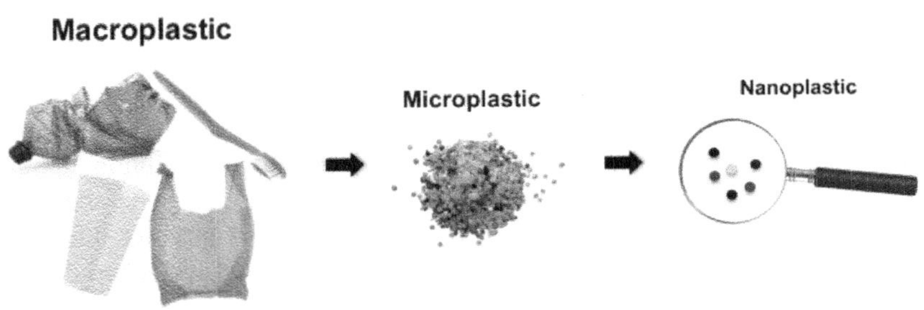

Figure 4.2. *Degradation of plastic into micro- and nano-plastic. For a color version of this figure, see www.iste.co.uk/singh/photodegradation.zip*

A narrow band gap tends to facilitate easier electron transfer, which can indeed enhance the efficiency of plastic degradation. This broader range of materials available for photocatalysts with narrower band gaps opens up more possibilities for efficient degradation methods.

Nano-plastics differ notably from micro-plastic in their behavior within environmental matrices. When present in natural waters, nano-plastic tends to acquire a coating of molecules, forming an eco-corona. These particles interact with dissolved organic molecules, ions and colloids like clay particles or organic matter, leading to potential aggregation through a process called hetroaggregation. The chemical nature of this corona influences particle stability and aggregation tendencies. For instance, organic matter coating colloids can create a negative charge, which prevents aggregation. In contrast, certain apolar organic structures or specific ions may induce aggregation by acting as bridges between particles. This intricate interplay of factors shapes the behavior and fate of nano-plastics in environmental settings (Boles et al. 2016).

Nano-plastic toxicity on freshwater crustaceans like *Macrobrachium nipponese* is crucial amidst rising concerns. Assessing survival rates, antioxidant activity, immune enzyme response and gene expression due to various nano-plastic concentrations in water can provide valuable insight into their impact on these organisms (Patra et al. 2022).

4.3. Carbon nanotube materials

CNTs have been examined for various scientific applications because of their distinctive characteristics but their use as photodegrading material is less common. However, their great surface area and excellent electrical conductivity could potentially aid in photocatalytic reactions for degrading certain pollutants or organic compounds under specific conditions, when combined with appropriate catalyst additives.

Incorporating oxygen functional groups onto the surface of CNTs not only influences their adsorption capacity but also enhances their hydrophilicity, making them more suitable for adsorption processes (Deline et al. 2020). The diverse nature of CNT allows them to interact with organic contaminants through mechanisms such as surface complexion, electrostatic interaction, exchange of ions, physical adsorption and precipitation. For organic pollutants, pi-pi interactions and electrostatic forces play significant roles alongside physical adsorption (Dubey et al. 2021). The versatility of CNTs makes them favorable adsorbents for different environmental uses. The scalability of CNT production through the chemical vapor deposition (CVD) process has contributed to their widespread use. Additionally, their ability to function as efficient electron sinks, coupled with high electrical conductivity and substantial electron storage capacity during photoexcitation, makes them promising for various electronic and energy storage capacity. The environmental application of CNTs presents promising and cost-effective solutions. Their superior adsorption capability and high efficiency properties not only offer structural and electrical advantages but also pave the way for widespread use across numerous industries globally (Zhang et al. 2022). Their versatility and effectiveness of CNTs in adsorption make them a valuable asset for tackling various environmental challenges. Analyzing the catalytic behavior under the light of carbon nanoparticle synthesized via solvent assisted hydrothermal carbonization in the breakdown of methylene blue with sunlight exposure must have involved a comprehensive characterization process. Techniques such as transmission electron microcopy (TEM) and scanning electron microscopy (SEM), ultraviolet visible spectroscopy (UV-VIS), FTIR stands for Fourier transform infrared spectroscopy (FTIR), fluorescence and X-ray photoelectron spectroscopy (XPS) offer valuable insight into the structural and chemical properties of the carbon nanoparticles (Kolhe et al. 2023). The impact of CNT loading on zinc oxide photocatalyst, revealing enhanced photocatalytic activity, attributed to reduced aggregation and blue shifted band gap energy. The modified sol gel synthesis method excluded organic additives and annealing treatments, promoting zinc oxide attachment to CNT surfaces. The zinc oxide/CNT photocatalyst exhibited superior performance in degrading methylene blue under sunlight with optimal results achieved at 10 wt.% CNTs loading (Phin et al. 2020). Addressing issues such as recombination and photo corrosion, this approach presents a promising solution for water pollution concerns.

4.4. Coating of carbon nanotube as photocatalytic degradation materials

CNTs are ideal as thermal coatings and functional fillers. They are among the best materials known for thermal conductivity. CNTs are one-dimensional nanomaterials with a large surface area and are recognized as some of the darkest materials in the world, with notable radiation coefficient is close to 1. Nano-fibrous CNTs exhibit superior thermal conductivity, forming effective networks in comparison to other particulate thermal radiation fillers. This characteristic contributes to a noticeable toughening effect in coatings, enhancing their overall performance. The coating of CNT as photocatalytic degrading material can significantly enhance their photocatalytic performance. The coated CNTs provide a larger surface area and improved stability, promoting efficient adsorption of pollutants and facilitating photocatalytic reactions (Mohamed et al. 2016). This enhanced photocatalytic capability makes them effective in degrading various organic pollutants under light irradiation. Coating CNTs onto surfaces can enhances their photocatalytic properties when coating occurs on CNT enhances surface area to volume ratio, providing more active sites for interaction with micro-plastic and nano-plastic (Li et al. 2012; Liu et al. 2005). Coated CNT has excellent light absorption properties across wide spectrum, including UV and visible light, which can facilitate the generation of ROS upon illumination. Upon exposure to light, CNT can generate ROS (such as hydroxyl radicals and singlet oxygen), which are highly reactive and can oxidize organic compounds present in micro-plastics and nano-plastics, leading to their degradation (Bacha et al. 2021). Coating CNTs onto their photocatalytic materials can create synergic effects, further enhancing the photodegrading efficiency by improving charge separation and reducing electron hole recombination.

4.4.1. *TiO_2 coating*

The expansion of industries and modern technology has led to increased chemical utilization for plastic-based products, resulting in significant environmental impact global. The utilization of titanium oxide for photocatalytic degradation of macro- and micro-plastic is a burgeoning field of research, offering potential solutions to migrate plastic pollution (Gaur et al. 2023). This research aims to provide a comprehensive overview of TiO_2 application in plastic degradation, including synthesis methods, modification techniques, degradation efficiency, system limitations and future research directions. The use of TiO_2-based photocatalysts has shown promise in disregarding micro-plastics and recent studies suggest that this approach could also be effective for nano-plastic degradation.

However, research in this area is still limited, and there are challenges such as determining optimal parameters and tracking intermediate compounds during degradation. Given the significance of this issue the present study proposed through the use of three different TiO_2 photocatalyst structures, this paper seek to determine the impact of photocatalyst morphology on the degradation rates of nano-plastics in aqueous media, with a specific focus on nano-sized polystyrene. CNT films can be prepared by using films on a glass substrate using TiO_2 powder and CNT composites via the doctor blade technique that enhances the photocatalytic degradation of CNT (Zeinali et al. 2019). The thickness of the film ranged from 3.75 to 5.00 μm.

Depending on the type of TiO_2 used, the composite films with varying amounts of CNT exhibited different spectroscopic and morphological characteristics. Surface roughness was assessed via atomic force microscopy (Oke and Jen 2022). Photocatalytic activity was evaluated by testing the films' ability to degrade methylene blue under near UV visible irradiation. TiO_2 based photocatalysis, particularly for its surface activity and photocatalytic efficiency under UV illumination, which enhances its effectiveness in water treatment applications. Additionally, TiO2 is stable, non toxic, and relatively inexpensive, making it an ideal candidate for large scale environmental remediation efforts (Ribao et al. 2019). Recent research, such as that by Ariza-Tarazona et al. (2019), indicates that the photocatalytic process holds promise for regarding micro-plastic. Ariza-Tarazona et al. demonstrated the degradation of HDPE MPs from commercial facial scrub using N-TiO_2-based photocatalysts under visible light in both solid and aqueous environments. Their finding suggests the efficiency of the photocatalysts: MPs interaction depends on factors such as the developed semiconductor materials surface area and its photocatalytic activity regarding micro- and nano-plastics. The fabrication and characterization of 3D urchin-like black TiO_2-x\CNT hetrostructures, highlights their unique features such as hetrojunction structure and Ti^{3+} doping into anatase TiO_2 lattice (Zhang et al. 2017). These structures demonstrate enhanced visible-light photocatalytic activity and electrochemical properties due to abundant active sites, improved light scattering capability, and narrowed band gap, which is shown in Figure 4.3. The remarkable photocatalytic degradation rate of 99.6% for methylene orange and a hydrogen production rate of 242.9 micromolh^{-1} g^{-1} are achieved due to enhanced properties imparted by Ti^{3+} doping and specific structure features (Shandilya et al. 2021).

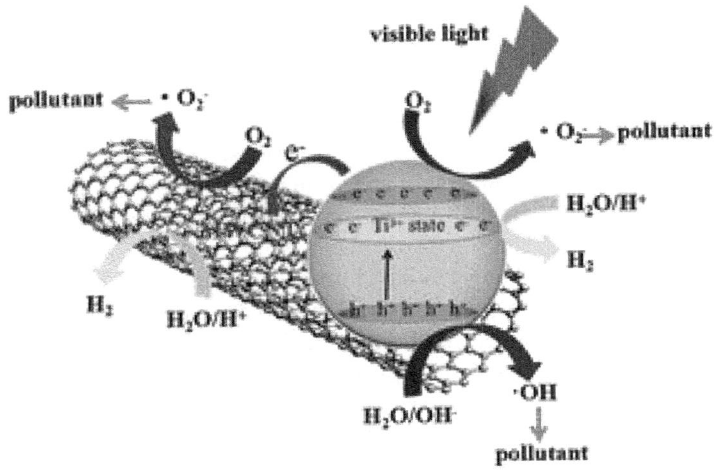

Figure 4.3. *Doping of TiO_2 on the surface of carbon nanotube. For a color version of this figure, see www.iste.co.uk/singh/photodegradation.zip*

4.4.2. *ZnO coating*

The resistance of micro-plastic and nano-plastic to decomposition under natural conditions underscores the severity of their environmental impact, leading to their classification as non-biodegradable plastics. The recognition by the UN environment program highlights the significance of addressing micro-plastic pollution as a top environmental concern. While our understanding of nano-plastics may not be as comprehensive due to methodological limitation, their ubiquitous presence and potential similarities in transport and fate pathways to MPs suggest they pose similar environmental risks. Photocatalytic and microbial degradation technologies show promise for transforming micro-plastic into harmless compounds. The ZnO nanostructure is using this to enhance photocatalytic degradation of organic pollutants, offering a potential solution to address environmental pollution caused by various industries (Amiri et al. 2022). The ability to manipulate the size and morphology of ZnO nanostructure contributes to the effectiveness' of ZnO-based photocatalysts. These photocatalysts demonstrate satisfactory performance in a solution with a pH range of 6.5–8. However, future advancements in mobilized ZnO-based photocatalysts should concentrate on addressing several key issues (Mumtaz et al. 2024). These include the use of flexible supporting substrate, increasing the specific area of both photocatalysts and supportive substrate, enhancing the lifespan of photo-generated charge carriers, ensuring stability, and optimizing applicability to industrial-scale operations. ZnO coating can increase the surface area of CNT, providing more effective sites for chemical reaction and potentially enhancing the degradation rates. ZnO is known for its photocatalytic properties, which can accelerate degradation reaction under UV or visible light irradiation; the ZnO coating may facilitate the degradation of organic contaminants adsorbed onto the CNT surface. ZnO coating may form chemical bonds, or interactions with the CNT are altering their chemical reactivity and degradation kinetics (Sanakousar et al. 2022). These interactions could either enhance or inhibit the degradation process, depending upon the specific condition. Alternatively, ZnO coating may act as a protective layer, shielding the underlying CNTs from degradation by environmental factors such as oxidation and UV radiation. The coating process typically involves chemical reactions between CNT and ZnO precursor solution (Azam et al. 2015). This reaction may release or absorb energy, depending on the specific reaction pathways and condition. The coating process often requires heating to promote the deposition of ZnO onto the CNTs. Energy is transferred from heating source to the reaction mixture, leading to temperature changes in the system.

4.5. Functionalized carbon nanotube as photocatalytic degradation materials

Functionalized CNTs have gained attention as versatile smart material for the photodegradation of micro- and nano-plastic due to having unique properties. Functionalized CNTs possess a high surface area that allows for efficient adsorption of MPs and NPs, enhancing their contact with a photodegrading agent and increasing degradation efficiency. Functionalized CNTs act as photocatalysts, facilitating the degradation of MPs and NPs under light irradiation (Agasti et al. 2022). Surface functionalization with specific groups can enhance their photocatalytic activity, promoting the generation of ROS for degradation. CNTs possess exceptional mechanical strength, which ensures their stability and durability during the photodegradation process. This enables continuous and long-term operation without significant degradation or loss of performance. Functionalized CNTs are generally biocompatible and environmentally friendly, minimizing potential adverse effects on ecosystems during the photodegradation process. Treatment with strong acids such as HNO_3 or H_2SO_4 can introduce oxygen containing functional groups (e.g. hydroxyl, carbonyl and carboxyl) onto the surface of CNTs (Yang et al. 2019). These functional groups will increase the force of attraction between CNTs and MPs\NPs, improving their photocatalytic degradation efficiency. Introduction of nitrogen atoms into the carbon lattice of CNTs through nitrogen doping can alter their electronic structure and enhance their photocatalytic activity. Nitrogen-doped CNTs exhibit improved charge separation and transfer properties, leading to enhanced degradation of MPs\NPs under light irradiation (Zou et al. 2019).

Functionalization with sensitizing agents such as organic dyes or quantum dots can enhance the absorption of light CNTs, thereby improving their photocatalytic activity. Sensitizing can extend the absorption range of CNTs into the visible region, where a significant portion of solar energy is available for photocatalysis (Younas et al. 2019). Overall, functionalized CNTs represent a promising avenue for the development of smart material for the photodegradation of micro- and nano-plastics, offering high efficiency, versatility and environmental sustainability.

4.5.1. *Single wall carbon nanotube*

Single-wall CNTs (SWCNTs) are tubular structures made from a single sheet of carbon atoms rolled into a seamless tube. They can be conceptualized as rolled up graphene sheets, where the diameter of the nanotube and its chirality determine the properties. SWCNTs demonstrate a unique combination of mechanical, strength, electrical and thermal conductivity, and distinctive optical properties, setting them apart as a versatile nanomaterial (Ali et al. 2023). They have high tensile strength, stiffness and flexibility, along with excellent electrical conductivity and thermal conductivity. Their unique electronic structure can range from metallic to semiconducting, depending on their chirality. SWCNTs have found applications in various fields such as electronics, energy storage, sensors, composite materials and biomedical devices due to their exceptional properties (Kharlamova and Kramberger 2021). They hold promise for advancing technology in areas such as nano-electronics, nano-composites and nano-medicine. However, challenges remain in large-scale production, purification and precise control over their properties for specification applications.

SWCNTs can be functionalized through various methods to enhance their photodegradation properties, particularly in environmental remediation applications. Chemical groups are covalently bonded to the surface of SWCNTs. These functional groups can include hydroxyl (–OH), carboxyl (–COOH), amino (–NH_2) or other reactive moieties (Khan and Alamry 2022) (Figure 4.4). Covalent functionalization can increase the dispersibility of SWCNTs, while molecules can be absorbed onto the surface SWCNTs through non-covalent interactions such as pi-pi stacking, Van der Waals forces or hydrogen bonding (Gulati et al. 2022). This method can improve the stability of SWCNT dispersions and allow for the attachment of photocatalytic or photosensitizing molecules. Combining SWCNTs with photocatalytic materials or photosensitizing molecules can create synergic effects, enhancing the overall photodegradation efficiency compared to individual components (Saleh 2013). Overall, functionalization SWCNTs offer great potential for improving photodegradation processes in environmental remediation, contributing to the removal of organic pollutants.

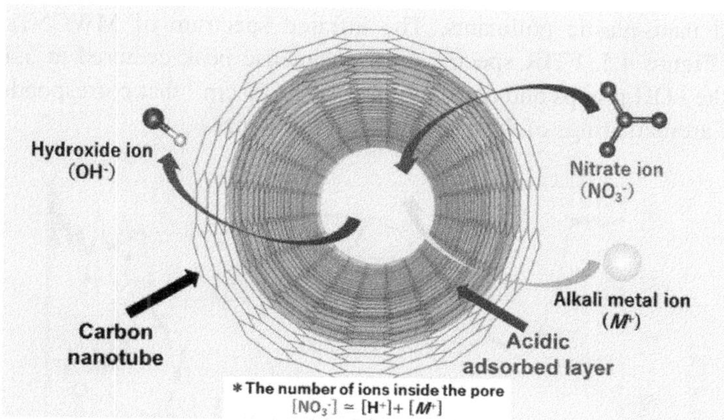

Figure 4.4. *Adsorption of different functional groups on the surface of CNT. For a color version of this figure, see www.iste.co.uk/singh/photodegradation.zip*

4.5.2. *Multiwall carbon nanotube*

A multiwall carbon nanotube (MWCNT), as opposed to SWCNTs, has a multi-layered structure with concentric cylinders of SWCNTs nested with one another. These result in large diameters, up to 30 nm, compared to typical SWCNTs which range from 0.7 to 2.0 nm. Both types of CNT exhibit unique properties that enable various applications and enhancements in existing technologies (Rathinavel et al. 2021). MWCNTs are typically prepared using techniques like chemical vapor deposition (CVD) or the arc discharge method. In CVD method, a substrate, usually made up of metal catalysts such as iron, nickel or cobalt, is heated at a high temperature (typically around 700–900°C) in the presence of a carbon containing precursor gas, such as methane ethylene. In the arc discharge method, a high electrical current is passed between two graphite electrodes in an inert atmosphere. The high temperature generated by the arc vaporizes the graphite, and carbon atoms condense to form CNT (Purohit et al. 2014). The resulting nanotubes are collected on the cathode and can be further purified and processed to obtain MWCNTs.

Functionalization of CNTs involves attaching functional groups or molecules to their surface, which can alter their properties and enhance their performance in various applications, including photocatalytic degradation of micro- and nano-plastic (Arora and Attri 2020). Functionalization modifies the electronic properties of CNTs, leading to improved charge separation and migration during photocatalytic reactions. This enhances the creation and extends the lifespan of ROS such as hydroxyl radicals (–OH) and superoxide radicals ($-O^{2-}$), which are responsible for degrading micro- and nano-plastic pollutants. The infrared spectrum of MWCNTs _Ag is present in Figure 4.5. FTIR spectra show an intense peak centered at 3,425 cm^{-1} related to the –OH groups and the sharp peak at 1,640 cm^{-1} that corresponds to C=C bonding of aromatic rings of the carbon Skelton structure.

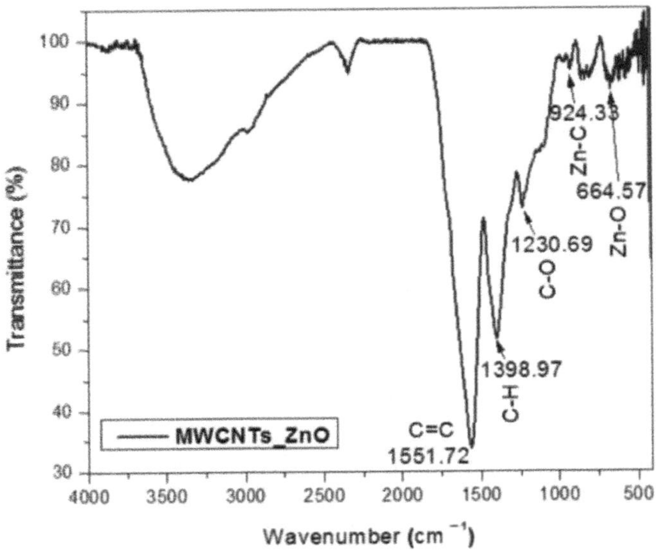

Figure 4.5. *FTIR spectra of MWCNTs_ZnO. For a color version of this figure, see www.iste.co.uk/singh/photodegradation.zip*

In MWCNTs, only the outer wall contributes to the conductivity of the composite structure. The inner walls of MWCNTs do not actively participate in conducting electricity. This is because the electronic properties of the inner walls are strongly influenced by the presence of the outer walls and the surrounding environments (Milowska et al. 2017), leading to a diminished contribution to overall conductivity. Three steps chemical modification can occur by oxidation, chlorination and grafting of polydiol, which aims to improve the dispersion of MWCNTs in the polymer matrices by enhancing the interaction between the filler and the polymer. This causes a more uniform distribution of MWCNTs throughout the polymer matrix, as revealed by SEM. Although nanoparticles enhance the shape fixity, they contribute weaker shape recovery, which can be attributed to the limited mobility of polymer chains caused by the presence of MWCNTs (Lofrano et al. 2016).

4.5.3. *Noncovalent endohedral and exohedral functionalization with surfactants*

Noncovalent exohedral functionalization of CNTs with surfactants can indeed enhance their properties as a smart material for the degradation of micro- and nano-plastic. Surfactants can attach to the surface of CNTs via non-covalent interactions such as Van der Waals forces, hydrogen bonding or pi-pi stacking (Dharmadhikari 2015). This functionalization enhances the disposability of CNTs in various solvents or polymers matrices, facilitating their incorporation into composite materials for plastic degradation application. Non-covalent endohedral functionalization is the encapsulation of small molecules, metal clustered or even other nanoparticles within the cavity of a CNT. This guest species can be stabilized inside the CNT structure through non-covalent interactions, offering protection from external environments and enabling unique properties or functionalities. Endo and exohedral functionalization is shown in Figure 4.6.

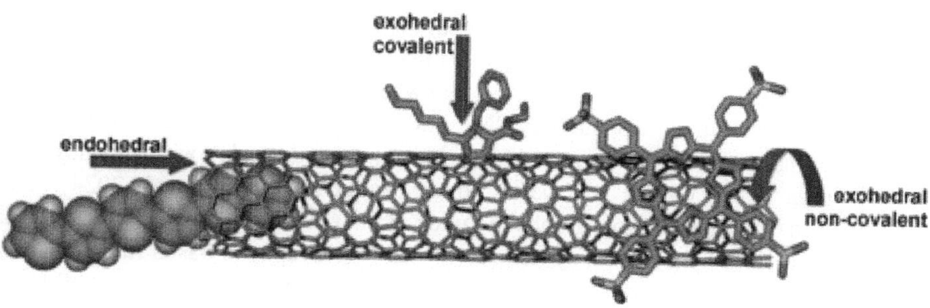

Figure 4.6. *Noncovalent endohedral and exohedral functionalization of CNT. For a color version of this figure, see www.iste.co.uk/singh/photodegradation.zip*

Certain molecules or catalysts encapsulated inside CNTs can act as an active site for catalytic degradation reactions, facilitating the breakdown of micro-plastic and nano-plastic into smaller, less harmful fragments. The hydrophobic functional groups on the CNT surface enhance the adsorption of hydrophobic plastic particles, bringing them closer to the active site, thereby facilitating the degradation reaction (Hassan et al. 2020). By combining endo- and exohedral functionalization strategies, XNT-based smart materials can be designed with enhanced capabilities for the degradation of micro- and nano-plastics. These functionalized CNTs can offer improved efficiency, selectivity and versatility in addressing the challenges posed by plastic pollution in the environment. Additionally, the use of CNTs as smart material for plastic degradation holds promise for developing sustainable and environmentally friendly solutions to mitigate the impact of plastic waste. Graphene broad absorption spectrum and efficient light harvesting capabilities can enhance the utilization of light energy for photocatalysis (Li et al. 2016). This ensures that the larger portion of the incident light is utilized for driving the degradation reaction of micro- and nano-plastic adsorbed on the surface of functionalized CNTS.

4.5.4. *Graphene-functionalized carbon nanotube*

Graphene-functionalized CNTs hold promise as smart material for degradation of micro- and nano-plastics due to their unique properties. By leveraging the high surface area and reactivity of graphene-functionalized CNTs, they can potentially adsorb and catalyze the degradation of plastic particles at the micro- and nano-scales (Goh et al. 2022). Additionally, their mechanical strength and conductivity could enable efficient collection and plastic detection. Graphene provides mechanical reinforcement of CNTs, making them more robust and resistant to structural damage during the degradation process. This ensures the longevity and effectiveness of the composite material for plastic degradation. The theoretically proposed graphene\SWCNT hybrid indeed holds great promise for extraordinary physical properties and various engineering application (Rao et al. 2018). However, the challenges arise in fabricating this hybrid as intended due to difficulties in precisely controlling the covalent C-C bonding between graphene and SWCNTs. The integration of nanomaterials, such as CNT, fullerene and graphene nano-sheets, into the recycling process shows promise for improving degradation efficiency and transforming waste plastic into value added products (Tewari et al. 2024). The combination of graphene or graphene oxide with CNT creates synergic effects, further enhancing the photocatalytic activity of CNT. Graphene based materials can serve as support structure for CNTs, improving their dispersion and accessibility to light and pollutants, thus amplifying the overall photocatalytic degradation. Exohedral functionalization with graphene can promote the generation of ROS such as hydroxyl irradiation. The ROS are highly reactive and play a crucial role in oxidizing micro- and nano-plastic, leading to their degradation. Initially, the micro- and nano-plastic particles are adsorbed on the surface of the functionalized CNTs through physical interaction such as the Van der Waals force (Chelombitko 2018). The large surface area and enhanced adsorption capacity of the graphene functionalized CNTs facilitate the efficient adsorption of plastic pollutants.

Under light irradiation, the functionalized CNTs adsorb photons and generate electron hole pairs. The graphene component enhances light absorption and promotes charge separation, leading to the formation of ROS. These ROS oxidize the adsorb micro- and nano-plastic particles, breaking down their chemical structure and initiating degradation. The functional group present on the surface of graphene, as well as defects and active sites on the CNTs, can participate in chemical reaction with the plastic pollutants (Khan 2023). This may involve the formation of a covalent bond or the transfer of electrons, leading to the degradation of the plastic molecules into smaller, more biodegradable fragments. Overall, the degradation of micro- and nano-plastic by graphene-functionalized CNT involves a combination of adsorption, photocatalytic oxidation and chemical modification, synergic effects, etc., leading to an efficient breakdown of plastic pollutants into smaller or less harmful fragments.

4.6. Hetero atom doping of carbon nanotube as photocatalytic degradation material

Doping of CNTs with heteroatoms such as nitrogen, boron or sulfur can enhance the photocatalytic properties by introducing new energy levels and increasing surface area, thus improving their efficiency in degrading pollutants through photocatalysis. This modification expands the range of wavelengths adsorbed, leading to better performance in environmental remedial application. Doping creates defects and introduces new energy levels, allowing for better adsorption of levels, and for better absorption of light across a wider spectrum (Bai et al. 2018). When used as a photocatalyst for degrading micro- and nano-plastic, these doped CNTs exhibit improved efficiency due to enhanced light absorption and increased active sites for catalytic reactions, thereby facilitating the depredating process more effectively than undoped CNT. The doping can be achieved through various methods such as CVD, arc discharge or laser ablation, where precursor gases containing the desired heteroatoms are introduced during nanotube growth (Gupta et al. 2019). The heteroatom replaces some carbon atoms in the nanotube structure (Figure 4.7), creating defects and introducing new energy levels with the band structure. These introduced defects and energy levels modify the electronic properties of the CNTs, enabling them to absorb to a broader range of light wavelength, including visible light, which is crucial for photocatalytic reaction.

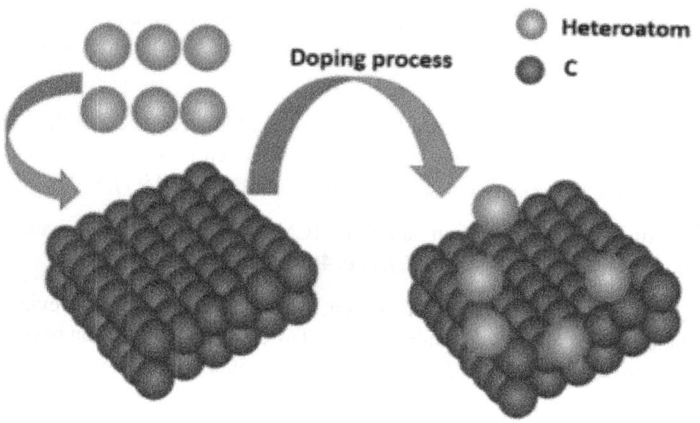

Figure 4.7. *Doping of heteroatoms. For a color version of this figure, see www.iste.co.uk/singh/photodegradation.zip*

4.7. Conclusion

CNTs exhibit immense potential as smart materials for photocatalytic degradation of micro- and nano-plastics. Their unique structural properties, such as high surface area, excellent electron transport properties and chemical stability, make them effective photocatalysts for breaking down plastic pollutants under light irradiation. Through photocatalysis, CNTs can generate ROS, such as hydroxyl radicals, which efficiently degrade plastic materials into smaller, less harmful compounds. Additionally, CNTs can be functionalized or doped with other materials to enhance their photocatalytic performance and selectivity toward specific types of plastics. Moreover, their versatility allows for various remediation systems, including filters, membranes and coatings, for efficient plastic waste management in both micro- and nano-cales. However, challenges remain in optimizing CNT-based photocatalytic systems for large-scale applications, including cost-effectiveness, stability, and scalability.

4.8. References

Agasti, N., Gautam, V., Pandey, N., Genwa, M., Meena, P.L., Tandon, S., Samantaray, R. (2022). Carbon nanotube based magnetic composites for decontamination of organic chemical pollutants in water: A review. *Applied Surface Science Advances*, 10, 100270.

Ali, A., Koloor, S.S.R., Alshehri, A.H., Arockiarajan, A. (2023). Carbon nanotube characteristics and enhancement effects on the mechanical features of polymer-based materials and structures: A review. *Journal of Materials Research and Technology*, 24, 6495–6521.

Amiri, M.R., Alavi, M., Taran, M., Kahrizi, D. (2022). Antibacterial, antifungal, antiviral, and photocatalytic activities of TiO_2 nanoparticles, nanocomposites, and bio-nanocomposites: Recent advances and challenges. *Journal of Public Health Research*, 11(2), 22799036221104151.

Ariza-Tarazona, M.C., Villarreal-Chiu, J.F., Barbieri, V., Siligardi, C., Cedillo-González, E.I. (2019). New strategy for microplastic degradation: Green photocatalysis using a protein-based porous N-TiO_2 semiconductor. *Ceramics International*, 45(7), 9618–9624.

Arora, B. and Attri, P. (2020). Carbon nanotubes (CNTs): A potential nanomaterial for water purification. *Journal of Composites Science*, 4(3), 135.

Azam, M.A., Zulkapli, N.N., Nawi, Z.M., Azren, N.M. (2015). Systematic review of catalyst nanoparticles synthesized by solution process: Towards efficient carbon nanotube growth. *Journal of Sol-Gel Science and Technology*, 73, 484–500.

Bacha, A.U.R., Nabi, I., Zhang, L. (2021). Mechanisms and the engineering approaches for the degradation of microplastics. *ACS ES&T Engineering*, 1(11), 1481–1501.

Bagheri, S., TermehYousefi, A., Do, T.O. (2017). Photocatalytic pathway toward degradation of environmental pharmaceutical pollutants: Structure, kinetics and mechanism approach. *Catalysis Science and Technology*, 7(20), 4548–4569.

Bai, S., Zhang, N., Gao, C., Xiong, Y. (2018). Defect engineering in photocatalytic materials. *Nano Energy*, 53, 296–336.

Boles, M.A., Engel, M., Talapin, D.V. (2016). Self-assembly of colloidal nanocrystals: From intricate structures to functional materials. *Chemical Reviews*, 116(18), 11220–11289.

Boyle, K. and Örmeci, B. (2020). Microplastics and nanoplastics in the freshwater and terrestrial environment: A review. *Water*, 12(9), 2633.

Campanale, C., Massarelli, C., Savino, I., Locaputo, V., Uricchio, V.F. (2020). A detailed review study on potential effects of microplastics and additives of concern on human health. *International Journal of Environmental Research and Public Health*, 17(4), 1212.

Chelombitko, M.A. (2018). Role of reactive oxygen species in inflammation: A minireview. *Moscow University Biological Sciences Bulletin*, 73, 199–202.

Deline, A.R., Frank, B.P., Smith, C.L., Sigmon, L.R., Wallace, A.N., Gallagher, M.J., Goodwin, D.G., Durkin, D.P., Fairbrother, D.H. (2020). Influence of oxygen-containing functional groups on the environmental properties, transformations, and toxicity of carbon nanotubes. *Chemical Reviews*, 120(20), 11651–11697.

Dharmadhikari, B.V. (2015). Computing the fundamental interactions between carbon nanotubes and pulmonary surfactant proteins. PhD Thesis, University of Bridgeport, Bridgeport.

Domínguez-Jaimes, L.P., Cedillo-González, E.I., Luévano-Hipólito, E., Acuña-Bedoya, J.D., Hernández-López, J.M. (2021). Degradation of primary nanoplastics by photocatalysis using different anodized TiO_2 structures. *Journal of Hazardous Materials*, 413, 125452.

Dubey, R., Dutta, D., Sarkar, A., Chattopadhyay, P. (2021). Functionalized carbon nanotubes: Synthesis, properties and applications in water purification, drug delivery, and material and biomedical sciences. *Nanoscale Advances*, 3(20), 5722–5744.

Eder, D. (2010). Carbon nanotube – Inorganic hybrids. *Chemical Reviews*, 110(3), 1348–1385.

Enfrin, M., Hachemi, C., Hodgson, P.D., Jegatheesan, V., Vrouwenvelder, J., Callahan, D.L., Lee, J., Dumée, L.F. (2021). Nano/micro plastics – Challenges on quantification and remediation: A review. *Journal of Water Process Engineering*, 42, 102128.

Gaur, J., Kumar, S., Pal, M., Kaur, H., Batoo, K.M., Momoh, J.O. (2023). Current trends: Zinc oxide nanoparticles preparation via chemical and green method for the photocatalytic degradation of various organic dyes. *Hybrid Advances*, 5, 100128.

Goh, P.S., Kang, H.S., Ismail, A.F., Khor, W.H., Quen, L.K., Higgins, D. (2022). Nanomaterials for microplastic remediation from aquatic environment: Why nano matters? *Chemosphere*, 299, 134418.

Gulati, S., Kumar, S., Goyal, K., Arora, A., Varma, R.S. (2022). Improving the air quality with functionalized carbon nanotubes: Sensing and remediation applications in the real world. *Chemosphere*, 299, 134468.

Guo, C., Ran, J., Vasileff, A., Qiao, S.Z. (2018). Rational design of electrocatalysts and photo (electro) catalysts for nitrogen reduction to ammonia (NH_3) under ambient conditions. *Energy and Environmental Science*, 11(1), 45–56.

Gupta, N., Gupta, S.M., Sharma, S.K. (2019). Carbon nanotubes: Synthesis, properties and engineering applications. *Carbon Letters*, 29, 419–447.

Hassan, M., Liu, Y., Naidu, R., Du, J., Qi, F. (2020). Adsorption of perfluorooctane sulfonate (PFOS) onto metal oxides modified biochar. *Environmental Technology and Innovation*, 19, 100816.

Khan, M.M. (2021). Principles and mechanisms of photocatalysis. In *Photocatalytic Systems by Design*, Sakar, M., Geetha Balakrishna, R., Do, T.-O. (eds). Elsevier, Amsterdam.

Khan, M.T. (2023). Unraveling the impact of graphene nanostructures passivation on the electrical properties of the perovskite solar cell. *Materials Science in Semiconductor Processing*, 153, 107172.

Khan, A. and Alamry, K.A. (2022). Surface modified carbon nanotubes: An introduction. In *Surface Modified Carbon Nanotubes Volume 1: Fundamentals, Synthesis and Recent Trends*, Aslam, J., Hussain, C.M., Aslam, R. (eds). American Chemical Society, Washington, D.C.

Kharlamova, M.V. and Kramberger, C. (2021). Applications of filled single-walled carbon nanotubes: Progress, challenges, and perspectives. *Nanomaterials*, 11(11), 2863.

Kolhe, P., Shah, M., Gandhi, S. (2023). Spectroscopic and microscopic characterizations of functionalized carbon nanostructures. In *Handbook of Functionalized Carbon Nanostructures: From Synthesis Methods to Applications*, Barhoum, A. and Deshmukh, K. (eds). Springer, Cham.

Lange, J.P. (2021). Managing plastic waste- sorting, recycling, disposal, and product redesign. *ACS Sustainable Chemistry and Engineering*, 9(47), 15722–15738.

Li, S., Zhang, X., Zhao, J., Meng, F., Xu, G., Yong, Z., Jia, J., Zhang, Z., Li, Q. (2012). Enhancement of carbon nanotube fibres using different solvents and polymers. *Composites Science and Technology*, 72(12), 1402–1407.

Li, L., Yu, L., Lin, Z., Yang, G. (2016). Reduced TiO_2-graphene oxide heterostructure as broad spectrum-driven efficient water-splitting photocatalysts. *ACS Applied Materials and Interfaces*, 8(13), 8536–8545.

Liu, M.S., Lin, M.C.C., Huang, I.T., Wang, C.C. (2005). Enhancement of thermal conductivity with carbon nanotube for nanofluids. *International Communications in Heat and Mass Transfer*, 32(9), 1202–1210.

Lofrano, G., Carotenuto, M., Libralato, G., Domingos, R.F., Markus, A., Dini, L., Gautam, R.K., Baldantoni, D., Rossi, M., Sharma, S.K. et al. (2016). Polymer functionalized nanocomposites for metals removal from water and wastewater: An overview. *Water Research*, 92, 22–37.

Matiyani, M., Pathak, M., Bohra, B.S., Sahoo, N.G. (2022). Noncovalent functionalization of carbon nanotubes. In *Handbook of Carbon Nanotubes*, Abraham, J., Thomas, S., Kalarikkal, N. (eds). Springer, Cham.

Mennani, M., Kasbaji, M., Benhamou, A.A., Boussetta, A., Kassab, Z., El Achaby, M., Grimi, N., Moubarik, A. (2024). The potential of lignin-functionalized metal catalysts – A systematic review. *Renewable and Sustainable Energy Reviews*, 189, 113936.

Milowska, K.Z., Ghorbani-Asl, M., Burda, M., Wolanicka, L., Ćatić, N., Bristowe, P.D., Koziol, K.K.K. (2017). Breaking the electrical barrier between copper and carbon nanotubes. *Nanoscale*, 9(24), 8458–8469.

Mohamed, A., Osman, T.A., Toprak, M.S., Muhammed, M., Yilmaz, E., Uheida, A. (2016). Visible light photocatalytic reduction of Cr(VI) by surface modified CNT/titanium dioxide composites nanofibers. *Journal of Molecular Catalysis A: Chemical*, 424, 45–53.

Mumtaz, F., Li, B., Al Shehhi, M.R., Feng, X., Wang, K. (2024). Treatment of phenolic-wastewater by hybrid technologies: A review. *Journal of Water Process Engineering*, 57, 104695.

Oke, J.A. and Jen, T.C. (2022). Atomic layer deposition and other thin film deposition techniques: From principles to film properties. *Journal of Materials Research and Technology*, 21, 2481–2514.

Ouda, M., Banat, F., Hasan, S.W., Karanikolos, G.N. (2023). Recent advances on nanotechnology-driven strategies for remediation of microplastics and nanoplastics from aqueous environments. *Journal of Water Process Engineering*, 52, 103543.

Patra, I., Huy, D.T.N., Alsaikhan, F., Opulencia, M.J.C., Van Tuan, P., Nurmatova, K.C., Majdi, A., Shoukat, S., Yasin, G., Margiana, R. et al. (2022). Toxic effects on enzymatic activity, gene expression and histopathological biomarkers in organisms exposed to microplastics and nanoplastics: A review. *Environmental Sciences Europe*, 34(1), 80.

Phin, H.Y., Ong, Y.T., Sin, J.C. (2020). Effect of carbon nanotubes loading on the photocatalytic activity of zinc oxide/carbon nanotubes photocatalyst synthesized via a modified sol–gel method. *Journal of Environmental Chemical Engineering*, 8(3), 103222.

Purohit, R., Purohit, K., Rana, S., Rana, R.S., Patel, V. (2014). Carbon nanotubes and their growth methods. *Procedia Materials Science*, 6, 716–728.

Radhika, N.P., Selvin, R., Kakkar, R., Umar, A. (2019). Recent advances in nano-photocatalysts for organic synthesis. *Arabian Journal of Chemistry*, 12(8), 4550–4578.

Rao, R., Pint, C.L., Islam, A.E., Weatherup, R.S., Hofmann, S., Meshot, E.R., Wu, F., Zhou, C., Dee, N., Amama, P.B. et al. (2018). Carbon nanotubes and related nanomaterials: Critical advances and challenges for synthesis toward mainstream commercial applications. *ACS Nano*, 12(12), 11756–11784.

Rathinavel, S., Priyadharshini, K., Panda, D. (2021). A review on carbon nanotube: An overview of synthesis, properties, functionalization, characterization, and the application. *Materials Science and Engineering*, 268, 115095.

Saleh, T.A. (2013). The role of carbon nanotubes in enhancement of photocatalysis. In *Syntheses and Applications of Carbon Nanotubes and Their Composites*, Suzuki, S. (ed.). IntechOpen, London, 479–493.

Sanakousar, F.M., Vidyasagar, C.C., Jiménez-Pérez, V.M., Prakash, K. (2022). Recent progress on visible-light-driven metal and non-metal doped ZnO nanostructures for photocatalytic degradation of organic pollutants. *Materials Science in Semiconductor Processing*, 140, 106390.

Shandilya, P., Raizada, P., Singh, P. (2021). Photocatalytic degradation of azo dyes in water. *Water Pollution and Remediation*, 57, 119–146.

Tewari, C., Tatrari, G., Sahoo, N.G., Mukhopadhyay, P. (2024). Plastics waste to carbon-based nanomaterials for water treatment and supercapacitor applications. In *Applied Plastics Engineering Handbook*, Kutz, M. (ed.). William Andrew Publishing, New York.

Xie, A., Jin, M., Zhu, J., Zhou, Q., Fu, L., Wu, W. (2023). Photocatalytic technologies for transformation and degradation of microplastics in the environment: Current achievements and future prospects. *Catalysts*, 13(5), 846.

Yang, X., Wan, Y., Zheng, Y., He, F., Yu, Z., Huang, J., Wang, H., Ok, Y.S., Jiang, Y., Gao, B. (2019). Surface functional groups of carbon-based adsorbents and their roles in the removal of heavy metals from aqueous solutions: A critical review. *Chemical Engineering Journal*, 366, 608–621.

Younas, M., Gondal, M.A., Dastageer, M.A., Harrabi, K. (2019). Efficient and cost-effective dye-sensitized solar cells using MWCNT-TiO_2 nanocomposite as photoanode and MWCNT as Pt-free counter electrode. *Solar Energy*, 188, 1178–1188.

Zeinali, M., Jaleh, B., Rashidian Vaziri, M.R., Omidvar, A. (2019). Study of nonlinear optical properties of TiO2–polystyrene nanocomposite films. *Quantum Electronics*, 49(10), 951.

Zhang, Y., Xing, Z., Zou, J., Li, Z., Wu, X., Shen, L., Zhu, Q., Yang, S., Zhou, W. (2017). 3D urchin-like black TiO_{2-x}/carbon nanotube heterostructures as efficient visible-light-driven photocatalysts. *RSC Advances*, 7(1), 453–460.

Zhang, S., Cheng, B., Jia, Z., Zhao, Z., Jin, X., Zhao, Z., Wu, G. (2022). The art of framework construction: Hollow-structured materials toward high-efficiency electromagnetic wave absorption. *Advanced Composites and Hybrid Materials*, 5(3), 1658–1698.

Zou, W., Gao, B., Ok, Y.S., Dong, L. (2019). Integrated adsorption and photocatalytic degradation of volatile organic compounds (VOCs) using carbon-based nanocomposites: A critical review. *Chemosphere*, 218, 845–859.

5

Environmental Justifications of MXene towards Photocatalytic Capture and Conversion of Micro- and Nano-Plastic

The environment and public health are seriously threatened by the exponential rise in plastic pollution, especially that which takes the form of micro- and nano-plastics (MNPs). In our daily lives, plastics are utilized in large quantities. Nonetheless, a substantial amount of plastic garbage is improperly recycled or reused, or it is released into the environment completely. MNPs are plastic particles that are either micro- or nanosized and are produced when plastic trash degrades. Plastic particles with a diameter of < 5 mm are referred to as micro-plastics (MPs), while those with a diameter of 1–100 or 1,000 nm are known as nano-plastics (NPs). The problems presented by these tiny particles are frequently too small for traditional approaches to handle plastic trash. In recent years, the development of cutting-edge materials and technology has opened up new possibilities for efficient plastic degradation. Among them, the special physicochemical characteristics of MXenes, a type of two-dimensional transition metal carbides/nitrides, such as their high surface area, conductivity and adjustable surface chemistry, have demonstrated great potential in a variety of applications. The use of MXene-based smart materials for photocatalytic degradation of MNPs is examined in this chapter.

Researchers across the globe have developed promising methods to reduce plastic pollution at the microscopic level by utilizing MXenes due to their remarkable adsorption abilities and photocatalytic activity. Because of this, composites based on MXene are synthesized with increased photoactivity and selectivity toward the destruction of plastic when exposed to light. The basic ideas behind the photocatalytic degradation of plastics with MXene-based materials are clarified. These include the mechanics of photocatalysis, the phenomena of adsorption–desorption and surface functionalization techniques to improve catalytic effectiveness. The synthesis and characterization of MXene-based photocatalysts for plastic degradation are also reviewed, with special emphasis on important factors including MXene composition, shape and heterostructure design. Furthermore, the potential environmental impact and scalability of MXene-based photocatalytic systems for plastic remediation are evaluated, along with challenges and future directions in this field. Overall, this chapter highlights the exciting potential of MXene-based smart materials as a cutting-edge and environmentally responsible way to

---

Chapter written by Sweta SINGH and Abhijeet KUMAR.

tackle the problem of MNP pollution, providing information on the development and application of sophisticated photocatalytic systems for environmental remediation.

5.1. Introduction

Plastic, once considered as materials of vast applications, which play pivotal role in life, started becoming a matter of great environmental concern due to its very slow degradation rate and fast accumulation. In today's world, plastic is so prevalent that the present era is considered to be the *Plasticene era* (Kumar et al. 2022b). The amount of plastics manufactured rises annually, because use plastics are suitable for a wide range of commercial purposes due to their low cost, ease of production, adaptability and hydrophobicity. According to the statistics, 370 Mt of plastic was produced in 2020 (Wang et al. 2021).

Sl. No.	Polymer	Chemical structures	Density (g/L)	Crystallinity (%)	Life span (years)	Recyclable?/ Recovery rate (%)
1.	Polystyrene (PS)	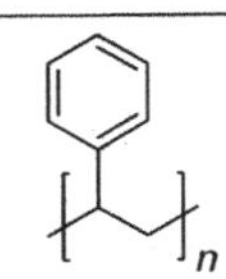	1.03–1.09	0	50–80	Sometimes/1
2.	Polyvinyl chloride (PVC)	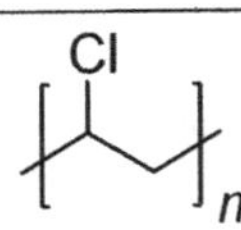	1.35–1.45	0	50–150	No/0
3.	Polyethylene terephthalate (PET)	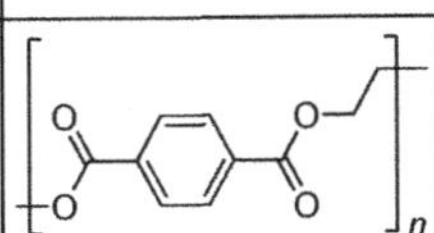	1.35	0–50	450	Yes/19.5
4.	Polypropylene (PP)	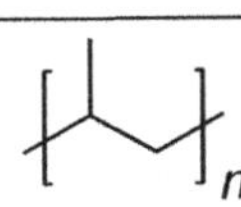	0.90–0.91	50	10–600	Sometimes/1
5.	Low-density polyethylene (LDPE)	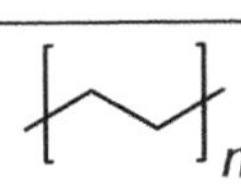	0.91–0.93	50	10–600	Mostly No/5
6.	High-density polyethylene (HDPE)	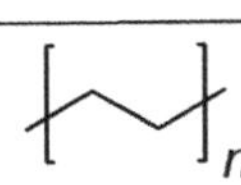	0.94–0.97	70	> 600	Yes/10

Table 5.1. *Examples and properties of some of the most commonly used polymers (Mohanan et al. 2020; Lee and Li 2021)*

Generally, we dispose of plastics after their use, which has lead to the generation of huge amount of plastic-based waste, accumulated over a period of time due to very slow rate of natural degradation. It is predicted that by the 2050, 12,000 Mt of plastic waste will have accumulated in the environment (Wang et al. 2021). Due to their very high production and very slow degradation, different types of plastics are considered to be the most severe forms of anthropogenic pollutants, which are threatening the life present on land as well as the aquatic ecosystem. Without appropriate post-treatment measures, plastics can contribute directly to severe the pollution of marine environments, the atmosphere and soil. According to assessments conducted by the United Nations Environment Programme (UNEP), it is estimated that there are between 75 million to 199 million tons of plastic present in the oceans alone, which accounts for the approximately 85% of the total weight of marine debris. In addition to this, more than eight million metric tonnes of plastics are added to the oceans each year. Such pollution poses serious threats to marine life, ecosystems and human health, highlighting the urgent need for effective waste-management strategies and global cooperation to mitigate the impacts of plastic pollution on the environment. Over a period of time, these plastic-waste materials start converting into smaller pieces by different modes of degradation such as physical, photochemical, biological and mechanical that occur naturally (Du et al. 2021). Based on the dimensions, these plastic pollutants could be termed as mega-plastics (> 1 m), macro-plastics (< 1 m), meso-plastics (< 2.5 cm), MPs (< 5 mm), and nano-plastics (< 1 μm). Among them, MPs, which are defined as "synthetic solid plastic particle or polymeric matrix with regular or irregular shapes and a size ranging from 1 mm to 5 mm" (Hamd et al. 2022), and NPs, which are < 1 μm in size, are considered to be some of the highly hazardous forms of the solid pollutants. MPs are considered as persistent pollutants due to their presence in various ecosystems worldwide, ranging from the depths of the oceans to the peaks of the mountains. High durability and a very slow degradation rate result in long half-lives, estimated to be up to 1,000 years or even longer (Table 5.1). Due to their smaller sizes, these could easily enter the bodies of vast ranges of organisms, starting from as small as zooplanktons to as big as human beings through direct exposure. For example, aquatic organisms may directly absorb plastic particles from the surrounding water. These particles can be suspended in the water column, and organisms may inadvertently ingest them during feeding or through other forms of contact. Similarly, these could also enter the body through indirect exposure via dietary exposure. These accumulation processes depend on various factors, including the exposure levels, uptake rates and excretion rates of organisms. Nevertheless, the accumulation of MNPs in aquatic organisms has been extensively documented, highlighting their bioaccumulation potential. One report indicates that a human being could consume more than 39,000 MPs per year through various sources that include food, beverages and tap and bottled water. Apart from the persistent nature and bioaccumulation through different mechanism, MNPs are also known to exert acute to chronic health effects. In view of the fast accumulation and slow degradation of plastics MNPs, in particular, are being categorized as PBT (persistent, bioaccumulative and toxic) pollutants similar to some of the nonsteroidal anti-inflammatory drugs such as diclofenac (Alava et al. 2023). Depending on the mode of action, these plastics could affect growth and developmental processes such as brain development, the reproductive process, genetic and protein expression, etc. Generally, plastic-waste materials are disposed according to following conventional methods (Li et al. 2023):

– *Landfill disposal*: this involves burying plastic-waste materials on land and leads to large areas of land with significant environmental concerns. For example, the leaching of toxic chemicals (which is commonly used as plasticizer, flame retardants (etc.) during the preparation of the plastic material), such as bisphenol, polybrominated diphenyl ether, etc., into the soil and groundwater, which leads to another kind of environmental health issue. It is known that almost 21%–42% of plastic-waste material is disposed through landfill (Ouyang et al. 2021).

– *Incineration*: plastics can be burned in specially designed facilities called incinerators. This process converts the plastics into heat, steam and ash. While it reduces the volume of waste and can generate energy, it also releases harmful emissions and greenhouse gases into the atmosphere if not properly controlled. For example, the incineration process may lead to the generation of poisonous gases such as CO, volatile organic compounds, dioxins and furans (Verma et al. 2016).

– *Recycling*: this method involves sorting, shredding, melting and reforming plastics into new products (Merrington 2017). It is a more environmentally friendly approach compared to landfill disposal and incineration because it promotes reuse and reduces the need for virgin plastic production. Mainly, four different approaches of recycling are adopted: *primary, secondary, tertiary and quaternary* recycling (Ignatyev et al. 2014). *Primary recycling*, which is also known as the *closed-loop recycling* process, involves directly using unpolluted plastic-waste materials to turn them into some new products. High costs involved in the selection and sorting of the unpolluted plastics among different plastic waste encourages us to adopt another approach.

Similarly, the *secondary recycling* process involves the recycling of contaminated or plastic wastes of heterogeneous nature that involves the dissolution followed by precipitation of mixed plastics (Achilias et al. 2009; Hadi et al. 2013). But this process leads to the reduction in mechanical properties and molecular weight of the resulting plastics. Thus, such recycled plastics are used for the preparation of less demanding plastic-based materials such as for the production of flooring tiles. In addition to the aforementioned conventional methods of plastic disposal, some of these lead to the generation of MNPs, which have emerged as matter of serious environmental concern. As per the data available in 2015, 12% of waste plastics were incinerated, while 79% were either landfilled or abandoned in the natural environment (Jan Kole et al. 2017; Li et al. 2023). These statistics highlight the urgent need for better waste management strategies, including increased recycling and the development of more sustainable alternatives to plastic products.

Although plastic-waste materials are increasing day by day and are known as *white pollution*, the type of plastics which are (basically) polymers rich in carbon could be treated as feedstock to generate diverse varieties of other valuable chemicals and commodities. Therefore, various approaches of the conversion of these plastics to other value-added products have also been developed through a *tertiary recycling* process, which is also known as the *chemical recycling process* that involves the conversion of plastic waste into other useful chemicals that could later be used for the production of new polymers and plastic (Zheng et al. 2023). For example: the conversion of PET into diols and dimethyl terephthalate through glycolysis using niobia-based catalyst (Shirazimoghaddam et al. 2023). For the plastic waste difficult to dispose of by any other means, the *quaternary recycling* process is used. This involves the incineration of complex plastic-waste materials to generate heat energy. Along with heat energy, some greenhouse and poisonous gases are also generated during this process. For example, the generation of tire-derived fuel (TDF) from tire waste is an example of quaternary recycling (Akbas and Yuhana 2021). In addition to the aforementioned conventional methods, various other novel strategies such as biochemical conversion, which basically converts plastics via enzyme-catalyzed conversion into other value-added products, have also been explored. For example, enzymes derived from microorganisms such as bacterial, fungi and algae are used for the degradation of polymers such as polylactic acid (PLA) and polyethene terephthalate (PET) (Tamoor et al. 2021). Similarly, microwave-assisted pyrolysis is another greener method for the conversion of plastic waste into useful products (Hu et al. 2023).

Likewise, compatibilization (Zervoudakis et al. 2022) and photoreforming (Gong et al. 2022) are also novel strategies for the conversion of plastic wastes into other useful products. Various strategies for the degradation and conversion of plastics into different chemicals and energies that could be executed either under mild- or violent-reaction conditions have also been developed (Table 5.2). For example, catalytic degradation of plastics under microwave-assisted conditions, pyrolysis, solvolysis, hydropyrolysis, etc. are examples of the methods that generally require violent-reaction conditions.

Similarly, catalytic methods such as photocatalysis, electrocatalysis and biocatalysis can be performed under milder conditions (Zheng et al. 2023).

<table>
<tr><th>Reaction Conditions</th><th>Methods</th><th>Temperature</th><th>Advantages</th><th>Disadvantages</th></tr>
<tr><td rowspan="4">Violent</td><td>Microwave catalysis</td><td>573–773 K</td><td rowspan="4">Efficiency of degradation is higher</td><td rowspan="4">Poor product selectivity, high power consumption and difficulty in mechanistic study</td></tr>
<tr><td>Pyrolysis</td><td>673–973 K</td></tr>
<tr><td>Solvolysis</td><td>< 473 K</td></tr>
<tr><td>Hydropyrolysis</td><td>473–523 K</td></tr>
<tr><td rowspan="3">Mild</td><td>Biocatalysis</td><td>303 K</td><td rowspan="3">Power consumption is low, product selectivity is high and the mechanism could be understood in a better way</td><td rowspan="3">Efficiency is low compared to degradation methods used under violent conditions</td></tr>
<tr><td>Electrocatalysis</td><td>Room temperature</td></tr>
<tr><td>Photocatalysis</td><td>Room temperature</td></tr>
</table>

Table 5.2. *Different methods for the degradation of plastics (Zheng et al. 2023)*

5.2. Nanomaterial catalyzed methods for the degradation of micro- and nano-plastics

In view of the emerging environmental and health concerns due to MNPs, several strategies have been adopted, among which catalysis such as photocatalysis, electrocatalysis and photo-electrocatalysis are some of the most prominent catalytic approaches.

In recent years, the catalytic properties of various types of nanomaterials (Figure 5.1) such as metal oxide like ZnO, TiO_2, etc., carbon-based materials such as CNT, MCNT, GO, etc., along with various other nanomaterials have also been exploited for the degradation of MNPs.

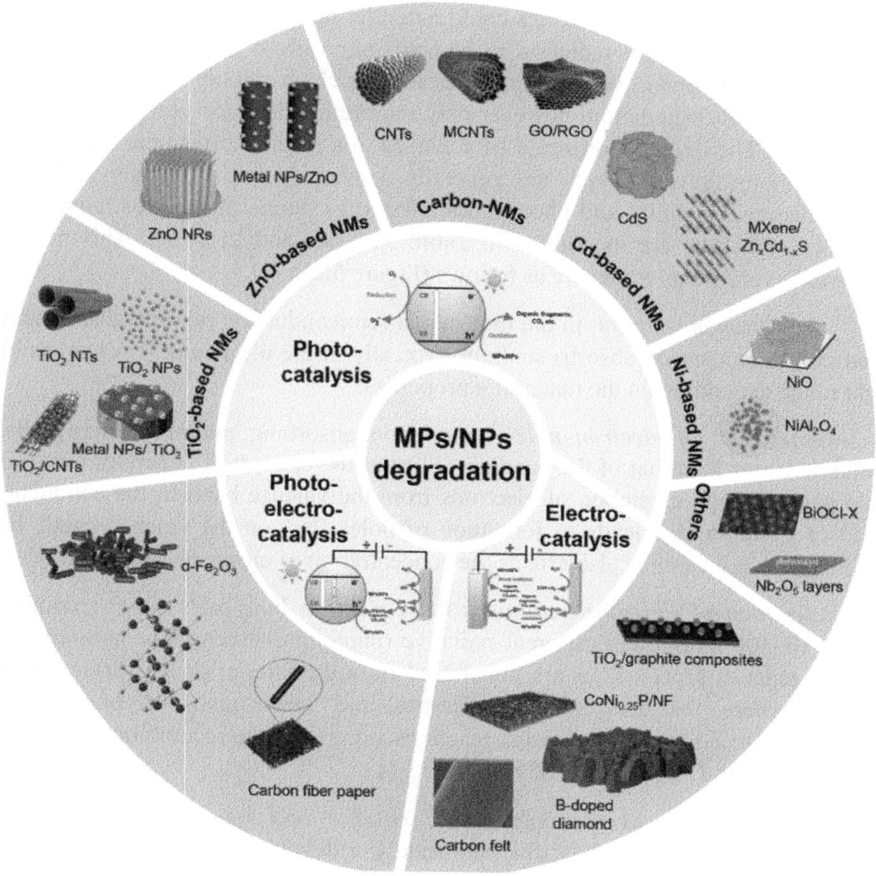

Figure 5.1. *Nanomaterials-based catalyst for the degradation of MPs and NPs (Du et al. 2022). Reproduced with permission from Du et al. (2022). Copyright (2022), Elsevier. For a color version of this figure, see www.iste.co.uk/singh/photodegradation.zip*

5.3. Photocatalytic degradation of micro- and nano-plastics

Compared to the aforementioned different catalytic methods used for the degradation of the plastics, photocatalyzed methods using different types of materials such as photocatalysts are emerging as environmentally benign and greener approaches, as previous methods generally required harsh-reaction conditions such as high temperatures, high pressures, along with high energy consumption in comparison with the photocatalytic methods, which generally use light energy such as UV light for the upcycling of the plastic-waste materials into useful value-added products (Chu et al. 2022). Plastics could naturally be degraded under the exposure to sunlight or ultraviolet radiation to convert into various smaller oligomers and other products involving various processes such as chain scission and the oxidation process. Natural degradation, however may require hundreds of years (Vivekanand et al. 2021). The degradation could be achieved in a much shorter time frame using suitable photocatalysts, which are generally semiconducting materials with high photo, chemical stability and photo-redox properties. Some of the most commonly used photocatalysts are mentioned in Table 5.2 (Lee and Li 2021). This process involves several steps which are as follows (Figure 5.2):

– *Absorption of sunlight*: in the first step, a semiconducting material that is being used as a photocatalyst absorbs sunlight, typically in the ultraviolet (UV) or visible light range depending on the material's properties.

– *Generation of electron–hole pairs*: upon absorbing photonic energy (hv), which is larger than that of the bandgap (Ebg) of the catalyst, it generates electron–hole pairs through excitation of electrons from the valence band to the conduction band (e^-), which also leads to formation of holes (h+) on the valence band. The nature of the light source also affects the concentration of electrons and holes.

– *Migration of electrons and holes*: these electron–hole pairs either recombine within 10–100 ns through different radiative (photoluminescence) or non-radiative (energy release) processes or are transferred from the interior to the surface of the catalyst where reactants such as plastics and other reactants interact and this leads to different types of redox reactions along with the recombination process (Qiu et al. 2016; Danish et al. 2021).

– *Redox reaction at the interface of catalyst and reactants*: two different strategies could be adopted for the photocatalytic degradation of plastics. The first one involves the photocatalytic reaction conducted in the oxygen-rich atmosphere, which is also known as oxygen evolution reaction (OER). This process leads to the generation of different types of reactive free radicals such as superoxide anion radicals and hydroxyl radicals, which initiates chain reactions that result in the degradation of the plastic into smaller fragments.

Due to the involvement of highly oxidizing radical species, the oxidation of MPs is nonselective (Shang et al. 2003; Hu et al. 2019). In different approach, photocatalytic reactions are performed under an inert atmosphere in the presence of gases such as N_2 and Ar. It is also known as H_2 evolution reaction (HER) as during the reaction hydrogen gas evolves (Takanabe and Shahid 2017). Herein, holes generated at the surface of photocatalyst by the absorption of solar energy leads to the conversion of (micro-)plastics to generate relatively selective value-added compounds, whereas photo-induced electron reacts with the proton to generate hydrogen gas (Melchionna and Fornasiero 2020). Reactions involved in these two approaches are summarized in Table 5.3.

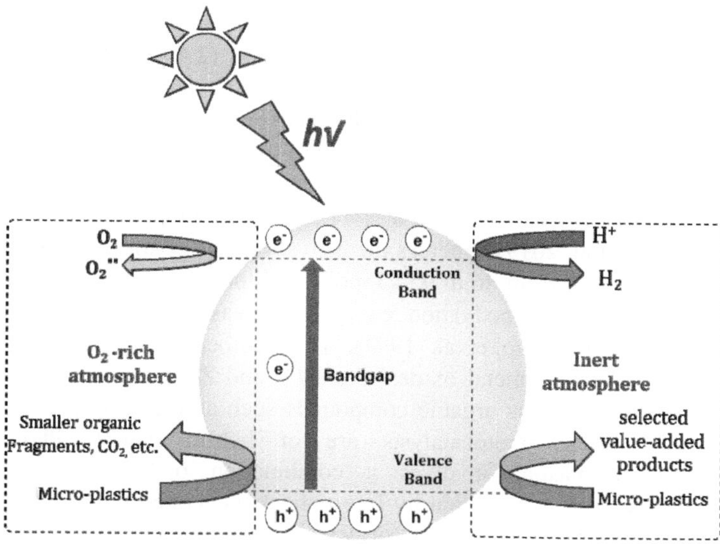

Figure 5.2. *Various steps involved in the photolytic degradation of MP into other products (Li et al. 2023). For a color version of this figure, see www.iste.co.uk/singh/photodegradation.zip*

Photochemical conversion of micro-plastics to products in oxygen-rich atmosphere	Photocatalytic reactions in inert atmosphere
Photocatalyst + $h\nu \longrightarrow h_{(VB)}^{+} + e_{(CB)}^{-}$ eq. 1 $O_2 + e_{(CB)}^{-} \longrightarrow O_2^{\bullet-}$ eq. 2 $H_2O + h_{(VB)}^{+} \longrightarrow {}^{\bullet}OH + H^{+}$ eq. 3 $O_2^{\bullet-} + 2H^{+} + e_{(CB)}^{-} \longrightarrow H_2O_2$ eq. 4 $H_2O_2 + h\nu \longrightarrow 2{}^{\bullet}OH$ eq. 5 $H_2O_2 + O_2^{\bullet-} \longrightarrow {}^{\bullet}OH + OH^{-} + O_2$ eq. 6 $H_2O_2 + e_{(CB)}^{-} \longrightarrow {}^{\bullet}OH + OH^{-}$ eq 7 ${}^{\bullet}OH$ or $O_2^{\bullet-}$ + (micro)plastics $\longrightarrow$ intermediates and products eq 8	Photocatalyst + $h\nu \longrightarrow h_{(VB)}^{+} + e_{(CB)}^{-}$ eq. 1 $H_2O + h_{(VB)}^{+} \longrightarrow {}^{\bullet}OH + H^{+}$ eq. 3 $H^{+} + e_{(CB)}^{-} \longrightarrow H_2$ eq. 9 $h_{(VB)}^{+}$ or ${}^{\bullet}OH$ + (micro)plastics $\longrightarrow$ valuable organic products eq 10

Table 5.3. *Photocatalytic reaction under two different environments (Li et al. 2023)*

Overall, the efficiency of the semiconductor as a photocatalyst mainly depends on the band gap, which would also affect the start and end of the reaction process. A catalyst with narrower band gap such as TiO_2 (3.2 eV) and ZnO (3.2 eV) (Kuehnel and Reisner 2018), C_3N_4 (2.7 eV) and CN-CNTs-NM (2.4 eV) (Gong et al. 2022) allows the degradation of (micro-)plastic in a facile manner. In addition to the outcome of the degradation process, the activation energy required for the degradation also varies depending on the nature of the polymers. For example, in case of HDPE, the activation energy for the formation of a carbonyl group is in the range of 85.6 ± 27.6 kJ mol^{-1}, while it is 71.5 ± 8.9 kJ mol^{-1} for the formation of vinyl (Fairbrother et al. 2019), whereas the activation energy for carbonyl formation in case of LDPE varies from 46 to 95 kJ mol^{-1} and for other products such as vinyl and hydroxyl groups, the activation energies decrease and lie in the range of 20–46 kJ mol^{-1} (Cruz-Pinto et al. 1994). The photocatalyst could be inorganic materials such as various metal oxides like TiO_2 and ZnO, metal halides such as $FeCl_3$ and nitride as well as organic compounds such as fluorenone and p-toluene sulfonic acids. Some photocatalysts are of hybrid nature. Photocatalytic decomposition of plastics involves a combination of direct oxidation by photoexcited holes and degradation by free radicals generated through the reaction of electrons and holes with oxygen or water molecules. This process offers a promising approach for mitigating plastic pollution by breaking down plastic waste into less harmful compounds under sunlight-driven conditions. In addition to that, it also allows the conversion of plastic waste into carbon-based fuels (Jiao et al. 2020).

Sl. No.	Catalyst type	Photocatalyst	Plastic	Light source	Time	Degradation efficiency (%)	
						Weight loss	Others
1.	Inorganic	TiO_2	PVC	Simulated sunlight	300 h	27	
2.		Ag-doped TiO_2	PE	Artificial light	300 h	14.28	
3.		TiO_2-MWCNTs	PE	UV light	180 h	35	
4.		ZnO	PS	UV light	2 h	16	
5.		C_3N_4	PS	300 W Xenon lamp (150°C)	24 h		60±4
6.	Organic	Fluorenone	PS	Blue LED	48 h		38±3
7.	Hybrid	FePc–TiO_2	PS	Sunlight	12 days	35	
8.		PPy/TiO_2	PE	Sunlight	240 h	35.4	

Table 5.4. *Examples of photocatalysts their application in degradation of different types of plastics (Li et al. 2023)*

5.4. MXene: a nanomaterial with diverse applications

In recent years, diverse varieties of 1D to 3D nanomaterials, having photocatalytic properties such as molybdenum sulfide (MOSx) (Rahman et al. 2022a), tungsten disulphide (WS_2) (Liu and Hua 2020), bismuth sulfide (BiSx) (Chao et al. 2016), silicon nitride (Sharma et al. 2020), titanium dioxide (TiO_2) (Schneider et al. 2014), ceric oxide (CeO_2) (Kusmierek 2020) and niobium (V) oxide (Nb_2O_5) (Ücker et al. 2021), are being discovered and developed, as well as their vast applications as photocatalysts in the purification of water and the degradation of organic dyes and polymers. However, due to their limited catalytic efficiency, the discovery of more efficient materials is an ongoing process. In this direction, the discovery of *MXenes* (Cui et al. 2023), which are basically two-dimensional (2D) carbides, nitrides or carbonitrides of early transition metals and having similarity with *Graphene*, by Michael Naguib in 2011 at Drexel University, Philadelphia, appeared as a milestone in the discovery of newer nano-sized 2D materials (Naguib et al. 2012).

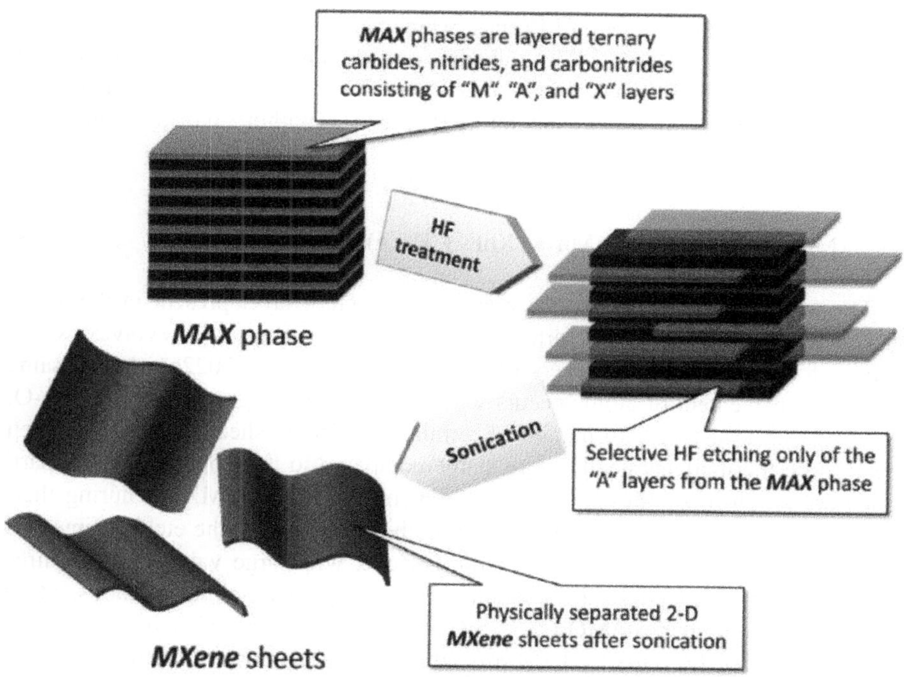

Figure 5.3. *Synthesis of MXene involving etching and exfoliation of MAX phases. Reproduced with permission from Naguib et al. (2012). Copyright (2012), American Chemical Society. For a color version of this figure, see www.iste.co.uk/singh/photodegradation.zip*

The MXene could be represented with general formula, such as $M_{n+1}X_nT_x$ (n = 1–3), where M stands for transition metals such as Sc, Ti, V, Cr, Zr, Mo, Hf and Nb, X = C and N, and T represents the hydrophilic surface terminating functional groups such as -O, -S, -NH, -OH, -F, -Br, and -Te and could easily be synthesized by the selective etching of elemental layer A using a suitable etchant such as HF from another 3D material known as MAX. This has the general formula $M_{n+1}AX_n$, where M and X are similar to MXene but A represents an element from group 13 or 14 of the periodic table such as Al, Si, In and Ga (Long et al. 2022). In general, MAX phases are chemically very stable materials due to having different types of bonding such as metallic, ionic and covalent. However, due to relatively weak bonding between M and A layer compared to M and X, at high temperatures, it decomposes according to the following reaction (equation [5.1]) to provide graphene like single layer sheets (Naguib et al. 2012).

$$M_{n+1}AX_n = M_{n+1}X_n + A \qquad [5.1]$$

Titanium carbide $Ti_3C_2T_x$ (T = -O, -OH, -F) is considered to be the first and one of the most studied examples of MXenes, which was first prepared by Naguib et al. (2012) by the exfoliation of Ti_3AlC_2, which is one of members of 3D-MAX material, which are layered hexagonal ternary metal carbides using hydrofluoric acid (HF) solution as the etchant at room temperature (Figure 5.3) within 2 h of reaction time.

$$Mn + 1AX_n + 3HF = Mn + 1Xn(s) + AF_3 + 1.5H_2\ (g) \qquad [5.2]$$

$$Mn + 1Xn(s) + 2HF = Mn + 1XnF_2 + H_2\ (g) \qquad [5.3]$$

The whole process involves the reaction between A layer present in the MAX phase with the fluoride ion to form salt AF and hydrogen gas also evolves, which helps exfoliate the different layers of MXene (Kumar et al. 2022a). At the same time, the replacement of atoms occurs with other functional groups such as -F, -O, and -OH that lead to the formation of terminated MXene sheet, which have been found to be thermodynamically more stable compared to their pristine counterpart, which are tough to synthesize due to the functionalization of MXene during their synthesis (Mishra et al. 2016). Depending on the number of n, the etching time and nature of etching would vary. In general, MXene, with large value of n, requires longer etching time and stronger etching. Following this etching methods, other MXene materials such as MO_2C, Ti_3SiC_2 and Cr_2TiC_2 have also been prepared (Kumar et al. 2022a). Due to the corrosive and hazardous nature of the HF, other less toxic salts such as LiF, FeF_3 and KF along with HCl have also been used as an etchant (Rahman et al. 2022b). In some cases, the combination of fluoride salt and HCl has also been used as an etchant for the formation of carbide MXenes such as $Ti_3C_2T_x$ (Wang et al. 2017).

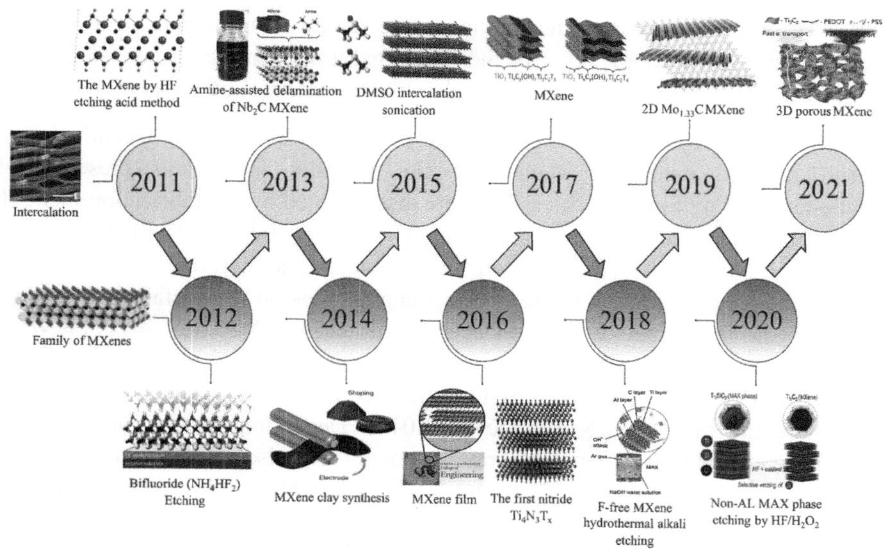

Figure 5.4. *Schematic diagram for the development of synthetic methods from 2011 to 2021 (Li et al. 2021). Reproduced with permission from Cui et al. (2023). Copyright 2022, Elsevier. For a color version of this figure, see www.iste.co.uk/singh/photodegradation.zip*

Likewise, various other approaches including *top-down approaches* such as solvothermal, microwave irradiation, ultrasonication, electrochemical (Sun et al. 2017) and ball milling, along with the *bottom-up approaches* such as molten-salt synthesis and pyrolysis have also been employed for the synthesis of different varieties of MXenes (Rahman et al. 2022b). In addition to the carbide MXenes, other nitride and carbonitride MXenes such as Ti_4N_3, Ti_3CN and Ti_2N have also been prepared and their wide applications have also been investigated (Soundiraraju and George 2017). Due to having better electrical conductivity in comparison to conventional graphene-based materials as well as other thermal, physical and chemical properties MXenes are gaining wide application in different sectors such as supercapacitors (Chen et al. 2022), batteries (Liu et al. 2022), photocatalysts (Ihsanullah 2022) and magnetic shielding (Iqbal et al. 2020). In addition to this, due to them having interesting electrical and biological properties, MXene-based sensors are also being developed (Xin et al. 2020).

5.5. Important properties of MXenes

Due to the excellent properties and characteristics, MXenes are considered to be *wonder materials* among different 2D nanomaterials such as graphenes (Li et al.

2021). These properties vary depending on the nature of transition metal (M), X (Cor N) and also on the presence of the functional groups at the surface along with various other factors such as use of etchants, doping material and temperature.

– *Electronic properties*: the electronic properties of MXene, mainly conductivity, depend on the nature of metals in the metallic layer and presence of surface-terminating functionalities. They are also affected by the fabrication method, extent of defects and nature of the etchants (Xin et al. 2020). Like their precursor MAX phase, all pristine MXenes are metallic in nature and therefore are conducting. However, depending on the nature of functional groups on the surface and also on the composition, some of the surface functionalized MXenes may be of a semiconducting or superconducting nature (Khazaei et al. 2017; Kamysbayev et al. 2020). For example, Ti_2CO_2, Zr_2CO_2 and Sc_2CT_2 (T = F, OH and O) are semiconducting in nature (Khazaei et al. 2017). Depending on the nature of the transition metals and functional groups, the energy gap also changes. For example, the energy gaps in case of Sc_2CT_2, where T = F, OH and O, are estimated to be 1.03, 0.45 and 1.8 eV, respectively. Similarly, by changing transition metals, this gap also changes. For example, the values in case of Ti_2CO_2, Zr_2CO_2 and Hf_2CO_2 are 0.24, 0.88 and 1.0 eV, respectively (Khazaei et al. 2017).

– *Mechanical properties*: the stability of differently functionalized MXenes varies depending on the nature of functional group present (Lai et al. 2016). For example, -O-terminated MXenes have a higher mechanical strength compared to their -F and -OH counterparts, which is also reflected in the difference in their lattice parameters, which have been found to be lower (maximum stiffness) in case of -O functionalized compared to later on (Zhan et al. 2020). Likewise surface functionalities, in case of functionalized MXenes, make it more flexible and stronger compared to their pristine counterpart. For example, the mechanical power of MXene composite with polyvinyl alcohol is found to be higher, that is, up to 91 ± 10 MPa compared to pristine MXene in which it is 22 ± 2 MPa (Sobolčiak et al. 2017).

– *Optical properties*: similarly, regarding optical properties, in particular, light absorption capacity and absorption performance, the reflectance of MXenes is also affected due to the presence and nature of different functionalization at their surface. For example, the MXene containing -O functional group at their surface exhibits higher absorption and reflectance in visible region compared to -F and -OH functionalized MXenes. Electrochemiluminescence and photoluminescence are some of the interesting optical properties associated with the MXene, which makes it suitable for the various applications as sensors such as photoelectrochemical sensor (Shi et al. 2023).

5.6. Application of MXene as photocatalyst

Compared to various other 2D material such as transition metal oxide, graphene, transition metal dichalcogenides, the presence of several active sites, higher electrical conductivity, thermal and chemical stability, large interlayer gap, and presence of functional groups at the surface and large specific surface area (SSA) make functionalized MXenes suitable for various applications and in particular for their use as photocatalysts. More specifically, the presence of large SSA and a large number of active sites at the surface contributes to excellent photocatalytic behavior of MXene-based materials. In addition to this, photocatalytic efficiency also depends on the electronic structure and more specifically on the band gap and also on the thickness of the materials as the electrical conductivity also varies with the thickness. For example, the electrical conductivity of MXene ranges from 850 to 9,880 $S.cm^{-1}$ when the thickness is between 1 and 5 nm. In addition to a considerable band gap, photocatalytic behavior also depends on their light-absorption efficiency, which varies with the thickness and functionalization of different ions. In general, the photocatalytic efficiency of the composite of the MXenes have exhibited better catalytic activity compared to the pristine MXene due to having an increase in SSA and a number of active sites (Sharma et al. 2022). In addition to this, the charges at the surface could be modified by tuning pH. For example, the point of zero charges (PZC), which is basically the pH at which the net charge of the surface of a particle is zero, of MXene ranges from 2 to 3. Below this, the surface could be positively charged, which would help it to bind with the anionic particles such as anionic dyes, whereas above PZC, the surface becomes negatively charged, which allows the positively charged particles, such as cationic dyes, to be absorbed on to it (Bury et al. 2023). Due to the aforementioned excellent properties associated with the MXene, it has been used in several areas such as for the development of wearable sensors and the removal of environmental contaminants and pollutants.

5.7. Application of MXene-based materials for the degradation of organic pollutants

Due to the excellent photocatalytic and adsorbent properties of MXene-based nanomaterials, they are being used for photocatalyzed degradation and the removal of various hazardous organic pollutants such as organic dyes and pharmaceuticals. For example, the application of $Ti_3C_2T_x$ MXene was demonstrated in the removal of different types of drugs such as amitriptyline (AMT), carbamazepine (CBM) and ibuprofen (IBP) from water in the presence of different types of other inorganic and organic materials.[44] MXenes with a higher functionalization at the surface and the presence of ions such as Na+ among different layers of MXene enhances the adsorption capacity. For example, $Ti_3C_2T_x$ MXene sheets intercalated with Na^+ (SI-$Ti_3C_2T_x$) demonstrated high adsorption efficiency due to an increase in inter spacing and the number of active sites, as this material adsorbed ciprofloxacin (CPX) up to 10 ppm for CPX, which could also be regenerated electrochemically at later stage (Ghani et al. 2021). Similarly, the application of Fe_3O_4@MXene nanocomposite was demonstrated for the removal of 17a-ethinylestradiol (EE2), a drug used for birth control, from water (Xu et al. 2021).

Apart from removal of different types of drugs, MXene-based nanomaterials have also been effectively used for the removal and degradation of different types of organic dyes. In particular, compared to the pristine MXene, the surface functionalized MXenes have displayed better adsorption capacity. For example, $Ti_3C_2T_x$ MXene, treated with different alkali such as NaOH, LiOH and KOH, have exhibited a better adsorption capacity, such as the adsorption capabilities of NaOH–$Ti_3C_2T_x$ and LiOH–$Ti_3C_2T_x$ for methylene blue are 189 and 121 mg g^{-1}, respectively (100 mg g^{-1}), due to enhanced inter-layer spacing, the number of active sites and also due to presence of -OH groups at the surface, interacting effectively with the ionic dyes (Wei et al. 2018). In addition to the removal of these organic pollutants by adsorption process, several examples are known where MXene-based materials have also been used as photocatalyst to catalyze the degradation of these pollutants. For example, Nb_2O_5/Nb_2CT_x composites was found to degrade organic dyes such as 98.5% of degradation of rhodamine B (RhB) within 120 min and 91.2% of tetracycline hydrochloride (TC-HCl) within 180 min (Cui et al. 2021). Similarly, the composite material developed with Bi_2WO_6 and Nb_2CT_x, that is, Bi_2WO_6/Nb_2CT_x exhibited enhanced photodegradation efficiency compared to bismuth tungstate (Bi_2WO_6), which is one of the most commonly used photocatalyst for the degradation of organic dyes due to its high chemical, photo and thermal stability. For example, the presence of MXene in Bi_2WO_6/Nb_2CT_x nanocomposite enhances the separation of electron–hole pairs generated under irradiation of light (Cui et al. 2020). Similarly, the photodegradation performance of TiO_2/Ti_3C_2 photocatalyst in degradation of methylene blue was also enhanced compared to the pristine TiO_2 (Peng et al. 2017).

5.8. MXene as photocatalyst for degradation of MPs and NPs

Apart from the metal oxide and sulfide-based photocatalysts used for the degradation of MPs and NPs, the current approach is to incorporate MXenes to enhance the electron–hole separation and thereby enhancing photocatalytic activity of the resulting materials. For example, $Zn_xCd_{1-x}S$ is considered to be a very good photocatalyst due to its better optical properties, band gap and also due to it having high utilization rate of visible light (Cui et al. 2015; Tang et al. 2018). The photocatalytic activity of such sulfides was enhanced by developing MXene-metal sulfides nanocomposites such as MXene/$Zn_xCd_{1-x}S$ nanocomposite. Cao et al. (2022) fabricated a series of MXene/$Zn_xCd_{1-x}S$ and also demonstrated their application as photocatalysts for the degradation of polymers such as polyethylene terephthalate (PET) along with the evolution of H_2 during this process.

5.9. Conclusion

In view of the emerging environmental and health concern due to plastic waste materials, mainly due to MNPs, this chapter discusses several strategies that include traditional, as well as modern methods for their degradation and utilization of these MNPs for their conversion into value added products and energy. In particular, the catalytic methods, mainly photocatalytic using different types of nanomaterials for the degradation of these plastic materials of smaller sizes, have been discussed. This chapter predominately focuses on the application of functionalized MXene, which has emerged as a smart 2D nanomaterial with excellent properties and other MXene-based nanocomposites as photocatalyst for the degradation of organic pollutants and in particular MNPs. Considering the pressing demand of effective catalytic methods and discovery of efficient materials to deal with emerging challenges due to MNPs, photocatalytic methods discussed using 2D nanomaterial-based photocatalysts in this chapter would be very useful for the scientific community working in this area.

5.10. References

Achilias, D.S., Giannoulis, A., Papageorgiou, G.Z. (2009). Recycling of polymers from plastic packaging materials using the dissolution-reprecipitation technique. *Polymer Bulletin*, 63, 449–465. doi: 10.1007/s00289-009-0104-5.

Akbas, A. and Yuhana, N.Y. (2021). Recycling of rubber wastes as fuel and its additives. *Recycling*, 6(4), 78.

Alava, J.J., Jahnke, A., Bergmann, M., Aguirre-Martínez, G.V., Bendell, L., Calle, P., Domínguez, G.A., Faustman, E.M., Falman, J., Kazmiruk, T.N. et al. (2023). A call to include plastics in the global environment in the class of persistent, bioaccumulative, and toxic (PBT) pollutants. *Environmental Science and Technology*, 57, 8185–8188. doi: 10.1021/acs.est.3c02476.

Bury, D., Jakubczak, M., Kumar, R., Ścieżyńska, D., Bogacki, J., Marcinowski, P., Jastrzębska, A.M. (2023). Cleaning the environment with MXenes. *MRS Bulletin*, 48, 271–282.

Cao, B., Wan, S., Wang, Y., Guo, H., Ou, M., Zhong, Q. (2022). Highly-efficient visible-light-driven photocatalytic H2 evolution integrated with microplastic degradation over MXene/ZnxCd1-xS photocatalyst. *Journal of Colloid and Interface Science*, 605, 311–319. doi: 10.1016/j.jcis.2021.07.113.

Chao, J., Xing, S., Zhao, J., Qin, C., Duan, D., Zhao, Y., He, Q. (2016). Bismuth sulfide nanoflowers as high performance near-infrared laser detectors and visible-light-driven photocatalysts. *RSC Advances*, 6, 55676–55681. doi: 10.1039/C6RA06339J.

Chen, Y., Yang, H., Han, Z., Bo, Z., Yan, J., Cen, K., Ostrikov, K.K. (2022). MXene-based electrodes for supercapacitor energy storage. *Energy and Fuels*, 36, 2390–2406.

Chu, S., Zhang, B., Zhao, X., Soo, H.S., Wang, F., Xiao, R., Zhang, H. (2022). Photocatalytic conversion of plastic waste: From photodegradation to photosynthesis. *Advanced Energy Materials*, 12(22), 2200435.

Cruz-Pinto, J.J.C., Carvalho, M.E.S., Ferreira, J.F.A. (1994). The kinetics and mechanism of polyethylene photo-oxidation. *Angewandte Makromolekulare Chemie*, 216, 113–133. doi: 10.1002/apmc.1994.052160108.

Cui, X., Zheng, Y.F., Yin, H.Y., Song, X.C. (2015). Novel $C_3N_4/Zn_{1-x}Cd_xS$ heterostructures with adjustment of the band gap and their visible light photocatalytic properties. *Physical Chemistry Chemical Physics*, 17, 29354–29362. doi: 10.1039/c5cp05464h.

Cui, C., Guo, R., Xiao, H., Ren, E., Song, Q., Xiang, C., Lai, X., Lan, J., Jiang, S. (2020). Bi_2WO_6/Nb_2CT_x MXene hybrid nanosheets with enhanced visible-light-driven photocatalytic activity for organic pollutants degradation. *Applied Surface Science*, 505. doi: 10.1016/j.apsusc.2019.144595.

Cui, C., Guo, R., Ren, E., Xiao, H., Lai, X., Qin, Q., Jiang, S., Shen, H., Zhou, M., Qin, W. (2021). Facile hydrothermal synthesis of rod-like Nb_2O_5/Nb_2CT_x composites for visible-light driven photocatalytic degradation of organic pollutants. *Environmental Research*, 193, 110587. doi: 10.1016/j.envres.2020.110587.

Cui, Y., Zhu, J., Tong, H., Zou, R. (2023). Advanced perspectives on MXene composite nanomaterials: Types synthetic methods, thermal energy utilization and 3D-printed techniques. *iScience*, 26, 105824.

Danish, M.S.S., Estrella, L.L., Alemaida, I.M.A., Lisin, A., Moiseev, N., Ahmadi, M., Nazari, M., Wali, M., Zaheb, H., Senjyu, T. (2021). Photocatalytic applications of metal oxides for sustainable environmental remediation. *Metals*, 11(1), 80. doi: 10.3390/met11010080.

Du, H., Xie, Y., Wang, J. (2021). Microplastic degradation methods and corresponding degradation mechanism: Research status and future perspectives. *Journal of Hazardous Materials*, 418, 126377.

Du, H., Wang, Q., Chen, G., Wang, J. (2022). Photo/electro-catalytic degradation of micro- and nano-plastics by nanomaterials and corresponding degradation mechanism. *TrAC Trends in Analytical Chemistry*, 157, 116815.

Fairbrother, A., Hsueh, H.C., Kim, J.H., Jacobs, D., Perry, L., Goodwin, D., White, C., Watson, S., Sung, L.P. (2019). Temperature and light intensity effects on photodegradation of high-density polyethylene. *Polymer Degradation and Stability*, 165, 153–160. doi: 10.1016/j.polymdegradstab.2019.05.002.

Ghani, A.A., Shahzad, A., Moztahida, M., Tahir, K., Jeon, H., Kim, B., Lee, D.S. (2021). Adsorption and electrochemical regeneration of intercalated Ti_3C_2Tx MXene for the removal of ciprofloxacin from wastewater. *Chemical Engineering Journal*, 421(Part 2), 127780. doi: 10.1016/j.cej.2020.127780.

Gong, X., Tong, F., Ma, F., Zhang, Y., Zhou, P., Wang, Z., Liu, Y., Wang, P., Cheng, H., Dai, Y., Zheng, Z., Huang, B. (2022). Photoreforming of plastic waste poly (ethylene terephthalate) via in-situ derived CN-CNTs-NiMo hybrids. *Applied Catalysis B*, 307. doi: 10.1016/j.apcatb.2022.121143.

Hadi, A.J., Najmuldeen, G.F., Bin, Y.K. (2013). Dissolution/reprecipitation technique for waste polyolefin recycling using new pure and blend organic solvents. *Journal of Polymer Engineering*, 33, 471–481. doi: 10.1515/polyeng-2013-0027.

Hamd, W., Daher, E.A., Tofa, T.S., Dutta, J. (2022). Recent advances in photocatalytic removal of microplastics: Mechanisms, kinetic degradation, and reactor design. *Frontiers in Marine Science*, 9, 885614.

Hu, C., Zhang, L., Gong, J. (2019). Recent progress made in the mechanism comprehension and design of electrocatalysts for alkaline water splitting. *Energy and Environmental Science*, 12, 2620–2645. doi: 10.1039/C9EE01202H.

Hu, X., Ma, D., Zhang, G., Ling, M., Hu, Q., Liang, K., Lu, J., Zheng, Y. (2023). Microwave-assisted pyrolysis of waste plastics for their resource reuse: A technical review. *Carbon Resources Conversion*, 6, 215–228. doi: 10.1016/j.crcon.2023.03.002.

Ignatyev, I.A., Thielemans, W., Vander Beke, B. (2014). Recycling of polymers: A review. *ChemSusChem*, 7, 1579–1593.

Ihsanullah, I. (2022). MXenes as next-generation materials for the photocatalytic degradation of pharmaceuticals in water. *Journal of Environmental Chemical Engineering*, 10(3), 107381. doi: 10.1016/j.jece.2022.107381.

Iqbal, A., Sambyal, P., Koo, C.M. (2020). 2D MXenes for electromagnetic shielding: A review. *Advanced Functional Materials*, 30(47), 2000883.

Jan Kole, P., Löhr, A.J., Van Belleghem, F.G.A.J., Ragas, A.M.J. (2017). Wear and tear of tyres: A stealthy source of microplastics in the environment. *International Journal of Environmental Research and Public Health*, 14(10), 1265.

Jiao, X., Zheng, K., Chen, Q., Li, X., Li, Y., Shao, W., Xu, J., Zhu, J., Pan, Y., Sun, Y., Xie, Y. (2020). Photocatalytic conversion of waste plastics into C_2 fuels under simulated natural environment conditions. *Angewandte Chemie International Edition*, 59, 15497–15501. doi: 10.1002/anie.201915766.

Kamysbayev, V., Filatov, A.S., Hu, H., Rui, X., Lagunas, F., Wang, D., Klie, R.F., Talapin, D.V. (2020). Covalent surface modifications and superconductivity of two-dimensional metal carbide MXenes. *Science*, 369, 979–983. doi: 10.1126/science.aba8311.

Khazaei, M., Ranjbar, A., Arai, M., Sasaki, T., Yunoki, S. (2017). Electronic properties and applications of MXenes: A theoretical review. *Journal of Materials Chemistry C*, 5, 2488–2503.

Kuehnel, M.F. and Reisner, E. (2018). Solar hydrogen generation from lignocellulose. *Angewandte Chemie*, 57, 3290–3296.

Kumar, J.A., Prakash, P., Krithiga, T., Amarnath, D.J., Premkumar, J., Rajamohan, N., Vasseghian, Y., Saravanan, P., Rajasimman, M. (2022a). Methods of synthesis, characteristics, and environmental applications of MXene: A comprehensive review. *Chemosphere*, 286, 131607. doi: 10.1016/j.chemosphere.2021.131607.

Kumar, R., Manna, C., Padha, S., Verma, A., Sharma, P., Dhar, A., Ghosh, A., Bhattacharya, P. (2022b). Micro(nano)plastics pollution and human health: How plastics can induce carcinogenesis to humans? *Chemosphere*, 298, 134267. doi: 10.1016/j.chemosphere.2022.134267.

Kusmierek, E. (2020). A CeO_2 semiconductor as a photocatalytic and photoelectrocatalytic material for the remediation of pollutants in industrial wastewater: A review. *Catalysts*, 10(12), 1435.

Lai, S., Jeon, J., Jang, S.K., Xu, J., Choi, Y.J., Park, J.H., Hwang, E., Lee, S. (2016). Surface group modification and carrier transport properties of layered transition metal carbides (Ti2CT(x), T: -OH, -F and -O). *Nanoscale*, 8(2), 1216. doi: 10.1039/c5nr90214b.

Lee, Q.Y. and Li, H. (2021). Photocatalytic degradation of plastic waste: A mini review. *Micromachines*, 12(8), 907.

Li, N., Peng, J., Ong, W.J., Ma, T., Arramel, Zhang, P., Jiang, J., Yuan, X., Zhang, C. (2021). MXenes: An emerging platform for wearable electronics and looking beyond. *Matter*, 4, 377–407.

Li, W., Zhao, W., Zhu, H., Li, Z.J., Wang, W. (2023). State of the art in the photochemical degradation of (micro)plastics: From fundamental principles to catalysts and applications. *Journal of Materials Chemistry A*, 11, 2503–2527.

Liu, J. and Hua, E. (2020). Photocatalytic properties and energy band offset of a tungsten disulfide/graphitic carbon nitride van der Waals heterojunction. *RSC Advances*, 10, 5260–5267. doi: 10.1039/c9ra10213b.

Liu, Y.H., Wang, C.Y., Yang, S.L., Cao, F.F., Ye, H. (2022). 3D MXene architectures as sulfur hosts for high-performance lithium-sulfur batteries. *Journal of Energy Chemistry*, 66, 429–439.

Long, Y., Tao, Y., Shang, T., Yang, H., Sun, Z., Chen, W., Yang, Q.H. (2022). Roles of metal ions in MXene synthesis, processing and applications: A perspective. *Adv. Sci.* 9, 2200296.

Melchionna, M. and Fornasiero, P. (2020). Updates on the roadmap for photocatalysis. *ACS Catalysis*, 10, 5493–5501. doi: 10.1021/acscatal.0c01204.

Merrington, A. (2017). Recycling of plastics. In *Applied Plastics Engineering Handbook: Processing, Materials, and Applications*, 2nd edition, Kutz, M. (ed.). William Andrew, Norwich, and Elsevier, Amsterdam.

Mishra, A., Srivastava, P., Mizuseki, H., Lee, K.R., Singh, A.K. (2016). Isolation of pristine MXene from Nb4AlC3 MAX phase: A first-principles study. *Physical Chemistry Chemical Physics*, 18, 11073–11080. doi: 10.1039/c5cp07609a.

Mohanan, N., Montazer, Z., Sharma, P.K., Levin, D.B. (2020). Microbial and enzymatic degradation of synthetic plastics. *Frontiers in Microbiology*, 11, 580709.

Naguib, M., Mashtalir, O., Carle, J., Presser, V., Lu, J., Hultman, L., Gogotsi, Y., Barsoum, M.W. (2012). Two-dimensional transition metal carbides. *ACS Nano*, 6, 1322–1331. doi: 10.1021/nn204153h.

Ouyang, Z., Yang, Y., Zhang, C., Zhu, S., Qin, L., Wang, W., He, D., Zhou, Y., Luo, H., Qin, F. (2021). Recent advances in photocatalytic degradation of plastics and plastic-derived chemicals. *Journal of Materials Chemistry A*, 9, 13402–13441.

Peng, C., Wang, H., Yu, H., Peng, F. (2017) TiO_2-x/Ti_3C_2: Synergy of active facets, interfacial charge transfer and Ti3+ doping for enhance photocatalytic activity. *Materials Research Bulletin*, 89, 16–25. doi: 10.1016/j.materresbull.2016.12.049.

Qiu, B., Deng, Y., Li, Q., Shen, B., Xing, M., Zhang, J. (2016). Rational design of a unique ternary structure for highly photocatalytic nitrobenzene reduction. *Journal of Physical Chemistry C*, 120, 12125–12131. doi: 10.1021/acs.jpcc.6b03800.

Rahman, A., Jennings, J.R., Tan, A.L., Khan, M.M. (2022a). Molybdenum disulfide-based nanomaterials for visible-light-induced photocatalysis. *ACS Omega*, 7, 22089–22110.

Rahman, U.U., Humayun, M., Ghani, U., Usman, M., Ullah, H., Khan, A., El-Metwaly, N.M., Khan, A. (2022b). MXenes as emerging materials: Synthesis, properties, and applications. *Molecules*, 27(15), 4909.

Schneider, J., Matsuoka, M., Takeuchi, M., Zhang, J., Horiuchi, Y., Anpo, M., Bahnemann, D.W. (2014). Understanding TiO_2 photocatalysis: Mechanisms and materials. *Chemical Reviews*, 114(19), 9919–9986.

Shang, J., Chai, M., Zhu, Y. (2003). Solid-phase photocatalytic degradation of polystyrene plastic with TiO_2 as photocatalyst. *Journal of Solid State Chemistry*, 174, 104–110. doi: 10.1016/S0022-4596(03)00183-X.

Sharma, G., Kumar, A., Sharma, S., Naushad, M., Dhiman, P., Vo, D.V.N., Stadler, F.J. (2020). Fe_3O_4/ZnO/Si_3N_4 nanocomposite based photocatalyst for the degradation of dyes from aqueous solution. *Materials Letters*, 278. doi: 10.1016/j.matlet.2020.128359.

Sharma, S.K., Kumar, A., Sharma, G., Vo, D.N., García-Peñas, A., Moradi, O., Sillanpää, M. (2022). MXenes based nano-heterojunctions and composites for advanced photocatalytic environmental detoxification and energy conversion: A review. *Chemosphere*, 291, 132923. doi: 10.1016/j.chemosphere.2021.132923.

Shi, Y., Li, T., Zhao, L., Liu, Y., Ding, K., Li, D., He, P., Jiang, D., Liu, J., Zhou, H. (2023). Ultrathin MXene nanosheet-based TiO_2/CdS heterostructure as a photoelectrochemical sensor for detection of CEA in human serum samples. *Biosensors and Bioelectronics*, 230, 115287. doi: 10.1016/j.bios.2023.115287.

Shirazimoghaddam, S., Amin, I., Faria Albanese, J.A., Shiju, N.R. (2023). Chemical recycling of used PET by glycolysis using niobia-based catalysts. *ACS Engineering Au*, 3, 37–44. doi: 10.1021/acsengineeringau.2c00029.

Sobolčiak, P., Ali, A., Hassan, M.K., Helal, M.I., Tanvir, A., Popelka, A., Al-Maadeed, M.A., Krupa, I., Mahmoud, K.A. (2017). 2D Ti3C2Tx (MXene)-reinforced polyvinyl alcohol (PVA) nanofibers with enhanced mechanical and electrical properties. *PLOS ONE*, 12, e0183705. doi: 10.1371/journal.pone.0183705.

Soundiraraju, B. and George, B.K. (2017). Two-dimensional titanium nitride (Ti_2N) MXene: Synthesis, characterization, and potential application as surface-enhanced Raman scattering substrate. *ACS Nano*, 11, 8892–8900. doi: 10.1021/acsnano.7b03129.

Sun, W., Shah, S.A., Chen, Y., Tan, Z., Gao, H., Habib, T., Radovic, M., Green, M.J. (2017). Electrochemical etching of Ti_2AlC to Ti_2CT:X (MXene) in low-concentration hydrochloric acid solution. *J. Mater. Chem. A*, 5, 21663–21668. doi: 10.1039/C7TA05574A.

Takanabe, K. and Shahid, S. (2017). Dehydrogenation of ethane to ethylene via radical pathways enhanced by alkali metal based catalyst in oxysteam condition. *AIChE Journal*, 63, 105–110. doi: 10.1002/aic.15447.

Tamoor, M., Samak, N.A., Jia, Y., Mushtaq, M.U., Sher, H., Bibi, M., Xing, J. (2021). Potential use of microbial enzymes for the conversion of plastic waste into value-added products: A viable solution. *Frontiers in Microbiology*, 12, 777727.

Tang, L., Kuai, L., Li, Y., Li, H., Zhou, Y., Zou, Z. (2018). $Zn_xCd_{1-x}S$ tunable band structure-directing photocatalytic activity and selectivity of visible-light reduction of CO_2 into liquid solar fuels. *Nanotechnology*, 29(6), 064003. doi: 10.1088/1361-6528/aaa272.

Ücker, C.L., Riemke, F.C., de Andrade Neto, N.F., Santiago, A.A.G., Siebeneichler, T.J., Carreño, N.L.V., Moreira, M.L., Raubach, C.W., Cava, S. (2021). Influence of Nb_2O_5 crystal structure on photocatalytic efficiency. *Chemical Physics Letters*, 764. doi: 10.1016/j.cplett.2020.138271.

Verma, R., Vinoda, K.S., Papireddy, M., Gowda, A.N.S. (2016). Toxic pollutants from plastic waste – A review. *Procedia Environmental Sciences*, 35, 701–708. doi: 10.1016/j.proenv.2016.07.069.

Vivekanand, A.C., Mohapatra, S., Tyagi, V.K. (2021). Microplastics in aquatic environment: Challenges and perspectives. *Chemosphere*, 282, 131151.

Wang, X., Garnero, C., Rochard, G., Magne, D., Morisset, S., Hurand, S., Chartier, P., Rousseau, J., Cabioc'h, T., Coutanceau, C., Mauchamp, V., Célérier, S. (2017). A new etching environment (FeF_3/HCl) for the synthesis of two-dimensional titanium carbide MXenes: A route towards selective reactivity vs water. *J. Mater. Chem. A*, 5, 22012–22023. doi: 10.1039/C7TA01082F.

Wang, T., Hu, M., Xu, G., Shi, H., Leung, J.Y.S., Wang, Y. (2021). Microplastic accumulation via trophic transfer: Can a predatory crab counter the adverse effects of microplastics by body defence? *Science of the Total Environment*, 754, 142099. doi: 10.1016/j.scitotenv.2020.142099.

Wei, Z., Peigen, Z., Wubian, T., Xia, Q., Yamei, Z., ZhengMing, S. (2018). Alkali treated Ti_3C_2Tx MXenes and their dye adsorption performance. *Materials Chemistry and Physics*, 206, 270–276. doi: 10.1016/j.matchemphys.2017.12.034.

Xin, M., Li, J., Ma, Z., Pan, L., Shi, Y. (2020). MXenes and their applications in wearable sensors. *Frontiers in Chemistry*, 8, 297.

Xu, M., Huang, C., Lu, J., Wu, Z., Zhu, X., Li, H., Xiao, L., Luo, Z. (2021). Optimizing adsorption of 17α-ethinylestradiol from water by magnetic MXene using response surface methodology and adsorption kinetics, isotherm, and thermodynamics studies. *Molecules*, 26. doi: 10.3390/molecules26113150.

Zervoudakis, A.J., Sample, C.S., Peng, X., Lake, D., Hillmyer, M.A., Ellison, C.J. (2022). Dihydroxy polyethylene additives for compatibilization and mechanical recycling of polyethylene terephthalate/polyethylene mixed plastic waste. *ACS Macro Letters*, 11, 1396–1402. doi: 10.1021/acsmacrolett.2c00601.

Zhan, X., Si, C., Zhou, J., Sun, Z. (2020). MXene and MXene-based composites: Synthesis, properties and environment-related applications. *Nanoscale Horizons*, 5, 235–258.

Zheng, K., Wu, Y., Hu, Z., Wang, S., Jiao, X., Zhu, J., Sun, Y., Xie, Y. (2023). Progress and perspective for conversion of plastic wastes into valuable chemicals. *Chemical Society Reviews*, 52, 8–29.

6

Metal–Organic Framework based on Functional Materials for Photocatalytic Degradation of Micro- and Nano-Plastic

Micro-plastics (MPs) and nano-plastics (NPs) are considered as ecotoxicological hazards due to their capability to collect and transport lethal metals, organic pollutants or medical products. The release of these MPs and NPs into the atmosphere has produced considerable concern related to health and the environment. These MPs and NPs emerge in aquatic surface water from the disintegration of large plastics, through chemical degradation processes (ultraviolet (UV) biodegradation, mechanical and photodegradation), as well as from the transport in the marine environment as synthetic microparticles and microbeads (e.g. integrated into cosmetics, or during synthetic clothes washing). Removing MPs and NPs from water remains a severe challenge that needs to be urgently solved. Therefore, it is necessary to explore smart functional materials for managing the MPs and NPs that pose a massive risk to organisms in the ecosystem. Metal–organic framework (MOF)-based materials have interpenetrated pore, uniformity and excellent durability, which can be employed the removal of MPs and NPs from water. The aim of this chapter is to explore several efficient MOFs for the photocatalytic degradation of MPs and NPs. This chapter offers a vision in fabricating efficient heterojunction photocatalysts and converting micro- and nano-pollutants into value-added chemicals, coupled with clean energy production.

6.1. Introduction

The area of MOF has been an emerging and extremely fashioned scientific domain of material science. MOFs deal with the organic–inorganic hybrid materials, in which metal ions or clusters are linked by organic ligands via coordination bonds into indefinite crystal structures (Figure 6.1).

Chapter written by Vinita, Madhu TIWARI, Pravesh Kumar YADAV, Arun Pratap VERMA, Chandrakala SINGH and Sudhakar PANDEY.

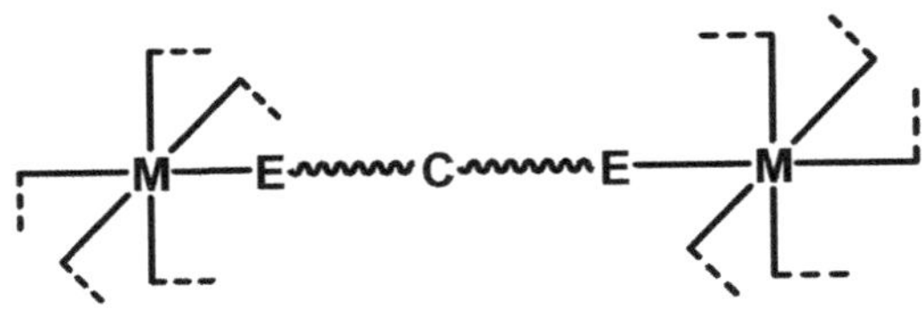

Figure 6.1. *Structural representation of metal–organic framework*

The MOFs show an attractive stability, polyfunctionality, adjustability and often has tunable porous structures with very large specific surface areas. The synthesis of MOFs can be regarded as "construction games" where the final architecture mainly depends on the molecular building modules of ligands, solvent and metal ions. Thus, the study of geometries with diversified interactions and optimized growth processes is known as crystal engineering. In particular, the building blocks are considered as a series of reproducible molecular recognition events, remarking that the overall module regulation can be imaginable along with the dimensionalities and topography resembling with single chemical entities of self–assembly and supra-molecular chemistry. The self-assembly of the MOF is based on the explicitness and complementary interactions of the molecular units to engineer the final building block architectures. The metal ions designated as nodes and ligands are labeled as linkers in molecular architectures.

6.2. Historical background and discovery of metal–organic frameworks

In the history of MOFs, Diesbach, a German pigment and dye producer from Berlin, prepared a red pigment of cochineal red lake from 1704 to 1705 (Ball 2001). Although the method was quite simple, using potash and iron sulfate, due to tiny mistake, it turned pale, which consequently resulted in dark blue when increasing concentration. Diesbach synthesized Prussian blue by using extensively cheap potash with animal oil impurities, and thus Prussian blue was the first man-made MOF (Herren et al. 1908; Buser et al. 1977). Within a short time period, it was commercially available as a valuable pigment and its recipe was continuously scrutinized. Later on, a Russian batch reported the chemical structures of $Cd(CN)_2$ and $Zn(CN)_2$ during Second World War. Furthermore, Rayner and Powell ascertained the chemical structures of Hoffmann clathrate as $[Ni(NH_3)_2Ni(CN)_4]\cdot 2C_6H_6$ (Powell and Rayner 1949). Thereafter, extensive studies followed by the Iwamoto group, further explored the research in this area. A Japanese researcher group accounted the structure of $[Cu(adiponitrile)]NO_3$ containing six penetrating diamond networks and the 1D chain structure of $Ag(pyrizine)NO_3$ and $Cu(pyrazine)(NO_3)_2$ was established in 1966 and 1970, respectively (Kinoshita et al. 1959). In the 1980s, the entire focus was on magnetic properties-based molecules, which ultimately resulted a research communication in 1989 and then an extension to a full article in 1990 (Hoskins and Robson 1990). Successively, Robson, Hoskins and their coworkers sketched an assembly-based approach to engineer the MOF. They described the crystal structures of MOFs in terms of networks and employed it to develop new MOFs, with fascinating topology and interesting properties that may be advantageous in several applications of catalysis and sensing (Wells 1977).

Hybrid inorganic and organic materials and MOFs have received much attention in recent years. These MOFs deal with the repeating unit of the organic ligand attached to a metal in an infinite array and the functioning of these materials can be controlled with the choice of their individual components. Here, the assembly directed in various dimensions and volitionally chosen molecular building blocks regulate the physicochemical properties and structure of the materials up to the molecular level (Fromm et al. 2005). Recently, the metal–ligand coordination bond has found versatile applications in the molecular organization of various supra-molecular architectures, introducing the strength and specific directionality to the coordination bonds affiliated with metals. Metal complexes or ions and their bridging organic ligands assembled throughout the materials with various dimensionalities (1D, 2D and 3D) are referred to as MOFs. The engineering of the design and synthesis of MOFs by the assembly of organic ligands as connecting nodes and metal ions as linkers have achieved significant attention in the family and supra-molecular chemistry in the last few decades (Robin et al. 2003).

Several MOFs with high porosity and surface area persisting even upon removal of the solvent molecules has attracted a special and appealing interest because of their frontier potential applications in the broad spectrum of technologies such as catalysis, drug delivery, ion exchange, gas storage and separation. Among the various molecular building blocks studied, the organic ligands with electronegative active atoms are found to be highly efficient and attracting toward the construction of MOFs due to their stability and relative ease of functionalization due to flexibility, lone pair of electrons on heteroatoms. These organic ligands can be functionalized to build preprogrammed building blocks leading to extended structures with outstanding properties and topologies (Moulton and Zaworotko 2001).

6.3. Bonding in metal–organic frameworks

The extensive research in MOFs is greatly influenced by two closely associated areas, that is, metallo–supramolecular chemistry and crystal engineering. Crystal engineering deals with the molecular packing and arrangement of molecules in the materials. This is important because molecular properties are specially governed by the spatial arrangement of molecules. Thus, the molecular properties can be regulated by regulating the arrangements of molecules. The interactions are stronger in MOFs ranging from strong hydrogen bonding to π–π interactions and van der Waals forces. The researchers are investigating the interactions that regulate the bonding architecture of MOFs and a net-based advancement for MOFs is found to be a significantly valid for molecular architectures associated with well-established interactions (Barnett and Champness 2003).

Numerous concepts such as van der Waals forces and hydrogen bonding are introduced for bonding interactions in the modeling of MOFs. The intermolecular forces that govern synthons and assembly arrangements in MOFs are hydrogen bonding and chelation. The designing of MOFs resembles the supra-molecular chemistry, which engineers the building blocks of ligands with metal ion tectons. Enormous reports have been contributed in the area of MOFs to define and probe the structural engineering of MOFs. Incorporating metal ions leads to the achievement of required properties by regulating the metal atom positions throughout the coordination network. The organic ligands and types of metals are capable of tuning the interactions in a MOF to acquire the desired functional solid material. The possible diversity in the tecton and nodes lighten the thinking of investigators for the new aspects in topologies and intriguing designing, since the crystal study is much easier with the advancement of X-ray and computational resolution techniques.

6.4. Dimensionality of metal–organic frameworks

The self-assembly of molecular building blocks led to the MOFs of variant dimensionality (e.g. 1D, 2D and 3D). Linearly organized organic ligands with two donor molecules (D), metal ions (M) and the spacer (S) are presented in black (Figure 6.2). Here, dimensionality is regulated by the nodal centres (metal ions). 1D MOFs are those in which metal ions are in coordination with two donor organic ligands in the chain. 2D MOFs have metal ions in coordination with three or four organic ligands, thus expanding the polymeric array in two directions. Similarly, 3D MOFs can be constructed by metal ion coordinating to higher ligands.

The repeating units have variations depending on the experimental conditions and elementary units of building blocks (Kitagawa and Noro 2004). The MOFs primarily have the metal–ligand interactions, although there are several crucial interactions existing in the fabrication of these building blocks but the final dimensionality of MOFs is only dependent on the metal coordination with ligands.

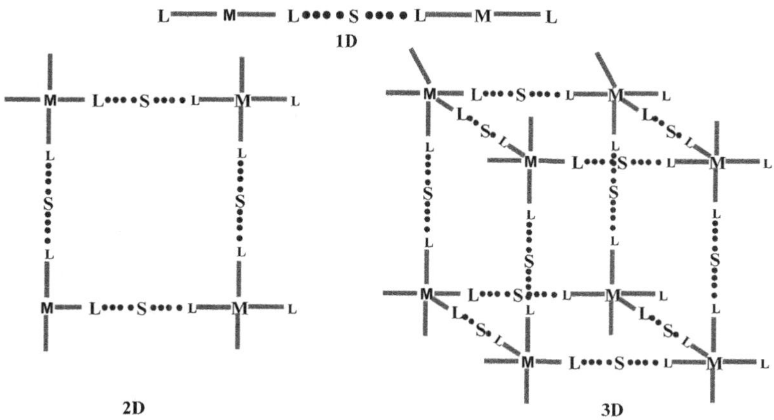

Figure 6.2. *MOFs with various dimensionalities: M: metal ions; S: spacer; and L: linker (Janiak 2003). For a color version of this figure, see www.iste.co.uk/singh/photodegradation.zip*

6.5. Methods for the synthesis of metal–organic frameworks

Several synthesis methods for MOFs have been illustrated, such as mechanochemical, microwave, solvothermal, sono-chemical and electrochemical. MOFs fill their pores with organic solvents, water and unreacted initial compounds since the dense structures are thermodynamically more stable. Therefore, construction of inorganic moiety and insertion of foreign molecules into the porous assembly play a crucial part in development of MOFs. A crucial step is post-synthetic dealing of MOFs purification and activation. Here, purification is a key factor for catalytic utilization of MOFs, as by-products and impurities may mimic stimulate and notably deteriorate the absorption potentials. Purification typically involves solvent treatment, often at high temperatures. Material activation infers pore evacuation, which is often a tedious job since the elimination of inclusions from a porous framework by heating may cause the breakdown of assembly, particularly if guest molecules and assembly is bounded by relatively robust bonds or if inclusion elimination needs high temperatures. There are diverse approaches in reducing complexity in activation of MOFs. The most popular is substituting more volatile solvent, facilitating the reduction in treated temperature. The reproducibility of synthesis outcomes is crucially important. The structure and properties of MOFs changes from experiment to experiment due to existence of pores and organic moieties in the complex structure. Similarly, variation in other properties such as thermal stability as a function of preliminary moiety ratio (Zr:BDC:DMF) and synthesis temperature has investigated for MOF UiO-66. Furthermore,

Hafizovic et al. (2007) investigated the assembly and properties of MOF-5 as a function of post-synthetic and synthesis reaction parameter. MOFs formulation is greatly influenced by solvent and various additives of reaction. Furthermore, Li et al. (1999) have disclosed that triethylamine addition encourages reaction of H_2BDC with $ZnNO_3$ due to increased pH, supportive for acid deprotonation in the synthesis of MOF-5. A tiny addition of H_2O_2 assists O^{2-} production, which is the focus of secondary metal–oxygen bunches (Shearer et al. 2014).

Several MOFs mentioned as MIL (Materials of Institute Lavoisier) have been formulated and synthesized at Lavoisier Institute in Versailles by solvothermal method. Their formulation and synthesis are listed in Table 6.1.

MOF	Formula	Time (h)	Temperature (°C)	Reference
MIL-122	$M_2(OH)_2[C_{14}H_4O_8]$ [M = Al, Ga, In] [$M(NO_3)_3$(M = Al, Ga or In), naphthalene-1,4,5, 8-tetracarboxylic acid ($C_{14}H_8O_8$)]	24	210	Volkringer et al. (2009)
MIL-116	$M_2(OH)_2[C_6(CO_2)_6]\cdot 2H_2O$ (M = Al, Ga or In) [$M(NO_3)_3$ (M = Al, Ga or In), benzenehexacarboxylic acid]	24	210	Volkringer et al. (2013)
MIL-74	$M_3Al_6(PO_4)_{12}\cdot 4(H_3NCH_2CH_2)_3N\cdot 17H_2O$ (M = Mg, Mn, Co) [$Mg(NO_3)_2\cdot 6H_2O$ [or $MnCl_2\cdot 4H_2O$, or $Co(NO_3)_2\cdot 6H_2O$], $Al(OH)_3$, H_3PO_4, $(H_2NCH_2CH_2)_3N$]	36	180	Loiseau et al. (2006)
MIL-69	$Al(OH)(O_2CC_{10}H_6CO_2)\cdot H_2O$ ($Al(NO_3)_3\cdot 9H_2O$, naphthalene-2,6-dicarboxylic acid ($HO_2CC_{10}H_6CO_2H$), KOH)	16	210	Loiseau et al. (2005)
MIL-64	$Al_{10}(PO_4)_4(HPO_4)_6F_9(H_2O)_4\cdot OH\cdot 2H_3N(CH_2)_3NH_3$ ($Al(OH)_3$, H_3PO_4, HF, $H_2N(CH_2)_3NH_2$)	720	180	Simon et al. (2006)
MIL-82	$[Fe^{III}(OH)[C_6H_2(CO_2)_2(CO_2H)_2]\cdot 0.88H_2O$ ($FeCl_2$, HF, $C_6H_2(CO_2H)_4$-1,2,4,5)	48	200	Sanselme et al. (2004)
MIL-49	$Fe^{III}(H_2O)(O_3P(CH_2)CO_2)$ (Fe, $(HO)_2(O)PCH_2CO_2H$, NaOH)	48	170	Sanselme et al. (2002)
MIL-38	$Fe_3^{II}(OH)_2(H_2O)_4(O_3P(CH_2)_2CO_2H)_2$ (Fe, $(OH)_2(O)P(CH_2)_2CO_2H$, NaOH)	48	170	Riou-Cavellec et al. (2002)
MIL-65	$[Fe_3^{III}(H_2O)_5(BTC)_2\cdot 3H_2O]$ (Fe, VCl_3, H_3BTC, NaOH)	48	180	Riou-Cavellec and Férey (2002)
MIL-37	$Fe(OH)(H_2O)(O_3P(CH_2)_2CO_2H$ ($FeCl_3\cdot 6H_2O$, $(OH)_2(O)P(CH_2)_2CO_2H$, NaOH)	24	170	Riou-Cavellec et al. (2000)

Table 6.1. *Example of some synthesized MOFs with the reaction condition*

6.5.1. *Ultrasonic synthesis*

The effect of ultrasonic waves on colloid and liquid systems is basically due to cavitation. First, resonant bubbles serve as an agitator, thus enhancing reactant's contact area. Additionally, thermal collision and differences in pressure lead to particle-aggregate breakdown, thus increasing the contact area. However, sonication has several other changes, including bond breaking in polymer chains, removal of solvation and adsorption shells from material surface and surface sites formation with uncompensated chemical and physical bonds. However, volatile solvents are not suitable for this synthesis as the high vapor pressure of solvent inside bubbles diminishes effects that lead to bubble implosion. First, $Zn_3(BTC)_2$ was synthesized in 2008 by the ultrasonication for 90 minutes of H_3BTC and zinc acetate in 20% ethanol (Qiu et al. 2008). Further synthesis of HKUST-1 was compared and the maximum yield with the least synthesis time was attained by MW radiation (Schlesinger et al. 2010).

6.5.2. *Electrochemical synthesis*

MOFs synthesis through electrochemical meaning resulted from electrochemical processes rather than solutions of metal ion salts. Metal ions are delivered in a mixture of synthon and electrolyte by the dissolution of anode, thus preventing the anion production and providing high yield of MOFs. First, green-blue precipitate of HKUST-1 was electrochemically synthesized in 150 minutes using a copper sheet as anode dipped in H_3BTC solution in methanol (Mueller et al. 2006). Later, several modifications were proposed by Van Assche et al. (2012).

6.5.3. *Mechanochemical synthesis*

In this method, the reaction between solids is started by mechanical energy, such as milling in ball mills. This technique is popular as it rapidly facilitates the reaction without solvent or with a tiny amount of solvent and a high yield of product obtained. The general theory for mechanochemical is not yet available. However, the magma ± plasma model and the hot-spot model have been attempted well. According to the former model, the temperature at the contact spot of particles goes beyond 104°C due to the instant plasma formation and huge energy release along free electron emissions (James et al. 2012). The other model illustrates that the localized distortion of the ductile substance with nearly 1 m^2 causes a spike in temperature of nearly 1,000°C for short time and brittle substances have hot spots, developed at the propagating cracks ending. It is assumed that the small duration may start reactions between milled parts in spite of their decomposition. Also, liquid presence in such reactions provide benefits of higher yield and easy crystallization.

The presence of a liquid component during mechanochemical reaction furnish additional benefits: facile crystallization, larger crystallinity of reaction product and larger yield due the larger mobility of reactants. Additionally, restless milling of components promotes diffusion and makes new surfaces with enhanced reactivity (Boldyrev 2006). First, MOF of iso-nicotinic acid and copper acetate, $Cu(INA)_2.xH_2O.yAcOH$, was prepared in 2006 in a ball mill. They acetic acid and water produced in reaction in the MOF's pores were eliminated by heating (Pichon et al. 2006). Later in 2010, HKUST-1 was synthesized using this method. H_3BTC and copper acetate both were used in solvent-free conditions in the aforesaid synthesis. Yuan et al. (2010) concluded that $Cu(INA)_2$ synthesized within 1 min milling and small addition of acetic acid sharply increases the rate of reaction, whereas HKUST-1 synthesis needs a longer time mixing, which may be due to the more porous nature of HKUST-1.

6.5.4. *Microwave synthesis*

The electrical component of microwaves (MWs) is involved in the synthesis of compounds. Heating resulted due to MW impact on free ions and polar molecules. Heating in electrolyte solutions is generated due to resistance of electric field. Resistance in polar molecules is due to hydrogen bonds that inhibit the dipole alignment with variable electric field. Free rotation, dipole moment, etc. are various factors affecting the heating efficiency. MW radiation affects the liquid state of water the strongest, rather than vapor with low concentration and no hydrogen bonds and thus no heating due to absence of resistance. In case of a solid state, MW radiation is not sufficient as the reorientation of firmly bounded ice molecules is not possible (Horikoshi and Serpone 2013).

First, MIL-100 was synthesized by MW radiation in 2005 in only 4 h with similar yield rather than 96 h by conventional hydrothermal method. The mixture of chromium, aqueous solution of HF and H_3BTC was mixed, kept in autoclave, and then gradually heated to 220°C. An isostructural iron analogue of same moiety was synthesized in 30 min using MW radiation (Taylor-Pashow et al. 2009).

6.6. Properties of metal–organic frameworks

MOFs are highly efficient sensors due to their high surface area and porous structure. Also, this high surface area, with the presence of functional groups, allows them to carry drugs proficiently and can be easily conjugated with active targeted ligands to further enhance drug accumulation in diseased locations and increase therapeutic efficacy. The coordination-assisted assembly strategy is expected to accelerate the development of new generation of photo-functional materials by rationally tuning optical properties of chromophores and properly selecting of the metal species (Zhang et al. 2009). Subsequent investigations revealed that a fluorescence switching between the nonfluorescent- and fluorescent-stretched compressed states of the supramolecular springs occur due to this dynamic conformational change. Such mechanical motions of the supramolecular spring offer a potential application in optical modulators, dynamic nano-devices and fluorescent thermometers (Kim et al. 2007). Ultrasonic degradation of polymers is quite a slow process but is highly selective, and the scission rate is chain-length dependent. N-heterocyclic carbene (NHC)-based MOF has an efficient ultrasonic scission and can be quantitatively cracked in minutes to provide highly reactive carbenes. Scission is irreversible in the presence of water. The mechano-chemical process that produce carbenes catalyzes organic transformations, for example, lactides polymerization and trans-esterification (Karthikeyan et al. 2008). Their helical MOF display temperature-regulated reversible extension–contraction motions. Furthermore, the aqueous solutions of MOF reversibly transform from transparent to translucent states above certain temperatures. Subsequent studies revealed that a fluorescence switching between the fluorescent-stretched and nonfluorescent-compressed states of the supramolecular helics occur due to this dynamic conformational change. Such mechanical motions of the supramolecular helical may be promising applications in the areas of dynamic nano-devices, optical modulators and fluorescent thermometers (Kim et al. 2007). MOFs can be employed as templates and sacrificial templates to engineer desired nanostructures (Moon et al. 2005).

6.7. Micro- and nano-plastics

MPs refer to non-soluble particles smaller than 5 mm, often imperceptible to the naked eye. The escalation of concern surrounding MPs stems from their potential threats to aquatic life and human health. These particles not only contribute to the accumulation of plastics in the environment but also facilitate the dissemination of micropollutants through absorption. MPs adsorb organic and metal contaminants, which are subsequently released into aquatic organisms upon ingestion, leading to bioaccumulation along the food chain. The natural degradation of MPs can span decades, during which a variety of hazardous organic compounds are released from the polymers and additives used in manufacturing, such as plasticizers and antioxidants (Tanaka and Takada 2016; Zhang et al. 2017).

The origins of MPs are diverse, ranging from inputs via wastewater and runoffs from paved surfaces to inadequate waste management practices. Although wastewater treatment can mitigate the influx of plastics, a considerable number of particles still find their way into rivers globally. Additionally, biosolids, residues of wastewater treatment, may be repurposed in agriculture as fertilizers, potentially introducing MPs into soils and valuable groundwater resources through pathways that remain insufficiently understood.

NPs refer to a subset of MPs with dimensions equal to or less than 100 nm in at least one dimension. Nanoparticles typically demonstrate distinct chemical and physical characteristics compared to larger objects composed of the same material. Consequently, factors such as surface area and other physicochemical properties of nanoparticles can significantly influence their behaviour and availability in the environment. Thus, the disparities between MPs and NPs are substantial, particularly concerning their interactions with the environment and organisms. An in-depth comprehension of how NPs interact with the environment, particularly living organisms, is imperative for assessing potential health risks, given that NP particles may exhibit unique reactions compared to their larger counterparts (Mintenig et al. 2018; Al-Thawadi 2020; Jung et al. 2021).

Typical materials of plastic, for example, polyethylene terephthalate (PET), high-density polyethylene (HDPE), low-density polyethylene (LDPE), polypropylene (PP), polyvinyl chloride (PVC) and polystyrene (PS), are extensively utilized and studied. Additionally, there are various other plastics, including nylon, acrylic, polylactic acid, polyurethanes, polycarbonate and biodegradable plastics.

Recent studies have reported the presence of NPs in the North Atlantic (ter Halle et al. 2017), alongside the discovery of styrene monomers and oligomers in coastal sands and waters across the globe. These findings indicate that the anthropogenic signature of the colloidal fraction of seawater is attributable to a mixture of plastic sources. Notably, the polymer composition varied across different-sized categories. Specifically, PVC, PET, PS and polyethylene (PE) were identified at both the micro- and nanoscales. Additionally, changes in the pyrolytic signals of PE were observed with decreasing debris size, suggesting structural alterations resulting from weathering.

The proliferation of MP and NP pollution poses a significant and escalating global hazard to both aquatic and terrestrial ecosystems. These minute plastic particles are categorized in environmental studies primarily according to their dimensions and molecular composition, falling into three main types: primary MPs, secondary MPs and NPs. Primary MPs refer to minuscule plastic particles deliberately manufactured for diverse commercial applications (Murphy et al. 2016; Auta et al. 2017).

6.7.1. *Photocatalytic degradation of micro- and nano-plastics*

Photocatalytic degradation involves photocatalysts, such as semiconductors to encourage the organic pollutants degradation including MPs/NPs under the influence of light. Upon irradiation with UV or visible light, photocatalysts generate electron–hole pairs, which can initiate oxidation and reduction reactions with surrounding molecules, such as water and dissolved oxygen (Al-Mamun et al. 2019; Zhang et al. 2019). These reactions produce highly reactive species, which can effectively break down the chemical bonds in MPs, ultimately converting them into less harmful by-products, such as carbon dioxide and water.

One of the main advantages of photocatalytic technologies is their potential to degrade a wide range of MP types, including PS (Bandyopadhyay and Basak 2007), PE (Venkataramana et al. 2021) and PP (Kamrannejad et al. 2014), typically found in the environment. However, photocatalytic degradation can be employed to provide a universal solution to MP pollution from several sources. Furthermore, photocatalytic degradation can definitely lead to the absolute mineralization of MPs.

The potency of photocatalytic degradation can vary based on several elements, for example, the type of plastic or catalyst used, its light source and environmental conditions. The pursuit of photocatalytic tools in resolving MP pollution is a vital and thus areas of research holding notable potential for reducing this universal environmental concern are increasing.

6.7.2. *Mechanism of photocatalytic degradation*

Photocatalysis stands out as an environmentally friendly and energy-efficient method for degrading MPs and NPs. When exposed to appropriate light sources, photocatalysts or semiconductors undergo excitation, resulting in the formation of excited electrons (e–) and holes (h+). In the photocatalytic degradation process of MPs/NPs, the highly oxidizing photogenerated h+ can directly oxidize plastics, while certain free radicals (e.g. •OH and O_2•–) generated from the reaction of e– and h+ with H_2O, OH– and O_2 also contribute to plastic degradation. Simultaneously, the highly reductive e– can reduce H_2O into molecular hydrogen (Figure 6.3) (Uheida et al. 2021; Liu et al. 2022).

Photocatalysts such as Ti-, Zn-, Sn-, Bi-, and Cu-based oxides/sulfides/MOFs are widely used for degrading MPs/NPs from water (Domínguez-Jaimes et al. 2021; Jiang et al. 2021), with physicochemical properties of the catalysts significantly influencing their catalytic performance. In the degradation of PS NPs, for instance, a TiO_2 catalyst with a multilayer structure (TiO_2/M) outperforms other TiO_2 structures due to improved contact between MPs and the TiO_2/M photocatalyst and a lower bandgap (Haris et al. 2021). Additionally, the development of heterostructures and magnetic composite catalysts, acting as micromotors/microrobots, is promising for efficient MP/NP degradation (Beladi-Mousavi et al. 2021; Hu et al. 2022).

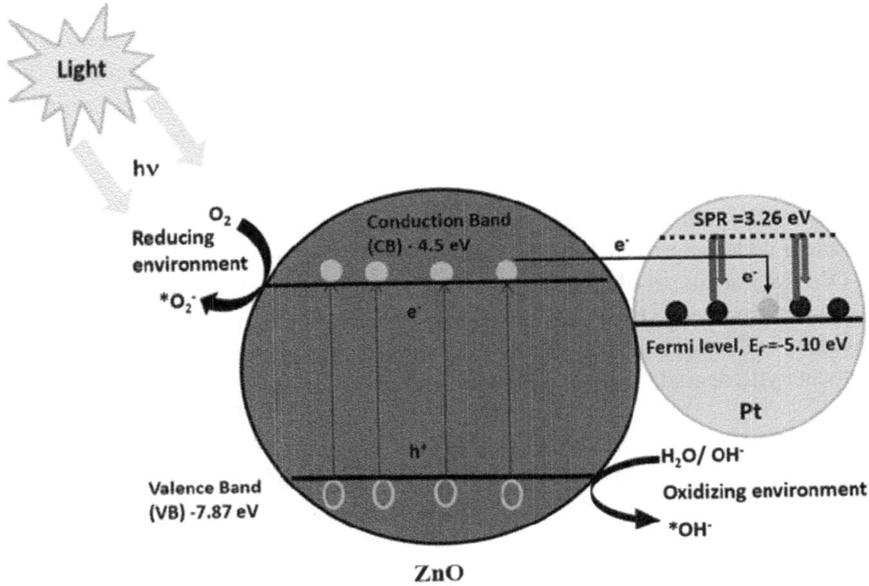

Figure 6.3. *Mechanism of photodegradation using ZnO–Pt photocatalyst (Tofa et al. 2019). For a color version of this figure, see www.iste.co.uk/singh/photodegradation.zip*

Another significant advancement in photocatalysis involves coupling hydrogen production with MP/NP degradation. For instance, MXene/ZnxCd1-xS photocatalysts can convert PET MPs into value-added organics while enabling hydrogen evolution from water (Cao et al. 2022). This process holds economic and environmental significance in addressing plastic pollution and energy challenges. Critical factors influencing the photocatalytic degradation of MPs/NPs include the properties of plastics and pH value.

Small-sized and dark-colored MPs/NPs exhibit higher degradation potential due to their high surface area and light absorption capacity, respectively. Furthermore, film-shaped MPs demonstrate lower degradation efficiency compared to MP beads due to a less oxygenated and illuminated reaction medium. pH value also plays a crucial role, with alkaline conditions favoring the hydrolysis of certain plastics like fibre-based PET MPs, while acidic conditions enhance the electronic interaction between catalysts and plastics like high-density PE MPs (Ariza-Tarazona and Villarreal-Chiu 2020).

6.7.3. *Changes in micro-/nano-plastics morphology in photocatalytic degradation*

The morphology of photocatalysts plays a critical character in the process of photo degradation. Variations in morphology may direct us to differences in porosity, surface area and accessibility to active sites, thereby affecting total photocatalytic activity. Understanding the interplay between influencing elements, photocatalytic mechanisms and catalyst morphology is crucial for optimizing performance in degradation process. For example, factors such as MP concentration, temperature and pH can impact morphology, subsequently affecting the photocatalytic mechanism and origin of reactive species. Optimization of catalyst morphology can enhance photocatalytic performance by facilitating improved MP adsorption, raised-reactive species generation and proficient implementation of catalytic sites.

To further investigate morphological changes during photodegradation, the surface of PP MPs was examined using scanning electron microscopy (SEM) (Uheida et al. 2021). Observable variations in the microstructure of MP surface can be credited to several elements, including the removal of photodegraded by-products, the identification of amorphous surface parts and an increase in crystalline parts, resulting in surface layer contraction and the development of cracks and cavities. The appearance of surface cracks and cavities can enhance the degradation process by facilitating deeper oxygen penetration into the sample, thereby promoting photooxidation. As the duration of photocatalytic treatment increases, the size and density of cavities consistently expand. Additionally, the formation of cavities may also be linked to the release of volatile degradation products from the polymer particle surface (Tofa et al. 2019).

6.8. Factors influencing photocatalytic degradation efficiency

In the photocatalytic degradation of MPs/NPs, various operational parameters play crucial roles (Ariza-Tarazona et al. 2020). Factors such as solution pH, termination rate and the generation of free radicals essential for extracting hydrogen atoms from polymer chains significantly impact the photocatalytic efficiency of MPs/NPs degradation. Understanding how these factors influence both the photocatalytic mechanism and the morphology of photocatalysts is essential for optimizing the degradation process and enhancing its effectiveness. For instance, a solution's pH affects the generation of reactive species (e.g. hydroxyl radicals and superoxide radicals) involved in the photocatalytic mechanism and influences the adsorption of MPs onto the photocatalyst surface. The temperature influences the reaction rate and the generation of reactive species, thereby affecting overall degradation efficiency. Additionally, MP concentration can influence the availability of active sites on the photocatalyst surface, thus impacting morphology and performance. By studying these factors and their effects on both mechanism and morphology, a comprehensive understanding of the photocatalytic degradation process can be achieved, leading to optimized processes with improved performance.

Moreover, the pH level can influence the surface charge of the photocatalyst and inter-particle electrostatic attraction (Zhao et al. 2007). MPs/NPs in alkaline solutions are more susceptible to degrade under acidic conditions caused by Coulomb repulsion, which prohibits disintegration by shielding MPs/ NPs.

The dimension of MPs/ NPs can affect their adsorption onto the photocatalyst's surface. While MPs of few nanometer diameters are probably adsorbed onto the semiconductor, longer-sized common MPs face challenges in adsorption, such as facial scrub MPs available on the market with an average particle size of 850 μm (Ariza-Tarazona et al. 2019), which is significantly longer than Degussa P25 TiO_2 particles (Kim and Choi 2002). As a result, a major issue in MP photocatalysis lies in directing their short or absent adsorption onto the semiconductor caused by their considerable particle size. Additionally, shape of MPs can affect their dispersion behavior and expose it to a semiconductor, necessitating examination in the development of photocatalytic systems aiming for full MP mineralization to CO_2 and H_2O. Llorente-García et al. (2020) investigated the catalysis of HDPE and LDPE MPs in visible light by mesoporous N-TiO_2 coating as promising solution. Their unraveling revealed that, "MP shape and size influence the photocatalytic disintegration by N-TiO_2 coating". Small MPs showed larger disintegration, while film-shaped MPs exhibited diminished disintegration. Enhancing selectivity is a significant challenge going forward, allowing micromotors to realize MPs thus improving MP removal proficiency.

6.9. Role of micromotors in photocatalytic degradation of MPs/NPs

Micromotors play a pivotal role in the photocatalytic degradation of MPs/NPs, offering a novel approach to address this environmental concern. These micromotors are engineered to convert energy from various sources, such as chemical reactions, external fields (like electric or magnetic forces), or light, into self-directed motion, permitting them to execute specific tasks. These micromotors exhibit diverse forms (Sun et al. 2021). Light-driven and chemical micromotors are extensively investigated in various applications for purification of water (Jurado-Sánchez and Wang 2018). Compared to stationary alternatives, micromotors induce micro-mixing by self-propulsion, enhancing mass transfer in their proximity, aiding in removal of desired pollutants. Orozco et al. (2013) demonstrated the beneficial impact of H_2O_2-directed micromotors on detoxification of organophosphate nerve agents. Besides their micro-stirring abilities, catalytic micromotors may serve as environment cleaners, as H_2O_2 is known for a sustainable oxidant for treatment of wastewater. Bio-framed tubular Fe_3O_4–MnO_2 micromotors have been shown to decompose H_2O_2, generating oxygen in self-propulsion and radicals of hydroxyl for proficient degradation of organic dyes (Chen et al. 2021).

The gathering and removal of minute MP particles with neutral surface impose sufficient challenges, often resulting in nonsuccess of current filtration technique to remove them from tap water. Hence, innovative methods for capturing and degrading these particles are essential. Beladi-Mousavi et al. (2021) unfolded visible light-directed microrobots able to catch and transport considerable number of MPs in complicated environments in sunlight exposure, successfully disintegrating them into tiny oligomers. Such energy-proficients procedures, solely governed by wireless origins, remove the use of pretreatment procedures or heavy mechanical stirrers commonly kept in conventional systems, providing a holistic remedy for MP withdrawal and in situ disintegration. Reviewing current research on sunlight-directed photocatalysts for MP disintegration, a novel approach represented in this study, including autonomous motion, permits proficient capture and disintegration of MPs, potentially laying the groundwork for smart systems useful in hard-to-access environments.

Similarly, Wang et al. (2019) introduced a novel passive particle removal technique comprising photocatalytic Au@Ni@TiO_2 micromotors. They investigated motion of sole particles and assembled chains in various fuels, particularly, H_2O and H_2O_2. They offered and accomplished two plans for removing MPs and suspended material that were employed to both suspended material and MPs extracted from individual care products and gathered from bare waters. This system works proficiently, operating effectively in dilute peroxide solutions and water, offering feasible advantages for MP ejection in true scenarios despite of fuel type. They also illustrated, first, that shape and matter shortcoming do not impede the elimination of material by light-driven micromotors. The primary issue ahead is enhancing selectivity to enable micromotors to recognize MPs more effectively, thus improving MP removal efficiency.

6.10. Photocatalytic water purification: removal of micro- and nano-plastics from water

Various plastic materials have different behaviors in aquatic environments, with some sinking, others floating and some persisting for extended periods, significantly impacting ecosystems. Due to the intricate nature of plastics, their degradation in the environment is not straightforward. Consequently, diverse processes, including physical, chemical and biogeochemical mechanisms, are employed to break down the denser polymer molecules into lighter components or MPs/NPs, facilitating their degradation. Upon discharge into industrial and municipal wastewater, these MPs/NPs fractions enter various aquatic ecosystems, such as ponds, lakes, rivers and oceans, directly or indirectly. This contamination occurs throughout the natural and anthropogenic hydrological cycles, leading to widespread concerns. Reports indicate a potential accumulation of MPs/NPs in both aquatic and terrestrial ecosystems, highlighting their significant threat (de Souza Machado et al. 2018).

Recent studies have identified scattered MP/NPs fragments, ranging from 0.001 to 5 mm, as detrimental environmental pollutants. Furthermore, traces of these MPs/NPs have been discovered in regions like shallow and deep-sea waters surrounding Antarctica in the Southern Ocean, indicating their widespread distribution. Researchers have warned about the potential hazards posed by MPs/NPs to aquatic flora and fauna, including microorganisms, zooplankton, phytoplankton and mammals, due to the various physicochemical damages they can inflict, which could subsequently affect human health through contaminated seafood consumption (Jaiswal et al. 2022).

The presence of MPs/NPs residues in urban sewage treatment plants also poses risks to aquatic microorganisms and, by extension, human health. Moreover, MPs are increasingly recognized as emerging persistent pollutants in terrestrial ecosystems. Soil, serving as the interface between lithosphere, hydrosphere, atmosphere and biosphere, acts as a habitat for various organisms crucial for agricultural ecosystems. MPs enter soil environments through various sources like sewage sludge landfill, plastic mulch, road runoff and atmospheric deposition, persisting and accumulating over time, potentially impacting biodiversity. Additionally, MPs can act as vectors for transporting pollutants, including plastic additives and other toxins, affecting agricultural crops, and posing risks to human health through the food chain.

Various reports have highlighted the adverse effects of MPs on aquatic ecosystems, including impaired enzyme assimilation and development in aquatic organisms resulting from MP ingestion (Tiwari et al. 2020). Despite their widespread use in various applications, such as grocery bags and plastic bottles, MPs are often released into marine or freshwater environments without consideration of their harmful consequences. Studies have confirmed that aquatic organisms, including fish, seabirds, sea turtles, invertebrates and marine mammals, ingest these MP wastes, causing detrimental effects on their health and ecosystems (Lin et al. 2019). Additionally, the adsorption of hydrophobic contaminants, like heavy metals, onto MPs poses risks to food chains.

6.10.1. *Photocatalytic degradation of polyethylene terephthalate nano-plastics*

PET stands out as one of the most common polymers found in both influent and effluent streams of wastewater treatment plants (WWTPs). PET, belonging to polyester family, is synthesized by reacting ethylene glycol and terephthalic acid at elevated temperatures and less vacuum pressures. Due to its high chemical resistance transparency, flexibility, low density and low cost, PET finds widespread use in the production of bottles, food containers and synthetic. Global PET generation reached 30.3 million tons in 2017 (Djapovic et al. 2021). The accumulation of plastics in the environment is primarily attributed to improper discard practices and their inherent resistance to disintegration. Over time, these plastics undergo fragmentation due to abiotic factors, for example, solar radiation or mechanical forces, resulting in formation of MPs, which pose risks to both human and animal health, potentially causing metabolic disorders and local inflammation.

For instance, in a study on the photocatalytic disintegration of PS NPs conducted using TiO_2-P25/β-SiC foam in under UV-A radiation (60 W/m^2; λmax = 365 nm) at pH 6.3, reducing the NPs' size from 508 to 140 nm resulted in 50% mineralization within 7 h (Allé et al. 2020). Similarly, Jiang et al. (2021) reported successful degradation of small HDPE MPs (200–250 μm) in visible light by a hydroxyl-rich BiOCl photocatalyst, achieving mass loss of 5.38% within 5 h. Other studies have also demonstrated the efficacy of photocatalytic degradation for PS NPs by immobilized TiO_2 with a multilayer structure, resulting in 23.5% elimination in UV irradiation (Domínguez-Jaimes et al. 2021) and 18–23% disintegration proficiency employing immobilized Cu_2O/CuO matters in visible light conditions (Acuña-Bedoya et al. 2021).

Recent investigations have highlighted relatively slow pace of photocatalytic disintegration when targeting fiber-based PET MPs, employing innovative setups such as the Bi_2O_3@N-TiO_2 heterojunction (Zhou et al. 2022b) and N-doped TiO_2 nanoflowers adorned with Pt nanoparticles (Zhou et al. 2022a), utilizing Xenon lamp for illumination. The attained disintegration proficiencies of 10.2% and 5.8% as weight loss after 48 h of reaction with these catalysts were attributed to the oxidation of PET surfaces through photocatalytic process.

Although titanium dioxide (TiO_2) stands as a commonly utilized photocatalyst in environmental contexts, its effectiveness is somewhat constrained by its huge band gap (Eg = 3.2 eV) and the propensity for rapid recombination of photoexcited electron–hole pairs. To address these limitations, researchers have explored the strategy of coupling TiO_2 with other materials, aiming to reduce its band gap, enhance absorption of visible light and facilitate detachment of photoexcited carriers.

Among these materials, MOFs have originated as particularly promising candidates. MOFs, composed of metal clusters and organic linkers, possess intriguing photocatalytic properties. In such structures, MOF metal-oxo clusters serve as semiconductor quantum devices, while organic coordinating ligands function as antennas, absorbing light and facilitating the transfer of photoexcited electrons to metal clusters via a ligand-metal charge transfer (LMCT) mechanism. In this composite setup, TiO_2 assumes the role of the primary catalyst, with the MOF serving as a co-catalyst, providing a platform for the efficient transportation of photogenerated electrons.

Among the diverse family of MOFs, MIL-100(Fe) stands out as a prominent member, featuring Fe(III) metal centers bonded with 1,3,5-benzene tricarboxylic acid (trimesic acid). MIL-100(Fe) exhibits several advantageous characteristics, including mesoporous voids, huge surface area and an appropriate HOMO-LUMO gap (ranging from 1.80 to 2.12 eV), enabling visible light photocatalytic responsiveness, large chemical and aqueous stability, and low toxicity. Typically, MIL-100(Fe) is synthesized via hydrothermal method, with reaction durations spanning from 12 to 70 h and temperatures maintained between 130 and 150°C or alternatively through mechanical stirring of reaction mixture for 12 h at room temperature (Rojas-Guerrero et al. 2023).

Composite structures involving TiO_2 and MIL-100(Fe) have demonstrated heightened photocatalytic efficacy in pollutant disintegration compared to separate components. For example, TiO_2/MIL-100(Fe) microspheres fabricated via layer-by-layer technique exhibited increased photocatalytic activity in methylene blue (MB) disintegration in visible light, getting 87% mineralization within 120 min (Hejazi et al. 2020). Furthermore, MIL-100(Fe) loaded onto N-doped TiO_2 nanoparticles attained remarkable removal rates of 99.1% and 93.5% for MB and rhodamine B dyes, respectively, in 180 min in visible irradiation (Huang et al. 2017).

Additionally, combinations of TiO_2 with other iron-based MOFs have been explored for degrading emerging pollutants. For example, Liu et al. (2021) investigated effectiveness of TiO_2/MIL-101(Fe) magnetic composites, synthesized via solvothermal method along with annealing at 350°C, in disintegration of tetracycline in sunlight, attaining a proficiency of 92.76% in 10 min. The photocatalytic activity of TiO_2/MIL-100(Fe) was estimated specifically in disintegration of PET NPs. Immobilizing the catalyst presents a promising approach to enhancing its recyclability and performance, facilitating a straightforward evaluation of PET NPs disintegration.

6.10.2. *Photodisintegration of emerging pollutants*

Emerging pollutants, for example, plasticizers, pharmaceuticals, personal care products and pesticides, have become a cause for concern in wastewater treatment due to their consistent presence in low concentrations. This persistence is attributed to ongoing human usage. The absence or incomplete regulation of discharge standards for these compounds raises significant danger to both environment and human health. The categories of emerging pollutants mentioned encompass pharmaceuticals, personal care products such as cosmetics, plasticizers used to enhance plastic flexibility and pesticides used in agriculture. These contaminants enter water sources through various pathways, such as human and animal excretion, industrial effluents and agricultural runoff.

Challenges associated with these pollutants include their low concentrations, which complicate detection and removal using conventional water treatment methods. Additionally, the cumulative and synergistic effects of multiple pollutants, the lack of specific regulations and the persistence and transformation of these compounds pose significant concerns. Regulatory frameworks often lag behind the identification of new pollutants, leading to uncertainties in acceptable concentration levels and treatment requirements.

To address these challenges, various water treatment strategies can be employed. Advanced oxidation processes (AOPs) such as photocatalysis and ozonation are effective in degrading a broad range of emerging pollutants. Activated carbon adsorption, membrane filtration technologies such as nanofiltration, reverse osmosis and biological treatment methods such as activated sludge treatment and constructed wetlands can also contribute to pollutant removal. Additionally, public awareness initiatives and drug take-back programs can play a crucial role in minimizing the introduction of pharmaceuticals into water systems.

Efficient photocatalysts play a crucial role in degrading emerging pollutants by absorbing visible radiation and generating charges necessary for creating radicals that oxidize contaminants. The chemical structure of these pollutants poses a greater challenge for oxidation compared to dyes. While research has focused on designing MOFs with photocatalytic activity, there is a limited body of work devoted to elimination of emerging pollutants, primarily centered around Fe-based MOFs.

In the realm of Fe-MOFs, specifically MIL-100, MIL-53 and MIL-101, efforts have been made to eliminate tetracycline from water. MIL-101 exhibited the highest activity, credited to its unique structure, achieving a 100% withdrawal of compound. Notably, 60% of tetracycline removal was attributed to the adsorption process, with only 40% eliminated through photocatalysis. Though there is a deficiency of data concerning spotting of by-products, a study on the radical mechanism revealed the involvement of O_2•−, h+ and •OH in disintegration of tetracycline.

The catalytic potential of MIL-100 and the derivative Pd@MIL-100 has been explored in disintegration of several pharmaceuticals and personal care products (PPCPs), for example, ibuprofen, bisphenol A and theophylline (Liang et al. 2015). However, their efficacy was somewhat bounded, necessitating inclusion of additional oxidants like H_2O_2 to get transformations exceeding 60%, essentially adopting a photo-Fenton type procedure. Huang et al. (2018) recently achieved more promising outcomes with a new Ag/AgCl@MIL-88A(Fe) composite. This composite achieved complete elimination of ibuprofen in visible light within 3 h, with a notable degree of mineralization, demonstrating a 90% reduction in total organic carbon (TOC) during that timeframe. Notably, the stability of this composite was exceptional, retaining its activity over four cycles without significant Fe and Ag leaching. The degradation pathway was investigated, highlighting the crucial roles played by h+, e− and O_2•− in photocatalytic disintegration process. The superior functioning of this composite can be attributed to the mutual effect between plasmon resonance of Ag nanoparticles, light absorption of both the MOF and AgCl, and efficient charge separation facilitated by interrelation among three components.

In another study, Fun et al. (2018) synthesized Ag/AgCl@ZIF-8 MOF for the elimination of acetaminophen under visible light irradiation. Recent research has also focused on disintegration of tetracycline antibiotics. Wang et al. (2016) fabricated a core–shell In_2S_3/MIL-125(Ti) heterostructure able to degrade 60% of tetracycline after 60 min under visible light. Despite this promising outcome, the mineralization rate was relatively low compared to other conventional photocatalysts, with only 17% of tetracycline converted into CO_2 and H_2O. Furthermore, stability of this heterostructure was not sufficiently high, as evidenced by a decrease in transformation rate to 45% after three consecutive cycles.

Higher degradation rates for tetracycline under sunlight were achieved by a new g-C_3N_4/ZIF-8 heterostructure (Panneri et al. 2017). Complete vanishing of tetracycline was observed after 1 h of reaction, though data regarding mineralization of antibiotic were not provided. The proposed mechanism suggests that ZIF-8 absorbs light, generating electrons that are subsequently relocated to g-C_3N_4. These electrons react with oxygen of medium, producing O_2•− radicals participated in tetracycline oxidation. Additionally, holes originated in ZIF-8 may react with target molecules. As observed in another heterojunctions, mutual action between both components enhances charge separation and prohibits unwanted recombination processes.

6.11. References

Acuña-Bedoya, J.D., Lúevano-Hipólito, J.D., Cedillo-González, E., Domínguez-Jaimes, E.I., Hurtado, L.P., Hernández-López, J.M.. (2021). Boosting visible-light photocatalytic degradation of polystyrene nanoplastics with immobilized Cu_xO obtained by anodization. *J. Environ. Chem. Eng.*, 9(5), 106208.

Al-Mamun, M.R., Kader, S., Islam, M.S., Khan, M.Z.H. (2019). Photocatalytic activity improvement and application of UV-TiO_2 photocatalysis in textile wastewater treatment: A review. *Journal of Environmental Chemical Engineering*, 7(5), 103248.

Al-Thawadi, S. (2020). Microplastics and nanoplastics in aquatic environments: Challenges and threats to aquatic organisms. *Arabian Journal for Science and Engineering*, 45, 4419–4440.

Allé, P.H., Garcia-Muñoz, P., Adouby, K., Keller, N., Robert, D. (2020). Efficient photocatalytic mineralization of polymethylmethacrylate and polystyrene nanoplastics by TiO_2/β-SiC alveolar foams. *Environmental Chemistry Letters*, 19, 1803–1808.

Ariza-Tarazona, M.C., Villarreal-Chiu, J.F., Barbieri, V., Siligardi, C., Cedillo-González, E.I. (2019). New strategy for microplastic degradation: Green photocatalysis using a protein-based porous N-TiO_2 semiconductor. *Ceramics International*, 45(7), 9618–9624.

Ariza-Tarazona, M.C. Villarreal-Chiu, J.F., Hernández-López, J.M., De la Rosa, J.R., Barbieri, V., Siligardi, C., Cedillo-González, E.I. (2020). Microplastic pollution reduction by a carbon and nitrogen-doped TiO_2: Effect of pH and temperature in the photocatalytic degradation process. *Journal of Hazardous Materials*, 395, 122632.

Auta, H.S., Emenike, C.U., Fauziah, S.H. (2017). Distribution and importance of microplastics in the marine environment: A review of the sources, fate, effects, and potential solutions. *Environment International*, 102, 165–176.

Ball, P. (2001). *Bright Earth: Art and the Invention of Color*. Farrar, Straus and Giroux, New York.

Bandyopadhyay, A. and Basak, G.C. (2007). Studies on photocatalytic degradation of polystyrene. *Materials Science and Technology*, 23(3), 307–314. doi: 10.1179/17432840 7X158640.

Barnett, S.A. and Champness, N.R. (2003). Structural diversity of building-blocks in coordination framework synthesis combining $M(NO_3)_2$ junctions and bipyridyl ligands. *Coordination Chemistry Reviews*, 246(1–2), 145–168.

Beladi-Mousavi, S.H., Hermanová, S., Ying, Y., Plutnar, J., Pumera, M. (2021). A maze in plastic wastes: Autonomous motile photocatalytic microrobots against microplastics. *ACS Appl. Mater. Interface*, 13(21), 25102–25110.

Boldyrev, V.V. (2006). Mechanochemistry and mechanical activation of solids. *Russian Chemical Reviews*, 75, 177–189. doi: 10.1016/0167-2738(93)90157-X.

Buser, H.J., Schwarzenbach, D., Petter, W., Ludi, A. (1977). The crystal structure of Prussian Blue: $Fe_4[Fe(CN)_6]_3.xH_2O$. *Inorganic Chemistry*, 16(11), 2704–2710.

Cao, B., Wan, S., Wang, Y., Guo, H., Ou, M., Zhong, Q. (2022). Highly-efficient visible light-driven photocatalytic H_2 evolution integrated with microplastic degradation over MXene/$Zn_xCd_{1-x}S$ photocatalyst. *Journal of Colloid and Interface Science*, 605, 311–319.

Chen, L., Yuan, H., Chen, S., Zheng, C., Wu, X., Li, Z., Liang, C., Dai, P., Wang, Q., Ma, X., Yan, X. (2021). Cost-effective, high-yield production of biotemplated catalytic tubular micromotors as self-propelled microcleaners for water treatment. *ACS Applied Materials and Interfaces*, 13(26), 31226–31235.

Djapovic, M., Milivojevic, D., Ilic-Tomic, T., Lješević, M., Nikolaivits, E., Topakas, E., Maslak, V., Nikodinovic-Runic, J. (2021). Synthesis and characterization of polyethylene terephthalate (PET) precursors and potential degradation products: Toxicity study and application in discovery of novel PETases. *Chemosphere*, 275, 130005.

Domínguez-Jaimes, L.P., Cedillo-González, E.I., Luévano-Hipólito , E., Acuña-Bedoya, J.D. Hernández-López, J.M. (2021). Degradation of primary nanoplastics by photocatalysis using different anodized TiO_2 structures. *J. Hazard. Mater.*, 413, 125452.

Fan, G., Zheng, X., Luo, J., Peng, H., Lin, H., Bao, M., Hong, L., Zhou, J. (2018). Rapid synthesis of Ag/AgCl@ZIF-8 as a highly efficient photocatalyst for degradation of acetaminophen under visible light. *Chemical Engineering Journal*, 351, 782–790.

Fromm, K.M., Gueneau, E.D., Robin, A.Y., Maudez, W., Sague, J., Bergougnant, R. (2005). Recent advances in the chemistry of 'clusters' and coordination polymers of alkali, alkaline earth metal and Group 11 compounds. *Zeitschrift für Anorganische und Allgemeine Chemie*, 631(10), 1725–1740.

Hafizovic, J., Bjørgen, M., Olsbye, U., Dietzel, P.D.C., Bordiga, S., Prestipino, C., Lamberti, C., Lillerud, K.P. (2007). The inconsistency in adsorption properties and powder XRD data of MOF-5 is rationalized by framework interpenetration and the presence of organic and inorganic species in the nanocavities. *Journal of the American Chemical Society*, 129(12), 3612–3620.

ter Halle, A., Jeanneau, L., Martignac, M., Jardé, E., Pedrono, B., Brach, L., Gigault, J. (2017). Nanoplastic in the North Atlantic subtropical gyre. *Environmental Science and Technology*, 51(23), 13689–13697.

Haris, M.F., Didit, A.M., Ibadurrohman, M., Setiadi, S. (2021). Silver doped TiO_2 photocatalyst for disinfection of *E. coli* and microplastic pollutant degradation in water. *Asian Journal of Chemistry*, 33(9), 2038–2042.

Hejazi, R., Mahjoub, A.R., Khavar, A.H.C., Khazaee, Z. (2020). Fabrication of novel type visible-light-driven TiO_2@MIL-100 (Fe) microspheres with high photocatalytic performance for removal of organic pollutants. *Journal of Photochemistry and Photobiology A*, 400, 112644.

Herren, F., Fischer, P., Ludi, A., Hälg, W. (1908). Neutron diffraction study of Prussian Blue, $Fe_4[Fe(CN)_6]_3.xH_2O$, location of water molecules and long–range magnetic order. *Inorganic Chemistry*, 19(4), 956–959.

Horikoshi, S. and Serpone, N. (2013). *Microwaves in Nanoparticle Synthesis: Fundamentals and Applications*. Wiley-VCH Press, Weinheim.

Hoskins, B.F. and Robson, R. (1990). Design and construction of a new class of scaffolding-like materials comprising infinite polymeric frameworks of 3D-linked molecular rods. A reappraisal of the zinc cyanide and cadmium cyanide structures and the synthesis and structure of the diamond-related frameworks $[N(CH_3)_4][Cu^IZn^{II}(CN)_4]$ and Cu^I[4,4′,4″,4‴-tetracyanotetraphenylmethane]$BF_4.xC_6H_5NO_2$. *Journal of the American Chemical Society*, 112(4), 1546–1554.

Hu, K.Z., Yang, P., Hall, Y., Nie, T., Yao, G., Duan, Y., X., Wang, S. (2022). Degradation of microplastics by a thermal Fenton reaction. *ACS ES&T Eng.*, 2(1), 110–120.

Huang, J., Song, H., Chen, C., Yang, Y., Xu, N., Ji, X., Li, C., You, J.A. (2017). Facile synthesis of N-doped TiO_2 nanoparticles caged in MIL-100(Fe) for photocatalytic degradation of organic dyes under visible light irradiation. *Journal of Environmental Chemical Engineering*, 5(3), 2579–2585.

Huang, W., Jing, C., Zhang, X., Tang, M., Tang, L., Wu, M., Liu, N. (2018). Integration of plasmonic effect into spindle-shaped MIL-88A(Fe): Steering charge flow for enhanced visible-light photocatalytic degradation of ibuprofen. *Chemical Engineering Journal*, 349, 603–612.

Jaiswal, K.K., Dutta, S., Banerjee, I., Pohrmen, C.B., Singh, R.K., Das, H.T., Dubey, S., Kumar, V. (2022). Impact of aquatic microplastics and nanoplastics pollution on ecological systems and sustainable remediation strategies of biodegradation and photodegradation. *Science of the Total Environment*, 806(Part 3), 151358.

James, S.L., Adams, C.J., Bolm, C., Braga, D., Collier, P., Friščić, T., Grepioni, F., Harris, K.D.M., Hyett, G., Jones, W. et al. (2012). Mechanochemistry: Opportunities for new and cleaner synthesis. *Chemical Society Reviews*, 41(1), 413–447.

Janiak, C. (2003). Engineering coordination polymers towards applications. *Dalton Transactions*, 14, 2781–2804.

Jiang, R., Lu, G., Yan, Z., Liu, J., Wu, D., Wang, Y. (2021). Microplastic degradation by hydroxy-rich bismuth oxychloride. *Journal of Hazardous Materials*, 405, 124247.

Jung, J.W., Park, J.W., Eo, S., Choi, J., Song, Y.K., Cho, Y., Hong, S.H., Shim, W.J. (2021). Ecological risk assessment of microplastics in coastal, shelf, and deep sea waters with a consideration of environmentally relevant size and shape. *Environmental Pollution*, 270, 116217. doi: 10.1016/j.envpol.2020.116217.

Jurado-Sánchez, B. and Wang, J. (2018). Micromotors for environmental applications: A review. *Environmental Science: Nano*, 5, 1530–1544.

Kamrannejad, M.M., Hasanzadeh, A., Nosoudi, N., Mai, L., Babaluo, A.A. (2014). Photocatalytic degradation of polypropylene/TiO_2 nano-composites. *Materials Research*, 17(4), 1039–1046.

Karthikeyan, S., Potisek, S.L., Piermattei, A., Sijbesma, R.P. (2008). Highly efficient mechanochemical scission of silver-carbene coordination polymers. *Journal of the American Chemical Society*, 130(45), 14968–14969.

Kim, S. and Choi, W. (2002). Kinetics and mechanisms of photocatalytic degradation of $(CH_3)_nNH_{4-n}^+$ $(0 \leq n \leq 4)$ in TiO_2 suspension: The role of OH radicals. *Environmental Science and Technology*, 36(9), 2019–2025.

Kim, H.J., Lee, E., Park, H.S., Lee, M. (2007). Dynamic extension – Contraction motion in supramolecular springs. *Journal of the American Chemical Society*, 129(36), 10994–10995.

Kinoshita, Y., Matsubara, I., Higuchi, T., Saito, Y. (1959). The crystal structure of bis(adiponitrilo)copper(I) nitrate. *Bulletin of the Chemical Society of Japan*, 32(11), 1221–1226.

Kitagawa, S. and Noro, S. (2004). Coordination polymers: Infinite systems. *Comprehensive Coordination Chemistry*, 35(47), 231–261.

Li, H., Eddaoudi M., O'Keeffe, M., Yaghi, O.M. (1999). Design and synthesis of an exceptionally stable and highly porous metal-organic framework. *Nature*, 402, 276–279.

Liang, R.L., Jing, S., Shen, F., Qin, L., N., Wu, L. (2015). A simple strategy for fabrication of Pd@MIL-100(Fe) nanocomposite as a visible-light-driven photocatalyst for the treatment of pharmaceuticals and personal care products (PPCPs). *Applied Catalysis B*, 176–177, 240–248.

Lin, W., Jiang, R., Wu, J., Wei, S., Yin, L., Xiao, X., Hu, S., Shen, Y., Ouyang, G. (2019). Sorption properties of hydrophobic organic chemicals to micro-sized polystyrene particles. *Science of the Total Environment*, 690, 565–572.

Liu, L., Liu, Y., Wang, X., Hu, N., Li, Y., Li, C., Meng, Y., An, Y. (2021). Synergistic effect of $BTiO_2$ and MIL-100(Fe) for high-efficiency photocatalysis in methylene blue degradation. *Applied Surface Science*, 561, 149969.

Liu, X., Duan, X., Bao, T., Hao, D., Chen, Z., Wei, W., Wang, D., Wang, S., Ni, B.J. (2022). High-performance photocatalytic decomposition of PFOA by $BiOX/TiO_2$ heterojunctions: Self-induced inner electric fields and band alignment. *Journal of Hazardous Materials*, 430, 128195.

Llorente-García, B.E., Hernández-López, J.M., Zaldívar-Cadena, A.A., Siligardi, C., Cedillo-González, E.I. (2020). First insights into photocatalytic degradation of HDPE and LDPE microplastics by a mesoporous $N-TiO_2$ coating: Effect of size and shape of microplastics. *Coatings*, 10(7), 658.

Loiseau, T., Mellot-Draznieks, C., Muguerra, H., Férey, G., Haouas, M., Taulelle, F. (2005). Hydrothermal synthesis and crystal structure of a new three-dimensional aluminum-organic framework MIL-69 with 2, 6-naphthalenedicarboxylate (ndc), Al (OH)(ndc)·H_2O. *Comptes rendus chimie*, 8(3–4), 765–772.

Loiseau, T., Beitone, L., Taulelle, F., Férey, G. (2006). Divalent metal incorporation in MIL-74, the super-sodalite aluminium phosphates $M_3Al_6(PO_4)_{12}$, 4 Tren, 17 H_2O (M= Mg, Mn, Co) and their gallium analogue $M'_3Ga_6(PO_4)_{12}$, 4 Tren, 17 H_2O (M'= Mg, Mn, Co, Fe, Zn). *Solid State Sciences*, 8, 346–352.

Mintenig, S.M., Bäuerlein, P.S., Koelmans, A.A., Dekker, S.C., Van Wezel, A.P. (2018). Closing the gap between small and smaller: Towards a framework to analyse nano-and microplastics in aqueous environmental samples. *Environmental Science: Nano*, 5(7), 1640–1649.

Moon, H.R., Kim, J.H., Suh, M.P. (2005). Redox-active porous metal–organic framework producing silver nanoparticles from AgI ions at room temperature. *Angewandte Chemie*, 44(8), 1261–1265.

Moulton, B. and Zaworotko, M.J. (2001). From molecules to crystal engineering: Supramolecular isomerism and polymorphism in network solids. *Chemical Reviews*, 101(6), 1629–1658.

Mueller, U., Schubert, M., Teich, F., Puetter, H., Schierle-Arndt, K., Pastré, J. (2006). Metal–organic frameworks – Prospective industrial applications. *J. Mater. Chem*, 16(7), 626–636.

Murphy, F., Ewins, C., Carbonnier, F., Quinn, B. (2016). Wastewater treatment works (WwTW) as a source of microplastics in the aquatic environment. *Environmental Science and Technology*, 50(11), 5800–5808.

Orozco, J., Cheng, G., Vilela, D., Sattayasamitsathit, S., Vazquez-Duhalt, R., Valdés-Ramírez, G., Pak, O.S., Escarpa, A., Kan, C., Wang, J. (2013). Micromotor-based high-yielding fast oxidative detoxification of chemical threats. *Angewandte Chemie International Edition*, 52(50), 13276–13279.

Panneri, S., Thomas, M., Ganguly, P., Nair, B.N., Mohamed, A.P., Warrier, K.G.K., Hareesh, U.S. (2017). C_3N_4 anchored ZIF 8 composites: Photo-regenerable, high capacity sorbents as adsorptive photocatalysts for the effective removal of tetracycline from water. *Catalysis Science and Technology*, 7(10), 2118–2128.

Pichon, A., Lazuen-Garay, A., James, S.L. (2006). Solvent-free synthesis of a microporous metal–organic framework. *CrystEngComm*, 8(3), 211–214.

Powell, H.M. and Rayner, J.H. (1949). Clathrate compound formed by benzene with an ammonia–nickel cyanide complex. *Nature*, 163, 566–567.

Qiu, L.G., Li, Z.Q., Wu, Y., Wang, W., Xu, T., Jiang, X. (2008). Facile synthesis of nanocrystals of a microporous metal–organic framework by an ultrasonic method and selective sensing of organoamines. *Chemical Communications*, 31, 3642–3644. doi: 10.1039/B804126A.

Riou-Cavellec, M. and Férey, G. (2002). A new three-dimensional iron trimesate $Fe^{III}_3(H_2O)_5(C_9O_6H_3)_2$, 3 H_2O or MIL-65. *Solid State Sciences*, 4(9), 1221–1225.

Riou-Cavellec, M., Sanselme, M., Grenèche, J.M., Férey, G. (2000). Hydrothermal synthesis, structural approach, magnetic and Mössbauer study of the layered Fe(III) carboxyethylphosphonate or [Fe(OH)(H_2O)(O_3P-$(CH_2)_2$-CO_2H] or MIL-37. *Solid State Sciences*, 2(7), 717–724.

Riou-Cavellec, M., Sanselme, M., Noguès, M., Grenèche, J.M., Férey, G. (2002). Synthesis, structure and metamagnetic behaviour of a three-dimensional Fe(II). *Solid State Sciences*, 4(5), 619–625.

Robin, A.Y., Fromm, K.M., Goesmann, H., Bernardinelli, G. (2003). Structural relationship of two coordination polymers of Cu(I) with the ligand ethanediylbis(isonicotinate). *CrystEngComm*, 5(71), 405–410.

Rojas-Guerrero, C.A., Villanueva-Rodríguez, M., Guzmán-Mar, J.L., Hernández-Ramírez, A., Cedillo-González, E.I., Longoria Rodríguez, F.E., Hinojosa-Reyes, L. (2023). Solar photocatalytic degradation of polyethylene terephthalate nanoplastics: Evaluation of the applicability of the TiO_2/MIL-100(Fe) composite material. *Journal of Environmental Chemical Engineering*, 11(5), 110415.

Sanselme, M., Riou-Cavellec, M., Grenèche, J.M., Férey, G. (2002). Hydrothermal synthesis, structure and magnetic properties of a layered Fe(III) carboxymethylphosphonate K[$M_3(BTC)_3$].5H_2O (M = Fe, Co) or MIL-49J. *Journal of Solid State Chemistry*, 164(2), 354–360.

Sanselme, M., Grenèche, J.M., Riou-Cavellec, M., Férey, G. (2004). The first ferric carboxylate with a three-dimensional hybrid open framework (MIL-82): Its synthesis, structure, magnetic behaviour and study of its dehydration by 57Fe Mössbauer spectroscopy. *Solid State Sciences*, 6(8), 853–858.

Schlesinger, M., Schulze, S., Hietschold, M., Mehring, M. (2010). Evaluation of synthetic methods for microporous metal–organic frameworks exemplified by the competitive formation of [$Cu_2(btc)_3(H_2O)_3$] and [Cu_2(btc)(OH)(H_2O)]. *Microporous and Mesoporous Materials*, 132(1–2), 121–127.

Shearer, G.C., Chavan, S., Ethiraj, J., Vitillo, J.G., Svelle, S., Olsbye, U., Lamberti, C., Bordiga, S., Lillerud, K.P. (2014). Tuned to perfection: Ironing out the defects in metal–organic framework UiO-66. *Chemistry of Materials*, 26(14), 4068–4071.

Simon, N., Marrot, J., Loiseau, T., Férey, G. (2006). Hydrothermal synthesis and crystal structures of two open-framework fluorinated aluminum phosphates templated by 1,3 diaminopropane (ULM-4 and MIL-64). *Solid State Sciences*, 8(11), 1361–1367.

de Souza Machado, A.A., Lau, C.W., Till, J., Kloas, W., Lehmann, A., Becker, R., Rillig, M.C. (2018). Impacts of microplastics on the soil biophysical environment. *Environmental Science and Technology*, 52(17), 9656–9665.

Sun, J., Tan, H., Lan, S., Peng, F., Tu, Y. (2021). Progress on the fabrication strategies of self-propelled micro/nanomotors. *JCIS Open*, 2, 100011.

Tanaka, K. and Takada, H. (2016). Microplastic fragments and microbeads in digestive tracts of planktivorous fish from urban coastal waters. *Scientific Reports*, 6, 34351.

Taylor-Pashow, K.M.L., Della Rocca, J., Xie, Z., Tran, S., Lin, W. (2009). Postsynthetic modifications of iron-carboxylate nanoscale metal–organic frameworks for imaging and drug delivery. *Journal of the American Chemical Society*, 131(40), 14261–14263.

Tiwari, N., Santhiya, D.S., Sharma, J.G. (2020). Microbial remediation of micro-nano plastics: Current knowledge and future trends. *Environmental Pollution*, 265(Part A), 115044.

Tofa, T.S., Ye, F., Kunjali, K.L., Dutta, J. (2019). Enhanced visible light photodegradation of microplastic fragments with plasmonic platinum/zinc oxide nanorod photocatalysts. *Catalysts*, 9(10), 819.

Uheida, A., Mejía, H.G., Abdel-Rehim, M., Hamd, W., Dutta, J. (2021). Visible light photocatalytic degradation of polypropylene microplastics in a continuous water flow system. *Journal of Hazardous Materials*, 406, 124299.

Van Assche, T.R.C., Desmet, G., Ameloot, R., De Vos, D.E., Terryn, H., Denayer, J.F.M. (2012). Electrochemical synthesis of thin HKUST-1 layers on copper mesh. *Microporous and Mesoporous Materials*, 158, 209–213.

Venkataramana, C., Botsa, S.M., Shyamala, P., Muralikrishna, R. (2021). Photocatalytic degradation of polyethylene plastics by $NiAl_2O_4$Spinels-synthesis and characterization. *Chemosphere*, 265, 129021.

Volkringer, C., Loiseau, T., Guillou, N., Férey, G., Elkaïm, E. (2009). RD and IR structural investigations of a particular breathing effect in the MOF-type gallium terephthalate MIL-53(Ga). *Solid State Sciences*, 11, 1507–1512.

Volkringer, C., Loiseau, T., Guillou, N., Férey, G., Popov, D., Burghammer, M., Riekel, C. (2013). Synthesis and structural characterization of metal–organic frameworks with the mellitate linker $M_2(OH)_2$ $[C1_2O_{12}H_2]\cdot 2H_2O$ (M = Al, Ga, In) MIL-116. *Solid State Sciences*, 26, 38–44.

Wang, H., Yuan, X., Wu, Y., Zeng, G., Dong, H., Chen, X., Leng, L., Wu, Z., Peng, L. (2016). In situ synthesis of In_2S_3@MIL-125(Ti) core-shell microparticle for the removal of tetracycline from wastewater by integrated adsorption and visible-light-driven photocatalysis. *Applied Catalysis B*, 186, 19–29.

Wang, L., Kaeppler, A., Fischer, D., Simmchen, J. (2019). Photocatalytic TiO_2 micromotors for removal of microplastics and suspended matter. *ACS Applied Materials and Interfaces*, 11(36), 32937–32944.

Wells, A.F. (1977). *Three-Dimensional Nets and Polyhedra.* Wiley Interscience, New York.

Yuan, W., Garay, A.L., Pichon, A., Clowes, R., Wood, C.D., Cooper, A.I., James, S.L. (2010). Study of the mechanochemical formation and resulting properties of an archetypal MOF: $Cu_3(BTC)_2$ (BTC = 1,3,5-benzenetricarboxylate). *CrystEngComm*, 12, 4063–4065.

Zhang, X., Chen, Z.K., Loh, K.P. (2009). Coordination-assisted assembly of 1-D nanostructured light–harvesting antenna. *Journal of the American Chemical Society*, 131(21), 7210–7211.

Zhang, K., Xiong, X., Hu, H., Wu, C., Bi, Y., Wu, Y., Zhou, B., Lam, P.K.S., Liu, J. (2017). Occurrence and characteristics of microplastic pollution in Xiangxi Bay of Three Gorges Reservoir, China. *Environmental Science and Technology*, 51(7), 3794–3801.

Zhang, F., Wang, X., Liu, H., Liu, C., Wan, Y., Long, Y., Cai, Z. (2019). Recent advances and applications of semiconductor photocatalytic technology. *Applied Sciences*, 9(12), 2489.

Zhao, X., Li, Z., Chen, Y., Shi, L., Zhu, Y. (2007). Solid-phase photocatalytic degradation of polyethylene plastic under UV and solar light irradiation. *Journal of Molecular Catalysis A: Chemical*, 268, 101–106.

Zhou, D., Luo, H., Zhang, F., Wu, J., Yang, J., Wang, H. (2022a). Efficient photocatalytic degradation of the persistent PET fiber-based microplastics over Pt nanoparticles decorated N-doped TiO_2 nanoflowers. *Advanced Fiber Materials*, 4, 1094–1107.

Zhou, D.W., Zhang, L., Wu, F., Wang, J., Yang, H., Feasible, J. (2022b). Feasible degradation of polyethylene terephthalate fiber-based microplastics in alkaline media with Bi_2O_3@N-TiO_2 Z-scheme photocatalytic system. *Adv. Sustain. Syst.*, 6(5), 2100516.

7

Carbon-based Materials for Photocatalytic Degradation of Micro- and Nano-plastics

In the universe, carbon is an abundant and multifaceted element. Carbon nanoparticles have proven to be quite beneficial as a class of nanomaterials (NMs). Carbon nanoparticles have been used in many consumer and commercial applications, ranging from carbon nanotubes (CNTs) and graphene to carbon nanodiamonds (NDs) and carbon dots. Due to their remarkable resistance to degradation and bioaccumulation, micro-plastics (MPs) and nano-plastics (NPs) constitute a new yet serious environmental concern. There is not enough technology available to eliminate MPs and NPs. Instead, these materials must be recycled or broken down. Furthermore, it has recently been determined that one major source of MPs in the environment is the fragmentation and degradation of poorly managed plastic wastes. Therefore, creating an environment free of MPs requires the development of efficient removal techniques as well as plastic recycling plans. This chapter provides a thorough overview of the methods already in use to remove MPs and NPs from the environment. It identifies two crucial components needed to accomplish this goal: (1) catalytic breakdown of NP and MPs into organics that are safe for the environment (carbon dioxide and water); (2) catalytic recycling and upcycling of plastic wastes into monomers, fuels and chemicals that can be used again. These techniques' mechanics, catalysts, viability and difficulties are also covered. Innovative catalytic techniques including photocatalysis, sophisticated oxidation processes and biotechnology are viable and sustainable options for converting plastic waste and MPs into useful and environmentally safe goods. The development of environmentally acceptable processes for the catalytic conversion of polymers into valuable goods with high selectivity, low cost and high efficiency under moderate circumstances is encouraged going forward.

7.1. Introduction

The scientific community has shown a great deal of interest in carbon-based functional NMs due to their exceptional combinations of favorable chemical and

---

Chapter written by Chandrakala SINGH and Devjani ADHIKARI.

physical properties, including superior electrical and heat conductivity, chemical stability, advanced optical properties and high mechanical strength (Li et al. 2019). Carbon-based nanomaterials (CBNs), such as fullerenes, graphene, CNTs and their derivatives, are widely utilized materials for a variety of applications, including biomedicine, because of their distinctive mechanical, electrical, thermal, chemical and optical capabilities. A few recent uses of CBNs in biomedicine include targeted drug delivery, biosensing, regenerative medicine, cancer therapy and imaging of cells and tissues. On the other hand, functionalization reduces CBNs' toxicity and increases their solubility in various solvents, including water, which is necessary for biological applications (Gupta et al. 2019). Deoxyribonucleic acid (DNA) is mostly made up of carbon, which is also responsible for the emergence of life on Earth. It might be offered in several formats. Due to significant electron configurations ($1s^2$, $2s^2$, $2p^2$), carbon can keep up in various crystalline and amorphous forms. Carbon is a very malleable material with a wide range of allotropes and configurations. Carbon hybridization (sp^3, sp^2 and sp^1) and its bonding across the atoms determine the kind of allotrope. For instance, sp2 hybridization forms graphite, whereas sp^3 hybridization forms diamond. Numerous technological uses for this carbon exist, including as the transportation of medications and artificial materials. The primary cause of this is that carbon can bond nearly all other elements as well as itself (Kumar and Kumbhat 2016). Since CBNs have unique structural dimensions and physicochemical features, they have garnered a lot of interest in a variety of fields (Rauti et al. 2019). CNTs, graphene, mesoporous carbon, NDs and fullerenes are the structural subtypes of CBNs. These materials have shown increased biocompatibility, decreased immunogenicity and higher drug-loading capacity due to their outstanding optical activity and large multifunctional surface area (Mohajeri et al. 2018). A lot of research has been done on the creation of biocompatible scaffolds and nanomedicines using functionalized CBNs. CBNs have been altered chemically to particular moieties (such as functional groups, compounds and polymers) and have been identified as appropriate for use in biological applications. Since the majority of the materials under CBNs have outstanding supramolecular π-π stacking, strong adsorption capacity and photothermal conversion ability, they have been widely researched for cancer therapy (Wang et al. 2020). Chemical functionalization combined with adaptable qualities can enhance therapy (Chen et al. 2013). All nanocarriers, however, have harmful or many side effects that limit their potential uses. Furthermore, CBNs' activity, biodistribution, pharmacokinetics, cellular absorption, toxicity and clearance may all be affected by the bio-corona that exists on their surface (Shannahan 2017). We categorized CBNs and emphasized their biological uses. The altered nanocarrier systems for improved health management can be made easier with the aid of the physicochemical characteristics of the CBNs discussed here. Here, the benefit of several operationalized methods for CBNs has been investigated. This section will assist in choosing the optimal operationalized method based on requirements or industrial scale-up. CBNs are utilized for targeted distribution in addition to being a carrier. Researchers will find this section useful in implementing unique targeted distribution strategies based on CBNs.

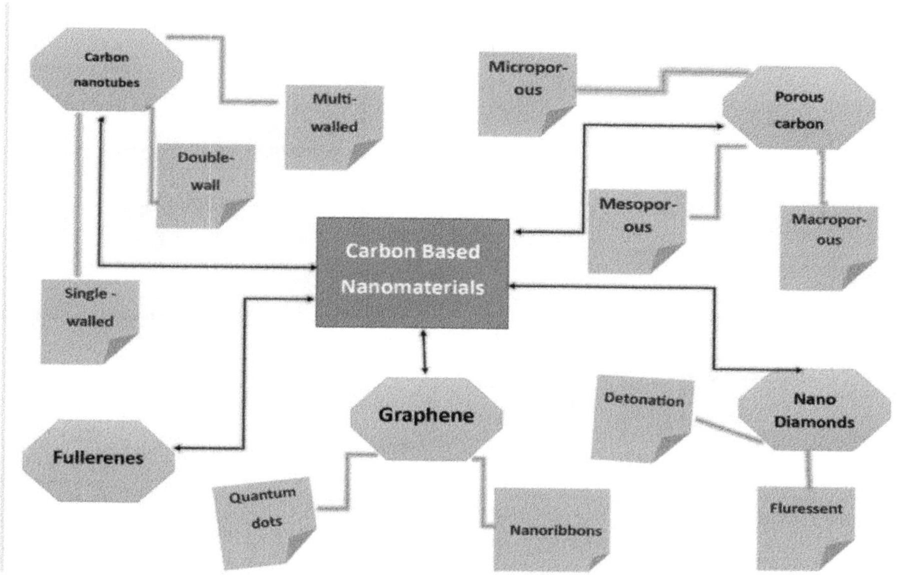

Figure 7.1. *CBN categorization based on the geometric shape and structure of the particles. For a color version of this figure, see www.iste.co.uk/singh/photodegradation.zip*

7.2. Classification of carbon-based nanomaterials

Carbon atoms are the main constituent of CBNs. They are categorized based on the geometric structure and shape of their bodies (Figure 7.1). The most often utilized carbon NMs include carbon dots, fullerenes (ellipsoidal or spherical), graphene, NDs and nanotubes.

7.2.1. *Carbon nanotubes*

The majority of CNTs have cylindrical or tube-like forms. Their diameters, lengths, chirality and number of layers can all vary. They display a lot of appealing qualities because of their special blend of stiffness, strength and flexibility. CNTs can be divided into single, double, multi-walled, and functionalized categories, based on differences in their structural configurations. A number of techniques, including chemical vapor deposition, arc discharge, laser ablation and high-pressure carbon monoxide disproportionation, are used in the production of CNTs.

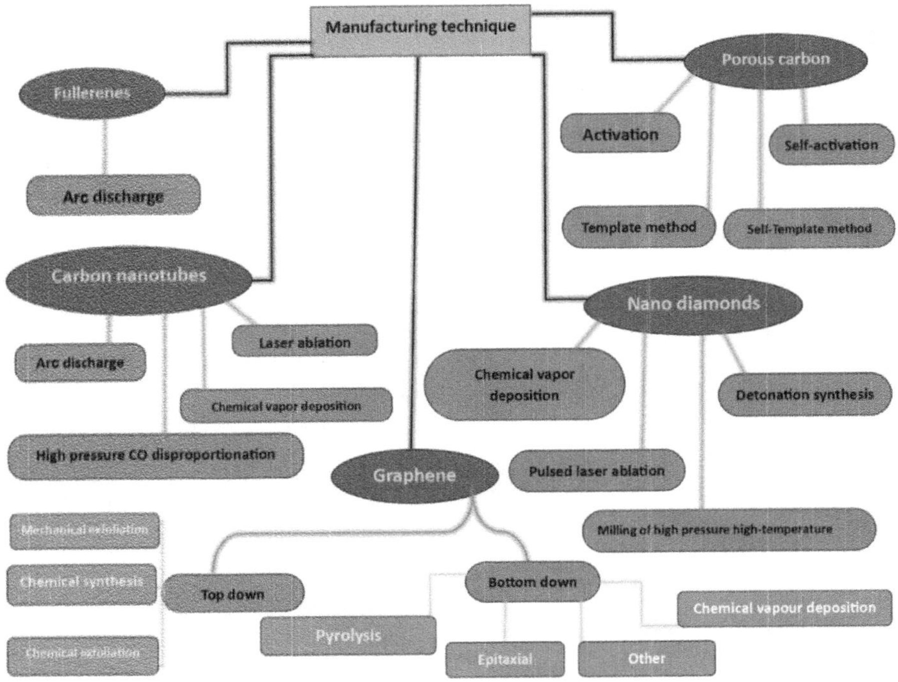

Figure 7.2. *Methods that can be easily implemented to produce CBNs. For a color version of this figure, see www.iste.co.uk/singh/photodegradation.zip*

7.2.2. *Single-walled carbon nanotubes*

Extended graphene sheets folded into tubes are called single-walled carbon nanotubes (SWCNTs). The graphene sheet's patterns are determined by its diameter and C-C orientation (Bepete and Coleman 2019). In this case, the ratio of length to diameter is almost 1,000:1. Their one-dimensional hollow and cylindrical forms made with sp2 hybridization are what define them. These CNTs can multiply thousands of times in length, with a diameter of approximately 1 nm. They are frequently employed in numerous nanotechnology applications. However, the primary barrier is the high cost of manufacturing (Raval et al. 2018). Therefore, more investigation is required to create a cost-effective synthesis method. Nanohorns are the revised figure of SWCNTs. There is just a single carbon ring with a hexagonal shape in it. They are used in cancer target drug compounds and medication delivery because of their special property, which is their increased diameter with increasing length (Beg et al. 2018). SWCNTs are widely employed in the biomedical industry as scaffolding for the growth of cell cultures and as diagnostic tools. Due to their strong π-π interaction, SWCNTs have a low solubility and are challenging to scatter in the aqueous phase. According to chemical modification, SWCNTs can be made more biomedical and pharmaceutically interesting by solubilizing and dispersing them in water (Liang and Chen 2010).

7.2.3. *Double-walled carbon nanotubes*

SWCNTs and DWCNTs (double-walled carbon nanotubes) have comparable morphologies and properties. It is essential to connect this feature to the nanotube to add new functionality (Bhatt et al. 2016). DWCNTs are made up of two CNTs and a double-layer structure made of two folded sheets of graphene. This is a newer category of CBNs. DWCNTs possess exceptional chemical resistance, thermal stability and mechanical strength, when juxtaposed with SWCNTs. Moreover, this class exhibits intriguing optical and electrical characteristics (Shen et al. 2011). Additionally, they are employed in the creation of biosensors. By grafting antibodies onto the surface of DWCNTs, an immunosensor for the obesity biomarker adiponectin was created in this field. In this fabrication, anti-salmonella was injected onto the surface using DWCNT bundles as an electrode (Punbusayakul et al. 2013). DWCNTs, as opposed to SiO_2 surfaces, offer a surface texture that is conducive to the growth of neuron cells in tissue engineering. Following the neuron growth on DWCNT, improved cell differentiation was seen (Béduer et al. 2012).

7.2.4. *Multi-walled carbon nanotubes*

MWCNTs (multi-walled carbon nanotubes) are made up of several graphene sheet layers and have more complex electrical characteristics. This class of nanotubes varies in diameter from 5 to 50 nm. With an interlayer dispersion of 3.4 Å, they are wrapped one outside of the other, creating structural complexity and variety in this class. MWCNTs are hence less well understood. A small structural change could make the desired material attributes less of it (Ganji and Kachapi 2015). The Russian Doll and Parchment model provides a good explanation for the structure of MWCNTs (Bhatt et al. 2016). They are discovered to have a better signal-to-noise ratio than DWCNT and SWCNT. However, the biological matrix's characteristics affect the sensitivity. MWCNTs were employed in the creation of pancreatic cancer cell scaffolds (Matta-Domjan et al. 2018). Verified materials encourage fiber- and sheet-based growth and proliferation, indicating a greater potential for destroying cancer cells.

7.2.5. *Fullerene*

The allotropic alteration of carbon known as fullerenes is sometimes referred to as a molecular form of carbon. The carbon atoms on a spherical surface make up the fullerene family, which has multiple atomic Cn clusters ($n > 20$). Pentagons and hexagons are formed by carbon atoms on the sphere's surface at its vertices. The sp^2-hybrid form of carbon atoms, which are joined by covalent bonds, can be found in fullerenes. The most prevalent and well-researched fullerene is fullerene C_{60} (Zaytseva and Neumann 2016). Chemical functionalization makes it feasible to increase fullerenes' water solubility. This method makes it possible to create novel materials or a bioactive redox medication (Hirsch 2010). Pure C60 enhances human health and exhibits no acute or chronic toxicity. C60 and its variants are a distinct class that continues to be of significant interest in the biological area (Moussa 2018). Functionalized fullerenes' applications in therapy and diagnosis are constantly expanding. With the aid of functionalized derivatives, their primary areas of interest are drug administration, targeted imaging and quenching reactive oxygen species (ROS) (Streicher et al. 2007).

7.2.6. *Nanodiamonds*

Three-dimensional cubic lattices formed by tetrahedrally linked carbon atoms are found in NDs. They have diamond-like qualities. NDs can be divided into two groups based on size and synthesis method: fluorescence NDs and detonation nanodiamonds (DNDs). DNDs are typically produced through shocks using hexogen and explosives similar to trinitrotoluene. They have a monodispersing size of about 5 nm. However, FNDs are made at high pressure and temperature and have a larger size distribution (Bondon et al. 2020). This kind of material exhibits the sp2/sp3 bonds' flexibility. NDs have an easy time sticking to many ligands, substances and medication molecules. Due to their large surface area and simplicity of doping and manufacture, these NDs have been examined in great detail (Reina et al. 2015). NDs' nitrogen–vacancy (NV) defect centers glow between 550 and 800 nm in wavelengths. This characteristic has been effectively utilized in the creation of fluorescent probes intended for tracking individual particles in a heterogeneous setting. These special optical and spectroscopic characteristics have led to the use of NDs in bio-imaging (25 NDs are additionally utilized in conjugation to deliver drugs, such as G9a inhibitors, that are low soluble) (Gu et al. 2019). It is determined that this combination is appropriate for intravenous delivery, and the drug's release exhibited pH responsiveness. As a result, NDs have been used to successfully transport drugs. After functionalization, this class outperforms CNTs in terms of biocompatibility (Patel et al. 2019). The discipline of theranostics is rapidly becoming more interested in NDs. They are also utilized in the creation of scaffolds, tissue engineering, biodegradable bone surgery and the transfer of genetic elements to cells (Turcheniuk and Mochalin 2017). Under vacuum, carbon

ND particles are converted into carbon nanoonions (CNOs). Smaller fullerenes in CNOs are shown as a cage in the cage structure (Plonska-Brzezinska 2019). These are graphitic layers with quasispherical and polyhedral shapes that are near to each other. These substances have distinct chemical and physical characteristics that are not found in any other kind of CBNs. They are discovered to be possible drug carriers (Bhinge 2017). Numerous in vivo, ex vivo and in vitro investigations have verified CNOs' biocompatibility. Compared to other forms of CBNs, they exhibit a few more advantages, such as ease of manufacture, high purity with a small polydispersibility index and improved stability, all of which negate the need for particular storage conditions. Furthermore, CNOs allow for both covalent and non-covalent functionalizations, which permit the attachment of additional functional moieties (Bartkowski and Giordani 2021).

7.2.7. *Carbon dots*

The newest and most distinctive component of CBNs are carbon dots (CDs). These carbon NMs are nearly spherical and have a particle size of less than 10 nm. These materials can destroy the infectious agent in the vicinity by producing ROS with light energy. A novel avenue for eliminating microorganisms, including the start of an infection, is provided by CDs-based photodynamic treatment or photocatalytic disinfection (Knoblauch and Geddes 2020). These substances frequently resemble the structure and activities of enzymes that are found in nature. They are frequently used in biosensing because of these special qualities (Garg and Bisht 2016). CDs are zero-dimensional spherical allotropes with several intriguing characteristics, including low toxicity, biocompatibility, low toxicity, conductivity, optical qualities and theranostics (Cohen et al. 2020). CDs have primarily been synthesized using the bottom-up method by hydrothermally treating different chemical precursors. Simple and targeted reactions are also used to create CDs. As template methods are simple to use and need readily available equipment, they are also used to synthesize spherical polymers (Tian and Yin 2019).

7.2.8. *Graphene*

A single layer of partially filled sp2-orbits carbon is what makes up graphene, both above and below the plane. It has a lot of surface area, great optical transparency, strong mechanical and elastic strength, and good heat conductivity. It is used as a semiconductor quite frequently. Once graphene and electrons interact, new quasi-particles are created. Quantum dots and graphene nanoribbons (GNRs) are the two main elements in this class. These particles exhibit ballistic transport, which allows electrons to move without scattering up to 1 mm (Crevillen et al. 2019). Good electrical conductivity is exhibited by graphene. It dissolves less readily in water. Consequently, several adjustments have been made to create layered graphene oxide (GO), chemically modified graphene and GO. In sol-gel chemistry, GO and its derivatives are widely employed and can be effective in the creation of biocompatible nanocomposites.

7.2.9. *Graphene nanoribbons*

Finite, planar and nearly one-dimensional graphene structures are known as GNRs. The graphene sheets are cut into finer, higher aspect ratio tiles in order to create GNRs. There is a set bandgap that is fewer than 10 nm broad. These materials behave differently as their width increases, transitioning from semiconductors to semimetals (Crevillen et al. 2019). Higher electrical and thermal conductivity is seen in GNRs. It is therefore regarded as a copper substitute for integrated circuits. These result in a bandgap that is tunable to the width and edge structure. While armchair GNRs can be either metallic or semiconductors, zigzag GNRs are metallic. GNRs can be created chemically by etching graphene using high-resolution electron-beam lithography or by exfoliating it in a liquid phase. Typically, they have a length of less than 1 μm and a width of about 10 nm (Hernandez et al. 2012). Functionalized GNRs with oxide have great potential uses in medicine, including drug delivery, cancer treatment and DNA applications. These substances hardly harm the environment or human health (Mousavi et al. 2019). GNRs are extensively utilized in biomedicine and diagnostics as an ideal replacement for silicon-based transistors because of their increased absorption by mammalian cells.

7.2.10. *Graphene quantum dots*

One way to produce graphene quantum dots (GQDs) is to slice graphene into pieces that are between 2 and 20 nm in size. When it comes to the environment, GQDs are less harmful than traditional quantum dots (Crevillen et al. 2019). These nanoparticles have zero dimensions. They are used in various biomedical applications because of their great photostability, high water solubility, ultra-small size and lack of toxicity. As GQDs offer robust photoluminescence, photo-stability and enhanced renal clearance, they solve the issues related to various fluorophores. They are therefore effectively applied in bioimaging. GQDs exhibit fluorescence. They are therefore utilized to create a fluorescent probe to observe cellular dynamics and tumor imaging, both in vivo and in vitro (Pan et al. 2010). Using GQDs, a redox fluorescent probe was created to assist in tracking the dynamic changes brought on by oxidative stress in real time (Li et al. 2016).

Plastic materials are vital to modern life, and as their effects on the environment and public health have come to light more and more, it is now widely acknowledged that improved management of their value chain is urgently needed. Plastics make life better for citizens, and because of all their uses, they cannot be outlawed. Simultaneously, adverse effects associated with their end-of-life, such as a deeper comprehension of possible chemical leaking throughout their use period, must be addressed immediately. The EU Member States are required by the Marine Strategic Framework Directive (MSFD) to make sure that marine litter, especially plastic litter, does not impair the coastal and marine ecosystems (Thiel et al. 2013). A variety of external factors, including biological metabolism, ultraviolet (UV) radiation and the mechanical strength of water, influence plastics as they enter the environment. This fragmentation process results in smaller plastic particles, known as MPs (sizes ranging from 1 μm to 5 mm) and NPs (sizes ranging from 1 nm to 1 μm). Particles that range in size from 1 nm to 1 mm and between 1 mm and 5 mm are referred to as MPs and NPs (Frias and Nash 2019). Even though the polymeric components of environmental MPs and NPs may be regarded as biochemically inert, numerous creatures can consume these. These all draw attention to the possible risks that MPs and NPs present to the ecosystem and biodiversity (Cole et al. 2013; Lambert et al. 2014). Two types of plastics, MP and NP waste, can break down into particles or small fragments in the environment as a result of weathering, biodegradation and solar exposure. MPs (<5 mm) and NPs (<1,000 nm) are the two categories into which small plastic fragments fall. There have been reports of microscale plastic particles found in the environment going back to the 1970s. The potential harm that MPs and NPs may do to living things has garnered increasing attention due to their widespread distribution in the environment (Hartmann et al. 2019; Sobhani et al. 2020). Primary MPs are made with the intention of falling into the micrometer size range. Primary MPs include plastic beads used in, for example, personal hygiene products (Rochman et al. 2015). The majority of MPs, found in the environment come from industrial and home items, such as drug vectors found in medications, cosmetics and face cleansers. Due to their enormous quantity, treated wastewater still contains a significant number of MPs, even though contemporary wastewater treatment methods can remove more than 90% of the MPs in wastewater (Carr et al. 2016). Larger plastic cracking produces secondary MPs. A combination of chemical and physical processes, such as hydrolytic degradation, photodegradation, oxidation and mechanical disintegration, are involved in this process (Lehner et al. 2019). High temperatures and UV radiation can cause chemical changes in plastics that increase their fragility and likelihood of breaking. Sunlight's UV rays have the ability to oxidize the polymer matrix, which breaks down chemical connections. Plastics can fracture due to a variety of mechanical stresses, including turbulence, abrasion and wave action. Comparatively speaking to research on other nanoparticles, however, NP research is still in its early stages. Research has demonstrated that plastic particles will continue to break down until they reach the nanoscale, at which point they will produce nanoparticles (Fu et al. 2020). Plastics and other pollutants degrade quite well when photocatalytic technology is used. Due to their complicated structure, limited water solubility and non-biodegradability, plastics are difficult to break down. Plastics and MPs have a lot of detrimental consequences on the environment. Plastic recycling has gained attention because "white pollution" wastes resources and pollutes the environment. Thus, to reduce the negative effects of plastic trash on the environment, appropriate management is required. Physical, chemical and cutting-edge technologies are the currently available techniques for removing MPs from water or wastewater. All of these techniques have drawbacks, and photocatalysis is the most successful technique for degrading plastic. The e^- in the valence band (VB) is moved into the conduction band (CB), leaving behind h^+, when the semiconductor absorbs light with photon energy (E) and gap energy (Eg). When these species encountered H_2O, OH^- and O_2 adsorbed on the semiconductor's surface, they produced ROS such as superoxide and hydroxyl radicals. These species start the breakdown of plastic, which eventually results in total mineralization into H_2O and CO_2, chain breaking, branching, and crosslinking (Ouyang et al. 2021). The potential harm that MPs/NPs pollution might do to human health and the environment has made it a global source of worry. Therefore, cleaning up MPs/NPs contamination in the environment is crucial. Reviews of the most recent MPs/NPs degrading processes over NMs are still lacking. As a result, the goal of this review is to examine and debate the different NMs used in the photo- and electro-catalysis of MPs and NPs.

Surface alteration, doping of elements and morphological dominion are carried out to improve performance. Nevertheless, a great variety of catalysts remain to be further studied. We also revisit the degradation mechanism concerning the discovery of intermediates and paths of degradation, density functional theory (DFT) calculations and ROS. During the breakdown of MPs/NPs, ROS are primarily responsible for the scission of the polymer chain and the production of by-products. The link between structure, mechanism and function still has to be investigated, though, by combining DFT computations. Furthermore, based on the inadequacies in the current research, some suggestions are made for future directions. As a result, this analysis offers a thorough resource for future directions in catalyst design and MPs/NPs degradation efficiency enhancements (Du et al. 2022). The possible threat posed by (micro-)plastic contamination to the biota has drawn attention from all across the world. It is still lacking in an overview of recent advances in photocatalytic breakdown of (micro-)plastics. Several bare and composite photocatalysts engaged in the photocatalytic degradation of (micro-)plastics have been covered in this work. There was also a discussion of the photocatalytic mechanisms and factors influencing the deterioration. Photocatalysts are made more effective by adding metal or non-metal dopants to their surface. Multifunctional composites are created by compounding these doped photocatalysts with a range of safe and nontoxic polymers. The photocatalytic degradation of (micro-)plastics is largely dependent on the production of ROS. Superoxide ions ($O_2^{\bullet-}$) and hydroxyl radicals ($^{\bullet}OH$) also contribute to this degradation process, which breaks the polymer chain and produces some intermediates. The majority of photocatalytic degradation technologies studied to date are unable to achieve the full mineralization of (micro-)plastics, even though satisfactory progress has been made in the photodegradation of (micro-)plastics. Furthermore, suggestions for future research directions have been made in light of the difficulties facing the current photocatalytic degradation technologies (Ge et al. 2022). Not even in the presence of peroxymonosulfate (PMS) and catalyst could MPs be broken down under the ambient conditions. This highlights how important it is for the HT condition and carbocatalysis to work together to develop highly reactive ROS to mineralize MPs and organic intermediates. Comparative MPs degradation experiments (Figure 7.3(A)) demonstrate that Mn@NCNTs perform better than MnO_2, N-doped CNTs (NCNTs) and GO. This suggests that manganese-carbides@carbon hybrids significantly boosted PMS activation and hastened MPs' catalytic breakdown. According to reports, metals can increase the oxidative chain scission of polyolefin and catalyze the dissociation of hydroperoxides.

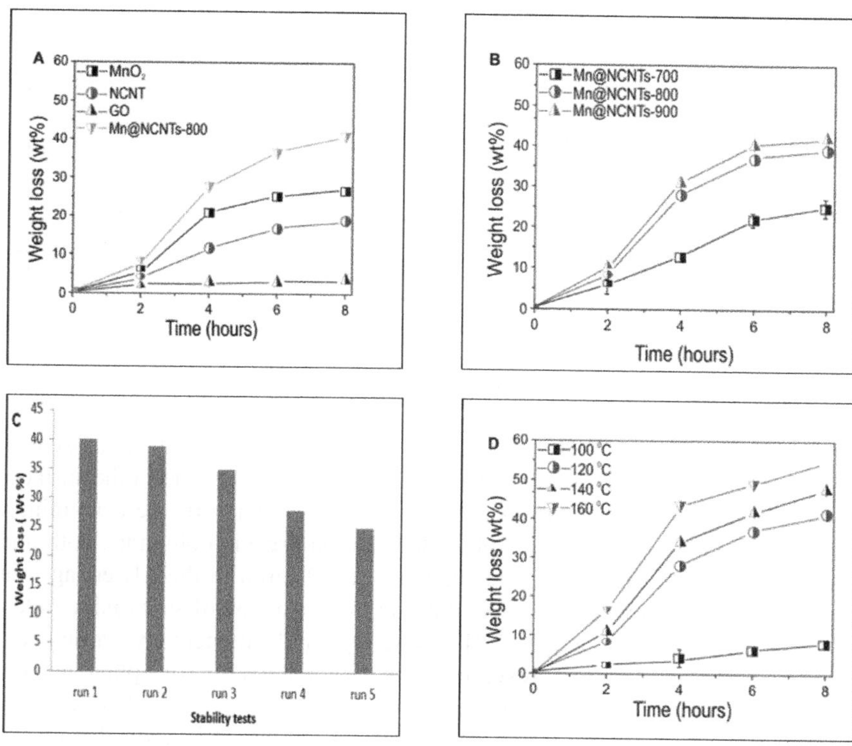

Figure 7.3. *Evaluation of the catalytic performances of Mn@NCNTs. (A) MPs removals in PMS activation over different catalysts. (B) Effects of pyrolysis temperature of catalysts on MPs removal. (C) Recyclability tests of Mn@NCNTs-800. (D) Effect of HT temperature on MPs removal. For a color version of this figure, see www.iste.co.uk/singh/photodegradation.zip*

Data are presented as mean ± SD; catalyst, 0.2 g/L; PMS, 6.5 mM; MPs, 5 g/L; and HT, 120°C (Hakkarainen and Albertsson 2004).

One well-known photocatalyst with strong photocatalytic activity is titanium dioxide (TiO_2). Its wide bandgap energy, great chemical stability, non-toxicity and abundance are the main factors contributing to its photocatalytic qualities. TiO_2 has the capacity to produce electron–hole pairs, which allows it to start a sequence of photochemical processes when exposed to UV light. When light irradiation moves an electron from the VB to the unoccupied CB, the photocatalytic activity begins. The energy of the incident photon is at least as big as the bandgap of TiO_2. The VB produces holes (h^+) as a result of this excitation. Equation [7.1] provides a summary of the entire response to light irradiation (He et al. 2020). Equation [7.2] illustrates the reaction that occurs when a water molecule comes into contact with a hole in the VB, producing hydroxyl radicals ($^•OH$). One of the most important oxidizing agents in photooxidation processes is the hydroxyl radical ($^•OH$) (Gligorovski et al. 2015; Proctor and Bartlett 2020).

$$TiO_2 \xrightarrow{h\nu} h^+_{VB} + e^-_{CB} \quad [7.1]$$

$$h^+_{VB} + H_2O \rightarrow {}^•OH + H^+ \quad [7.2]$$

$$O_2 + e^-_{CB} \rightarrow O_2^{•-} \quad [7.3]$$

$$O_2^{•-} + H^+ \rightarrow HOO^• \quad [7.4]$$

$$2HOO^• \rightarrow O_2 + H_2O_2 \quad [7.5]$$

$$H_2O_2 \rightarrow 2OH^• \quad [7.6]$$

The hydroxyl radical ($^•OH$) generated by the photocatalyst exhibits non-selective reactivity toward adsorbed contaminants that are attached to (or near) the surface of the photocatalyst. This reactivity results in the mineralization of the contaminants. TiO_2 is particularly effective in the photodegradation organic pollutants, both under UV and visible light irradiation. The photoexcited electron in the CB engages in a reaction with an oxygen molecule, resulting in the formation of superoxide radicals ($O_2^{•-}$) as described by equation [7.3]. These $O_2^{•-}$ radicals actively participate in oxidation reactions, while preventing the recombination of photogenerated electron–hole pairs. This process helps maintain the electron neutrality within the semiconductor (Gligorovski et al. 2015). The generated $O_2^{•-}$ radical undergoes protonation to form hydroperoxyl radical ($HO_2^•$), as depicted in equation [7.4]. The hydroperoxyl radical ($HO_2^•$) then reacts to produce hydrogen peroxide (H_2O_2) according to equation [7.5]. Subsequently, H_2O_2 dissociates to yield $^•OH$ radicals, as shown in equation [7.6]. To facilitate comprehension, the overall reaction to plastic degradation is illustrated through equations [7.7]–[7.9]:

$$h^{+}_{VB} + MPs \rightarrow \text{oxidize products} \quad [7.7]$$

$$^{\bullet}OH + MPs \rightarrow CO_2 + H_2O \quad [7.8]$$

$$e^{-} + MPs \rightarrow \text{reduced products} \quad [7.9]$$

Additionally, MPs are somewhat directly degraded by TiO_2 under visible light. When visible light photons enter this mechanism, MPs are excited from their ground state to an excited state. In this excited state, the polymers inject electrons into the catalyst's CB, producing semi-oxidized cation radicals. After that, the trapped electrons combine with the dissolved oxygen to generate $O_2^{\bullet -}$. The target pollutants decompose because of the $^{\bullet}OH$ radicals that are produced as a result of these $O_2^{\bullet -}$ radicals. Plastic decomposition is more common than the reaction brought on by visible light, indicating that the indirect process of plastic degradation predominates over the direct mechanism.

Furthermore, the reactions initiated by visible light are considerably slower compared to those triggered by UV light. The $OH^{\bullet}$ and $O_2^{\bullet -}$ radicals produced additionally initiate degradation at vulnerable points along the long polymeric chains of low-density polyethylene (LDPE), resulting in the formation of low molecular weight polyethylene alkyl radicals ($(\text{-}CH_2\text{-}CH_{\bullet}\text{-})n$). Furthermore, it has been shown that organic molecules and the photogenerated hole (h+) can react to make organic free radicals (Proctor and Bartlett 2020).

7.3. An overview of photocatalysts' breakdown of MPs and NPs

Photocatalyst	Kind of MPs/ NPs	Light source	Breakdown time (h)	Breakdown efficiency (weight losses) (%)	Limitations	Ref.
Protein-based porous TiO_2	HDPE	Visible light	20	6.40	Time-consuming, costly and shows incomplete breakdown	Ariza-Tarazona et al. (2019)
TiO_2 NPs	PS	UV light	12	98.40	Require MP extraction and breakdown experiments were carried out under UV irradiation	Nabi et al. (2020)
Au@Ni@TiO_2	PS	UV light	40 s	67	High cost for materials synthesis	Wang et al. (2019b)
C, N-TiO_2	HDPE	Visible light	50	70	Further study to explore other influencing factors	Ariza-Tarazona, Villarreal-Chiu & Hern
Ag/TiO_2	PE	UV light	1.5	100	Costly method and works only under UV light	Maulana et al. (2021)
Mesoporous N-TiO_2	HDPE; LDPE	Visible light	50	4.65	Insignificant breakdown	Llorente-García et al. (2020)
Degussa P25	Polyamide 66	UV light	n/a	97	Work only under the UVC system	Lee et al. (2020)
Anodized TiO_2	PS-NPs	UV light	50	23.5	Limited decomposition efficiency	Domínguez-Jaimes et al. (2021)

$Ag/TiO_2/RGO$	PE	UV light	n/a	76	Photocatalytic slurry needs 6 h irradiation that increases the synthesis cost	Fadli et al. (2021)
TiO_2@amylose/polyiodide/ hydroxyethyl cellulose	LDPE	UV light	n/a	15	Costly techniques for materials synthesis need further study on reaction intermediates and their toxicity	Zhao et al. (2021)
Copper phthalocyanine -sensitized TiO_2	PS	UV and visible light	250	n/a	Little formation of toxic by-products	Shang et al. (2003)
TiO_2-β silicon carbide alveolar foam	Poly (methyl methacrylate)	UV light	7	n/a	The reaction is dependent on the solution pH and flow rate	Allé et al. (2021)
ZnO NRs immobilized on a glass substrate	PP	Visible light	456	>65	More research is necessary to examine the determining elements	Uheida et al. (2021)
ZnO NRs	LDPE	Visible light	175	n/a	More research is necessary to examine the determining elements	Tofa et al. (2019a)
Polyacrylamide-ZnO	LDPE	UV light	50–200	<25	Only operate in UV	Kamalian et al. (2020)
ZnO-Pt	LDPE	Visible light	175	n/a	Did not display a significant loss	Tofa et al. (2019b)
Cu_xO	PS-NPs	Visible light	50	23	More research on intermediates' toxicity	Acuña-Bedoya et al. (2021)
ZnO NPs	PP	UV light	6	7.8	Restricted efficiency of decomposition	Razali et al. (2020)
CdS	PVC	UV light	250	4	Restricted efficiency of decomposition	Zhang et al. (2004)
CdS/CdO_x quantum dots	PVC, PE, PLA, PET, PUR	Visible light	22	n/a	Did not display a significant mass loss	Uekert et al. (2018)
$NiAl_2O_4$ spinels	LDPE	Visible light	5	12.5	Additional investigation to examine the intermediaries and mechanisms of degradation	Venkataramana et al. (2021)
MXene/Zn_xCd1-xS	PET	Visible light	4	n/a	Laborious material synthesis	Cao et al. (2022)
TiO_2-MWCNTs	PE	UV light	180	35	Insufficient details regarding the breakdown mechanism	An et al. (2014)
MoS_2/RGO/cotton	PE	Sunlight	1	32	Expensive materials synthesis methods require further research on the toxicity of reaction intermediates	Meng et al. (2021)
Nb2O5 atomic layers	PP, PVC, PE	Visible light	40, 60, 90	PP: 100 in 60 h; PVC: 100 in 90 h, PE: 100 in 40 h	Expensive methods for synthesizing materials	Jiao et al. (2020)
BiOCl-X	PE	Visible light	5	5.38	More improvement in the degrading efficiency is required.	Jiang et al. (2021)

* HDPE, high-density polyethylene; LDPE, low-density polyethylene; PS, polystyrene; PE, polyethylene; PVC, polyvinyl chloride; PP, polypropylene; PLA, polylactic acid; PUR, polyurethane; PET, polyethylene terephthalate.

Table 7.1. *Nanomaterials for the breakdown of micro- and nano-plastics*

7.4. Carbonaceous nanomaterials

7.4.1. *Graphene, RGO (reduced graphene oxide) and GO*

Graphene is a nanostructured material composed of one layer of carbon that has undergone sp^2 hybridization. Owing to its special qualities, it has been used in many reaction systems, including environmental remediation, as an affordable and efficient substitute (Peng et al. 2022). It successfully constructed layered double oxide (G@LDO) from synthetic graphene-like carbon. The G@LDO demonstrated a strong removal performance for PSMPs throughout a broad pH range and an efficient capacity for removing PS-MPs. However, in order to implement graphene-based catalysts, only a small number of studies have detailed the PC breakdown of MPs/NPs. For a variety of practical uses, pure graphene must be chemically altered due to its inert nature and low disintegration rate (Hegab and Zou 2015). GO, an oxidative graphene derivative, is also needed to create other GO-based natural products. The addition of oxygen functional groups to GO functionalizes it and improves its PC activity and adsorption (Bei et al. 2019). The use of GO-based NMs in environmental applications has been commonplace (Balarak et al. 2021). However, a lot more work has been done to treat MPs/NPs contamination, with a particular emphasis on removing MPs/NPs from water (Sun et al. 2020). Given the benefits of GO, moreover, sophisticated GO-based catalysts for PC degradation of MPs/NPs must be developed. Numerous active functional groups, including hydroxyl (-OH), epoxy (-C-O-C) and carboxylic (COOH) groups, can be added to RGO. RGO is a positively charged non-metal because it contains π-electrons (Jun et al. 2020). Furthermore, the reintegration of photo-induced electron–hole pairs is severely restricted by these features. Due to their superior functions, RGO-based catalytic materials have thus demonstrated themselves as an alternate catalyst for PC degradation of pollutants (Verma et al. 2017; Wang et al. 2019a) and created a TiO_2-RGO nanocomposite for PC deterioration of PP film under sun radiation by employing an easy solution casting approach. Increased degrading activity was demonstrated by this photocatalyst. Moreover, the produced polymer fragments help to clean up environmental contaminants because they are biodegradable. In addition to RGO-based NMs, plastic film was used to degrade MPs (Meng et al. 2021) and a 3D MoS_2/RGO/cotton photocatalyst generated from biomass that breaks down PE-MPs. Under the assault of ROS, the PE-MPs broke their C-C link and mineralized into H_2O and CO_2. PE-MPs lost 12% of their weight after one hour of sun exposure.

7.4.2. *Carbon nanotubes*

The controlled dimensional structure, programmable physicochemical properties and environmentally friendly nature of metal-free nanocarbon materials, such as graphene and CNTs, have made them attractive candidates to replace metal components in advanced oxidation processes. Kang et al. (2019) produced the Mn@N-doped CNTs using a one-pot method. The obtained outcome demonstrates that MPs are efficiently broken down and mineralized using this oxidation mechanism and that carbon catalysts can activate PMS, which then produces reactive radicals. Furthermore, toxicity tests demonstrate that the organic by-products are not harmful to the growth of algae. Since MWCNTs are the electron accumulation, they are likely to absorb excited photo-electrons from mixes or nanocomposite materials. This will prevent charge carriers from recombining to form further radicals and oxidants (Woan et al. 2009). Additionally, it has been claimed that MWCNTs can function as a photosensitizer to increase TiO_2's PC activity, which causes a redshift in catalyst activity (Wang et al. 2005). With the previously described properties, MWCNTs show promise for improving photocatalyst performance. Meng et al. (2021) explained a new acid vapor process to create functionalized MWCNTs, and then uniformly loaded anatase TiO_2 nanoparticles onto their surface. The outcome showed that a concentration of MWCNTs had a significant impact on the degrading activity of photocatalysts. The photocatalytic degradation efficiency of PE film was enhanced by the addition of MWCNTs. The weight loss (35%) of the TiO_2-MWCNTs (20 wt%)-PE sample was accomplished under 180 hours of UV light exposure. The synthesis of further MWCNT-based NMs could be done as part of the previously described research to inspire real-world environmental contamination, especially for MPs/NPs degradation.

7.4.3. *Nano-graphite*

Polyvinyl chloride (PVC) plastic was co-deposited with nano-graphite (Nano-G) and TiO_2 photocatalysts to create a unique photodegradable Nano-G/TiO_2/PVC composite film. The infrared (FT-IR), gel permeation chromatography (GPC), scanning electron microscopy (SEM), ultraviolet-visible spectroscopy (UV-vis), and physical and mechanical properties were used in the characterization tests. The photodegradability of the Nano-G/TiO_2/PVC composite film was higher than that of the TiO_2/PVC, Nano-G/PVC and PVC films. After 30 hours of UV irradiation, the weight loss rates of the TiO_2/PVC (TiO_2 1 wt%), Nano-G/PVC (Nano-G 1 wt%) and Nano-G/TiO_2/PVC (Nano-G 1 wt%, TiO_2 1 wt%) films were 8.94%, 7.68% and 17.24% respectively, while PVC's share was 2.12%. The ideal doping concentration of Nano-G for the Nano-G/TiO_2/PVC composite film is 1 weight percent. The Nano-G/TiO_2/PVC (Nano-G 1 wt%, TiO_2 1 wt%) composite film's Mw and Mn drop rates were 12.93% and 61.97%, respectively, which, according to the figures above, were far higher than those of PVC, and the photodegradability of the Nano-G/TiO2/PVC composite film is good. The PVC composite film's degrading efficiency was impacted by the Nano-G doping level. The dispersion degree increased with Nano-G doping increasing from 0 to 1 wt%. The maximum dispersion degree and the highest rates of Mw and Mn degradation were observed at

12.93% and 61.97%, respectively, with 1 wt% doping of Nano-G. However, the dispersion and molecular weight degradation rate progressively dropped as Nano-G was added in greater amounts. This could be because of the mixture's high concentration, which lowers the photocatalytic effect by causing the Nano-G to build up on the surface of TiO_2 and act as the charge carrier recombination center. After illumination, the high molecular weight of the Nano-G/TiO_2/PVC (1 wt%) composite film changed to a low molecular weight, the dispersion zone widened and the peak intensity decreased. The ideal amount of mixing is therefore 1 wt% of Nano-G/TiO_2/PVC, and the amount of Nano-G can be adjusted to control the rate of photodegradation (Zhang et al. 2020).

7.4. Conclusion

As plastic items were used extensively, there was a significant contamination of MPs and NPs, which had a negative impact on the environment. As a result, cleaning up MPs/NPs pollution has become a top priority. NMs have not yet been thoroughly studied on a bench scale for the cleanup of MPs/NPs contamination. The latest developments in PC/EC degradation of MPs/NPs using functional NMs, such as carbon-based NMs (e.g. graphene, GO, RGO and CNTS), are covered in this chapter. Furthermore, real-world environmental applications continue to present significant obstacles, with current research barely reaching the experimental level. Additionally, the DFT calculations and the detection and production of ROS were used to state the MPs/NPs degradation mechanism. It was concluded that ROS was essential to the breakdown of MPs and NPs. The structural efficiency connection and the ROS-induced degradation routes that were verified by DFT calculations were still, however, unknown. We are optimistic that sophisticated characterizations and computational techniques will reveal the ROS-dominated degradation mechanism and the structure mechanism–function link.

7.5. References

Acuña-Bedoya, J.D., Luévano-Hipólito, E., Cedillo-González, E.I., Domínguez-Jaimes, L.P., Hurtado, A.M., Hernández-López, J.M. (2021). Boosting visible-light photocatalytic degradation of polystyrene nanoplastics with immobilized CuxO obtained by anodization. *J. Environ. Chem. Eng.*, 9, 106208.

Allé, P.H., Garcia-Muñoz, P., Adouby, K., Keller, N., Robert, D. (2021). Efficient photocatalytic mineralization of polymethylmethacrylate and polystyrene nanoplastics by TiO_2/β-SiC alveolar foams. *Environmental Chemistry Letters*, 19, 1803–1808.

An, Y., Hou, J., Liu, Z., Peng, B. (2014). Enhanced solid-phase photocatalytic degradation of polyethylene by TiO_2-MWCNTs nanocomposites. *Materials Chemistry and Physics*, 148, 387–394.

Ariza-Tarazona, M.C., Villarreal-Chiu, J.F., Barbieri, V., Siligardi, C., Cedillo-González, E.I. (2019). New strategy for microplastic degradation: Green photocatalysis using a protein-based porous N-TiO_2 semiconductor. *Ceramics International*, 45, 9618–9624.

Ariza-Tarazona, M.C., Villarreal-Chiu, J.F., Hernández-López, J.M., Rivera De la Rosa, J., Barbieri, V., Siligardi, C., Cedillo-González, E.I. (2020). Microplastic pollution reduction by a carbon and nitrogen-doped TiO_2: Effect of pH and temperature in the photocatalytic degradation process. *J. Hazard Mater.*, 395, 122632.

Balarak, D., Mengelizadeh, N., Rajiv, P., Chandrika, K. (2021). Photocatalytic degradation of amoxicillin from aqueous solutions by titanium dioxide nanoparticles loaded on graphene oxide. *Environmental Science and Pollution Research International*, 28, 49743–49754.

Bartkowski, M. and Giordani, S. (2021). Carbon nano-onions as potential nanocarriers for drug delivery. *Dalton Transactions*, 50, 2300–2309. doi: 10.1039/d0dt04093b.

Béduer, A., Seichepine, F., Flahaut, E., Loubinoux, I., Vaysse, L., Vieu, C. (2012). Elucidation of the role of carbon nanotube patterns on the development of cultured neuronal cells. *Langmuir*, 28, 17363–17371. doi: 10.1021/la304278n.

Beg, S., Rahman, M., Jain, A., Saini, S., Hasnain, M. S., Swain, S., Imam, S., Kazmi, I., Akhter, S. (2018). Emergence in the functionalized carbon nanotubes as smart nanocarriers for drug delivery applications. In *Fullerens, Graphenes and Nanotubes: A Pharmaceutical Approach*, Grumezescu, A.M. (ed). Elsevier, Amsterdam. doi: 10.1016/B978-0-12-813691-1.00004-X.

Bei, H.P., Yang, Y., Zhang, Q., Tian, Y., Luo, X., Yang, M., Zhao, X. (2019). Graphene-based nanocomposites for neural tissue engineering. *Mol*, 24, 658.

Bepete, G. and Coleman, K.S. (2019). Carbon nanotubes: Electronic structure and spectroscopy. In *Comprehensive Nanoscience and Nanotechnology*, Andrews, D.L., Lipson, R.H., Nann, T. (eds). Elsevier, Amsterdam. doi: 10.1016/B978-0-12-803581-8.11401-8.

Bhatt, A., Jain, A., Gurnany, E., Jain, R., Modi, A., Jain, A. (2016). Carbon nanotubes: A promising carrier for drug delivery and targeting. In *Nanoarchitectonics for Smart Delivery and Drug Targeting*, Holban, A.M. and Grumezescu, A.M. (eds). Elsevier, Amsterdam. doi: 10.1016/B978-0-323-47347-7.00017-3.

Bhinge, S.D. (2017). Carbon nano-onions – An overview. *J. Pharm. Chem. Chem. Sci.*, 1, 1–2 [Online]. Available at: www.alliedacademies.org/journalpharmaceutical-chemistry-chemical-science/Editorial. [Accessed 30 January, 2021].

Bondon, N., Raehm, L., Charnay, C., Boukherroub, R. Durand, J.O. (2020). Nanodiamonds for bioapplications, recent developments. *Journal of Materials Chemistry. B*, 8, 10878–10896. doi: 10.1039/d0tb02221g.

Cao, B., Wan, S., Wang, Y., Guo, H., Ou, M., Zhong, Q. (2022). Highly-efficient visible-light-driven photocatalytic H_2 evolution integrated with microplastic degradation over MXene/ZnxCd1-xS photocatalyst. *Journal of Colloid and Interface Science*, 605, 311–319. doi: 10.1016/j.jcis.2021.07.113.

Carr, S.A., Liu, J., Tesoro, A.G. (2016). Transport and fate of microplastic particles in wastewater treatment plants. *Water Research*, 91, 174–182.

Chen, Y.C., Huang, X.C., Luo, Y.L., Chang, Y.C., Hsieh, Y.Z., Hsu, H.Y. (2013). Non-metallic nanomaterials in cancer theranostics: A review of silica and carbon-based drug delivery systems. *Science and Technology of Advanced Materials*, 14, 044407. doi: 10.1088/1468-6996/14/4/044407.

Cohen, E.N., Kondiah, P.P.D., Choonara, Y.E., du Toit, L.C., Pillay, V. (2020). Carbon dots as nanotherapeutics for biomedical application. *Current Pharmaceutical Design*, 26, 2207–2221. doi: 10.2174/1381612826666200402102308.

Cole, M., Lindeque, P., Fileman, E., Halsband, C., Goodhead, R., Moger, J., Galloway, T.S. (2013). Microplastic ingestion by zooplankton. *Environmental Science and Technology*, 47, 6646–6655.

Crevillen, A.G., Escarpa, A., Garcia, C.D. (2019). Carbon-based nanomaterials in analytical chemistry. In *Carbon-Based Nanomaterials in Analytical Chemistry*, Crevillen, A.G., Escarpa, A., García, C.D. (eds). Royal Society of Chemistry, Cambridge. doi: 10.1039/9781788012751-00001.

Domínguez-Jaimes, L.P., Cedillo-Gonzalez, E.I., Luévano-Hipólito, E., Acuña-Bedoya, J.D., Hernández-López, J.M. (2021). Degradation of primary nanoplastics by photocatalysis using different anodized TiO_2 structures. *Journal of Hazardous Materials*, 413, 125452.

Du, H., Wang, Q., Chen, G., Wang, J. (2022). Photo/electro-catalytic degradation of micro- and nano-plastics by nanomaterials and corresponding degradation mechanism. *TrAC Trends in Analytical Chemistry*, 157, 116815. doi: 10.1016/j.trac.2022.116815.

Fadli, M.H., Ibadurrohman, M., Slamet, S. (2021). Microplastic pollutant degradation in water using modified TiO_2 photocatalyst under UV-irradiation. *IOP Conference Series: Materials Science and Engineering*, 1011, 012055.

Frias, J.P.G.L. and Nash, R. (2019). Microplastics: Finding a consensus on the definition. *Marine Pollution Bulletin*, 138, 145–147.

Fu, Y., Zeng, G., Lai, C., Huang, D., Qin, L., Yi, H., Liu, X., Zhang, M., Li, B., Liu, S. et al. (2020). Hybrid architectures based on noble metals and carbon-based dots nanomaterials: A review of recent progress in synthesis and applications. *Chemical Engineering Journal*, 399, 125743.

Ganji, D.D. and Kachapi, S.H.H. (2015). Semi nonlinear analysis in carbon nanotube. In *Application of Nonlinear Systems in Nanomechanics and Nanofluids*, Ganji, D.D. and Kachapi, S.H.H. (eds). Elsevier, Amsterdam. doi: 10.1016/b978-0-323-35237-6.00002-9.

Garg, B. and Bisht, T. (2016). Carbon nanodots as peroxidase nanozymes for biosensing. *Molecules*, 21, 1653. doi: 10.3390/molecules21121653.

Ge, J., Zhang, Z., Ouyang, Z., Shang, M., Liu, P., Li, H., Guo, X. (2022). Photocatalytic degradation of (micro)plastics using TiO_2-based and other catalysts: Properties, influencing factor, and mechanism. *Environmental Research*, 209, 112729. doi: 10.1016/j.envres.2022.112729.

Gligorovski, S., Strekowski, R., Barbati, S., Vione, D. (2015). Environmental implications of hydroxyl radicals (_OH). *Chemical Reviews*, 115, 13051–13092.

Gu, M., Toh, T.B., Hooi, L., Lim, J.J., Zhang, X., Chow, E.K.H. (2019). Nanodiamond-mediated delivery of a G9a inhibitor for hepatocellular carcinoma therapy. *ACS Applied Materials and Interfaces*, 11, 45427–45441. doi: 10.1021/acsami.9b16323.

Gupta, T.K., Budarapu, P.R., Chappidi, Sudhir Sastry, Y.B., Paggi, M., Bordas, S.P. (2019). Advances in carbon based nanomaterials for bio-medical applications. *Current Medicinal Chemistry*, 26(38), 6851–6877. doi: 10.2174/0929867326666181126113605.

Hakkarainen, M. and Albertsson, A.C. (2004). Environmental degradation of polyethylene. *Advances in Polymer Science*, 169, 177–200.

Hartmann, N.B., Hüffer, T., Thompson, R.C., Hassellöv, M., Verschoor, A., Daugaard, A.E., Rist, S., Karlsson, T., Brennholt, N., Cole, M. et al. (2019). Are we speaking the same language? Recommendations for a definition and categorization framework for plastic debris. *Environmental Science and Technology*, 53, 1039–1047.

He, X., Wang, A., Wu, P., Tang, S., Zhang, Y., Li, L., Ding, P. (2020). Photocatalytic degradation of microcystin-LR by modified TiO_2 photocatalysis: A review. *Science of the Total Environment*, 743, 140694.

Hegab, H.M. and Zou, L. (2015). Graphene oxide-assisted membranes: Fabrication and potential applications in desalination and water purification. *Journal of Membrane Science*, 484, 95–106.

Hernandez, Y., Pang, S., Feng, X., Müllen, K. (2012). Graphene and its synthesis. In *Polymer Science: A Comprehensive Reference, 10 Volume Set,* Matyjaszewski, K. and Moller, M. (eds). Elsevier, Amsterdam. doi: 10.1016/B978-0-444-53349-4.00216-8.

Hirsch, A. (2010). The era of carbon allotropes. *Nature Materials*, 9, 868–871. doi: 10.1038/nmat2885.

Jiang, R., Lu, G., Yan, Z., Liu, J., Wu, D., Wang, Y. (2021). Microplastic degradation by hydroxy-rich bismuth oxychloride. *Journal of Hazardous Materials*, 405, 124247. doi: 10.1016/j.jhazmat.2020.124247.

Jiao, X., Zheng, K., Chen, Q., Li, X., Li, Y., Shao, W., Xu, J., Zhu, J., Pan, Y., Sun, Y. et al. (2020). Photocatalytic conversion of waste plastics into C2 fuels under simulated natural environment conditions. *Angewandte Chemie*, 59, 15497–15501.

Jun, B.M., Elanchezhiyan, S.S., Yoon, Y., Wang, D., Kim, S., Muthu Prabhu, S., Park, C.M. (2020). Accelerated photocatalytic degradation of organic pollutants over carbonate-rich lanthanum-substituted zinc spinel ferrite assembled reduced graphene oxide by ultraviolet (UV)-activated persulfate. *Chemical Engineering Journal*, 393, 124733.

Kamalian, P., Khorasani, S.N., Abdolmaleki, A., Karevan, M., Khalili, S., Shirani, M., Neisiany, R.E. (2020). Toward the development of polyethylene photocatalytic degradation. *Journal of Polymer Engineering*, 40, 181–191.

Kang, J., Zhou, L., Duan, X., Sun, H., Ao, Z., Wang, S. (2019). Degradation of cosmetic microplastics via functionalized carbon nano springs. *Matter*, 1, 745–758.

Knoblauch, R. and Geddes, C.D. (2020). Carbon nanodots in photodynamic antimicrobial therapy: A review. *Materials*, 13, 4004. doi: 10.3390/ma13184004.

Kumar, N. and Kumbhat, S. (2016). Carbon-based nanomaterials. In *Essentials in Nanoscience and Nanotechnology*. Wiley, New York. doi: 10.1002/9781119096122.ch5.

Lambert, S., Sinclair, C., Boxall, A. (2014). Occurrence, degradation, and effect of polymer-based materials in the environment. *Reviews of Environment Contamination and Toxicology*, 227, 1–53.

Lee, J.M., Busquets, R., Choi, I.C., Lee, S.H., Kim, J.K., Campos, L.C. (2020). Photocatalytic degradation of polyamide 66; Evaluating the feasibility of photocatalysis as a microfibre-targeting technology. *Water*, 12, 3551.

Lehner, R., Weder, C., Petri-Fink, A., Rothen-Rutishauser, B. (2019). Emergence of nanoplastic in the environment and possible impact on human health. *Environmental Science and Technology*, 53, 1748–1765.

Li, N., Than, A., Sun, C., Tian, J., Chen, J., Pu, K., Dong, X., Chen, P. (2016). Monitoring dynamic cellular redox homeostasis using fluorescence-switchable graphene quantum dots. *ACS Nano*, 10, 11475–11482. doi: 10.1021/acsnano.6b07237.

Li, Z., Wang, L., Li, Y., Feng, Y., Feng, W. (2019). Carbon-based functional nanomaterials: Preparation, properties and applications. *Composites Science and Technology*, 179, 10–40. doi: 10.1016/j.compscitech.2019.04.028.

Liang, F. and Chen, B. (2010). A review on biomedical applications of single-walled carbon nanotubes. *Current Medicinal Chemistry*, 17, 10–24. doi: 10.2174/092986710789957742.

Llorente-García, B.E., Hernández-López, J.M., Zaldívar-Cadena, A.A., Siligardi, C., Cedillo-González, E.I. (2020). First insights into photocatalytic degradation of HDPE and LDPE microplastics by a mesoporous N-TiO_2 coating: Effect of size and shape of microplastics. *Coatings*, 10, 658.

Matta-Domjan, B., King, A., Totti, S., Matta, C., Dover, G., Martinez, P., Zakhidov, A., La Ragione, R., Macedo, H., Jurewicz, I. et al. (2018). Biophysical interactions between pancreatic cancer cells and pristine carbon nanotube substrates: Potential application for pancreatic cancer tissue engineering. *Journal of Biomedical Materials Research. Part B, Applied Biomaterials*, 106, 1637–1644. doi: 10.1002/jbm.b.34012.

Maulana, D.A., Ibadurrohman, M., Slamet, S. (2021). Synthesis of nano-composite Ag/TiO_2 for polyethylene microplastic degradation applications. *IOP Conference Series: Materials Science and Engineering*, 1011, 012054.

Meng, X., Peng, X., Xue, J., Wei, Y., Sun, Y., Dai, Y. (2021). A biomass-derived, all-day round solar evaporation platform for harvesting clean water from microplastic pollution. *Journal of Materials Chemistry A*, 9, 11013–11024.

Mohajeri, M., Behnam, B., Sahebkar, A. (2018). Biomedical applications of carbon nanomaterials: Drug and gene delivery potentials. *Journal of Cellular Physiology*, 234, 298–319. doi: 10.1002/jcp.26899.

Mousavi, S.M., Soroshnia, S., Hashemi, S.A., Babapoor, A., Ghasemi, Y., Savardashtaki, A., Amani, A.M. (2019). Graphene nano-ribbon based high potential and efficiency for DNA, cancer therapy and drug delivery applications. *Drug Metabolism Reviews*, 51, 91–104. doi: 10.1080/03602532.2019.1582661.

Moussa, F. (2018). Fullerene and derivatives for biomedical applications. In *Nanobiomaterials: Nanostructured Materials for Biomedical Applications*, Narayan, R. (ed.). Elsevier, Amsterdam. doi: 10.1016/B978-0-08-100716-7.00005-2.

Nabi, I., Bacha, A.U.R., Li, K., Cheng, H., Wang, T., Liu, Y., Ajmal, S., Yang, Y., Feng, Y., Zhang, L. (2020). Complete photocatalytic mineralization of microplastic on TiO_2 nanoparticle film. *iScience*, 23, 101326.

Ouyang, Z., Yang, Y., Zhang, C., Zhu, S., Qin, L., Wang, W., He, D., Zhou, Y., Luo, H., Qin, F. (2021). Recent advances in photocatalytic degradation of plastics and plastic-derived chemicals. *Journal of Materials Chemistry A*, 9, 13402–13441.

Pan, D., Zhang, J., Li, Z., Wu, M. (2010). Hydrothermal route for cutting graphene sheets into blue-luminescent graphene quantum dots. *Advanced Materials*, 22, 734–738. doi: 10.1002/adma.200902825.

Patel, K.D., Singh, R.K., Kim, H.W. (2019). Carbon-based nanomaterials as an emerging platform for theranostics. *Materials Horizons*, 6, 434–469. doi: 10.1039/C8MH00966J.

Peng, G., Xiang, M., Wang, W., Su, Z., Liu, H., Mao, Y., Chen, Y., Zhang, P. (2022). Engineering ^{3}D graphene-like carbon-assembled layered double oxide for efficient microplastic removal in a wide pH range. *Journal of Hazardous Materials*, 433, 128672.

Plonska-Brzezinska, M.E. (2019). Carbon nano-onions: A review of recent progress in synthesis and applications. *ChemNanoMat*, 5, 568–580. doi: 10.1002/cnma.201800583.

Proctor, A.D. and Bartlett, B.M. (2020). Hydroxyl radical suppression during photoelectrocatalytic water oxidation on WO3|FeOOH. *Journal of Physical Chemistry C*, 124, 17957–17963.

Punbusayakul, N., Talapatra, S., Ajayan, P.M., Surareungchai, W. (2013). Label-free as-grown double wall carbon nanotubes bundles for *Salmonella typhimurium* immunoassay. *Chemistry Central Journal*, 7, 102. doi: 10.1186/1752153X-7-102.

Rauti, R., Musto, M., Bosi, S., Prato, M., Ballerini, L. (2019). Properties and behavior of carbon nanomaterials when interfacing neuronal cells: How far have we come? *Carbon*, 143, 430–446. doi: 10.1016/j.carbon.2018.11.026.

Raval, J.P., Joshi, P., Chejara, D.R. (2018). Carbon nanotube for targeted drug delivery. In *Applications of Nanocomposite Materials in Drug Delivery*, Inamuddin, Asiri, A.M., Mohammad, A. (eds). Elsevier, Amsterdam. doi: 10.1016/B978-0-12-813741-3.00009-1.

Razali, N., Wan Abdullah, W.R., Mohd Zikir, N. (2020). Effect of thermo-photocatalytic process using zinc oxide on degradation of macro/micro-plastic in aqueous environment. *Journal of Sustainability Science and Management*, 15, 1–14.

Reina, G., Orlanducci, S., Cairone, C., Tamburri, E., Lenti, S., Cianchetta, I., Rossi, M., Terranova, M.L. (2015). Rhodamine/nanodiamond as a system model for drug carrier. *Journal of Nanoscience and Nanotechnology*, 15, 1022–1029. doi: 10.1166/jnn.2015.9736.

Rochman, C.M., Kross, S.M., Armstrong, J.B., Bogan, M.T., Darling, E.S., Green, S.J., Smyth, A.R., Veríssimo, D. (2015). Scientific evidence supports a ban on microbeads. *Environmental Science and Technology*, 49, 10759–10761.

Shang, J., Chai, M., Zhu, Y. (2003). Solid-phase photocatalytic degradation of polystyrene plastic with TiO_2 as photocatalyst. *Journal of Solid State Chemistry*, 174, 104–110.

Shannahan, J. (2017). The biocorona: A challenge for the biomedical application of nanoparticles. *Nanotechnology Reviews*, 6, 345–353. doi: 10.1515/ntrev-2016-0098.

Shen, C., Brozena, A.H., Wang, Y. (2011). Double-walled carbon nanotubes: Challenges and opportunities. *Nanoscale*, 3, 503–518. doi: 10.1039/c0nr00620c.

Sobhani, Z., Zhang, X., Gibson, C., Naidu, R., Megharaj, M., Fang, C. (2020). Identification and visualisation of microplastics/nanoplastics by Raman imaging (i): Down to 100 nm. *Water Research*, 174, 115658.

Streicher, R.M., Schmidt, M., Fiorito, S. (2007). Nanosurfaces and nanostructures for artificial orthopedic implants. *Nanomedicine*, 2, 861–874. doi: 10.2217/17435889.2.6.861.

Sun, C., Wang, Z., Chen, L., Li, F. (2020). Fabrication of robust and compressive chitin and graphene oxide sponges for removal of microplastics with different functional groups. *Chemical Engineering Journal*, 393, 124796.

Thiel, M., Hinojosa, I.A., Miranda, L., Pantoja, J.F., Rivadeneira, M.M., Vásquez, N. (2013). Anthropogenic marine debris in the coastal environment: A multi-year comparison between coastal waters and local shores. *Marine Pollution Bulletin*, 71, 307–316.

Tian, X.T. and Yin, X.B. (2019). Carbon dots, unconventional preparation strategies, and applications beyond photoluminescence. *Small*, 15, e1901803. doi: 10.1002/smll.201901803.

Tinwala, H. and Wairkar, S. (2019). Production, surface modification and biomedical applications of nanodiamonds: A sparkling tool for theranostics. *Materials Science and Engineering. C, Materials for Biological Applications*, 97, 913–931. doi: 10.1016/j.msec.2018.12.073.

Tofa, T.S., Kunjali, K.L., Paul, S., Dutta, J. (2019a). Visible light photocatalytic degradation of microplastic residues with zinc oxide nanorods. *Environmental Chemistry Letters*, 17, 1341–1346.

Tofa, T.S., Ye, F., Kunjali, K.L., Dutta, J. (2019b). Enhanced visible light photodegradation of microplastic fragments with plasmonic platinum/zinc oxide nanorod photocatalysts. *Catalysts*, 9, 819.

Turcheniuk, K. and Mochalin, V.N. (2017). Biomedical applications of nanodiamond. *Nanotechnology*, 28, 252001. doi: 10.1088/1361-6528/aa6ae4.

Uekert, T., Kuehnel, M.F., Wakerley, D.W., Reisner, E. (2018). Plastic waste as a feedstock for solar-driven H_2 generation. *Energy and Environmental Science*, 11, 2853–2857.

Uheida, A., Mejía, H.G., Abdel-Rehim, M., Hamd, W., Dutta, J. (2021). Visible light photocatalytic degradation of polypropylene microplastics in a continuous water flow system. *Journal of Hazardous Materials*, 406, 124299.

Venkataramana, C., Botsa, S.M., Shyamala, P., Muralikrishna, R. (2021). Photocatalytic degradation of polyethylene plastics by $NiAl_2O_4$ spinels-synthesis and characterization. *Chemosphere*, 265, 129021.

Verma, R., Singh, S., Dalai, M.K., Saravanan, M., Agrawal, V.V., Srivastava, A.K. (2017). Photocatalytic degradation of polypropylene film using TiO_2-based nanomaterials under solar irradiation. *Materials and Design*, 133, 10–18.

Wang, W., Serp, P., Kalck, P., Faria, J.L. (2005). Visible light photodegradation of phenol on MWNT-TiO_2 composite catalysts prepared by a modified sol–gel method. *Journal of Molecular Catalysis A*, 235, 194–199.

Wang, D., Saleh, N.B., Sun, W., Park, C.M., Shen, C., Aich, N., Peijnenburg, W.J.G.M., Zhang, W., Jin, Y., Su, C. (2019a). Next-generation multifunctional carbon-metal nanohybrids for energy and environmental applications. *Environmental Science and Technology*, 53, 7265–7287.

Wang, L., Kaeppler, A., Fischer, D., Simmchen, J. (2019b). Photocatalytic TiO_2 micromotors for removal of microplastics and suspended matter. *ACS Applied Materials and Interfaces*, 11, 32937–32944.

Wang, S.Y., Hu, H.Z., Qing, X.C., Zhang, Z.C., Shao, Z.W. (2020). Recent advances of drug delivery nanocarriers in osteosarcoma treatment. *Journal of Cancer*, 11, 69–82. doi: 10.7150/jca.36588.

Woan, K., Pyrgiotakis, G., Sigmund, W. (2009). Photocatalytic carbon-nanotube-TiO_2 composites. *Advanced Materials*, 21, 2233–2239.

Zaytseva, O. and Neumann, G. (2016). Carbon nanomaterials: Production, impact on plant development, agricultural and environmental applications. *Chemical and Biological Technologies in Agriculture*, 3, 17. doi: 10.1186/s40538-016-0070-8.

Zhang, K., Cao, W., Zhang, J. (2004). Solid-phase photocatalytic degradation of PVC by tungstophosphoric acid–A novel method for PVC plastic degradation. *Applied Catalysis A*, 276, 67–73.

Zhang, Y., Sun, T., Zhang, D., Shi, Z., Zhang, X., Li, C., Wang, L., Song, J., Lin, Q. (2020) Enhanced photodegradability of PVC plastics film by codoping nano-graphite and TiO_2. *Polymer Degradation and Stability*, 181, 109332. doi: 10.1016/j.polymdegradstab.2020.109332.

Zhao, Y., Zhang, F., Zhang, J., Zou, K., Zhang, J., Chen, C., Long, M., Zhang, Q., Wang, J., Zheng, C. et al. (2021). Preparation of composite photocatalyst with tunable and self-indicating delayed onset of performance and its application in polyethylene degradation. *Applied Catalysis B*, 286, 119918.

8

Graphene-based Materials for Photodegradation of Micro- and Nano-Plastics

Micro-plastics (MPs, ≤5mm) and nano-plastics (NPs, ≤1 mm) are a significant environmental concern due to their ubiquitous presence in oceans, waterways and soil. These tiny plastics can harm aquatic life, contaminate the food chain and even affect human health. Graphene has unique properties such as large surface area, outstanding electrical conductivity and exceptional mechanical strength that make it an ideal candidate for photodegradation applications. Graphene-based materials (GBMs) have emerged as a promising solution for the photo degradation of MPs and NPs. Researchers have investigated various GBMs, such as graphene oxide (GO), reduced graphene oxide (rGO), and graphene nanosheets, for the photodegradation of MPs and NPs. The photodegradation mechanism of GBMs involves the absorption of light energy, which generates reactive oxygen species (ROS) that can break down the plastic molecules. The high surface area of graphene allows for efficient interaction with the plastic particles, facilitating the degradation process. The advantages of GBMs for photodegradation of MPs and NPs include their high efficiency, low toxicity and potential for scalability. However, further research is required to optimize the materials properties, investigate their long-term stability and explore their application in real-world scenarios. Studies have demonstrated the effectiveness of GBMs in degrading various types of plastics including polyethylene, polypropylene and polystyrene. The degradation rates and efficiencies can be tuned by adjusting the graphene material's properties, such as its surface functionalization, layer thickness and composite structure. Overall, GBMs hold significant potential for tackling the urgent problem of MP and NP pollution. This chapter highlights the challenge of plastic pollution, the overview of MPs and NPs, the increasing presence of MPs and NPs in environments, the composition of GBMs and the applications of GBMs in the detection and degradation of various pollutants, with an emphasis on MPs and NPs.

Chapter written by Geeta SINGH and Preeti GUPTA.

8.1. Introduction

Plastics are used to create synthetic fibers such as polyester, nylon and acrylic. These fibers are used in everything from clothing to upholstery, providing durability and ease of care. Plastic materials are lightweight, durable, convenient, corrosion and flame-resistant, and have also become famous over a short period of time because they have improved the quality of life of millions of people across the globe. But we also see the consequences of plastic usage when they enter into the environment (Thompson et al. 2004). However, it is important to balance these benefits with considerations around environmental impact (Bratovcic et al. 2015). Synthetic plastics have indeed revolutionized many industries due to their versatility and cost-effectiveness. Their widespread use can be dependent on many factors such as the manufacturing industry, textile industry and food packaging industry, where plastics are integral in producing a wide range of products, from automotive parts to household goods, and in transportation. They offer a combination of strength, flexibility, and lightweight properties that are often superior to traditional materials like metal and glass.

8.1.1. *Overview of micro-plastics*

MPs are tiny plastic particles smaller than 5 mm in size. They result from the breakdown of larger plastic materials or are intentionally manufactured at this small scale for use in products like cosmetics and clothing. "Nano-plastics" are plastic particles with a length that ranges between 1 nm and 1 μm. MPs may stem from two different origins: primary MPs which are commercially produced plastics (e.g. some industrial abrasives or cosmetic scrubs), whereas secondary MPs originate from the degradation of larger plastic items (e.g. bottles and packaging materials), either by mechanical forces or mainly by photo-degradation induced by UV-light (ISO 2020; Braun et al. 2020). MPs are chemically and structurally stable and resist biodegradation for decades, hence, they tend to accumulate in the environment, presenting an increasing threat to global ecosystems (Shi et al. 2015; Bajt 2021; Sorasan et al. 2022). MPs are a complex class of pollutants with a multifaceted nature, requiring careful analysis on several parameters. For example, their size variation (1 μm–1 mm) affects their behavior, transport and interaction with organisms and environments. Secondly, they are composed of various polymer types, including both conventional synthetic plastics and biodegradable biopolymers, and these polymers differ in their chemical composition, structure and density, which can influence how they break down, their environmental impact and how organisms may interact with them. MPs often contain a range of chemical additives such as antioxidants, light stabilizers, plasticizers, flame retardants and pigments.

These chemicals enhance the functional properties of plastics, but can leach into the environment and potentially cause harm. Additionally, the particles may absorb contaminants from their surroundings, such as persistent organic pollutants, antibiotics and heavy metals (Koelmans et al. 2019; Nguyen et al. 2019; Pico et al. 2019). The tremendous and increasing production of plastic materials has resulted in the appearance of MPs, which have been described as contaminants of emerging concern.

8.1.2. *Overview of nano-plastics*

Recently, NPs, which refer to plastic particles smaller than 1 μm, are emerging as a significant environmental concern. These tiny particles are generated unintentionally through the breakdown of larger plastics, such as MPs, or from industrial processes. Their small size makes them particularly challenging to study, monitor and remove from the environment. NPs can infiltrate ecosystems more deeply than larger plastic particles, entering water systems, soils and even the atmosphere (Shi et al. 2015; Nguyen et al. 2019). To understand the impacts of NP particles on the environment and human health requires detailed information about several key characteristics like chemical composition (polymer type, additives, absorbed contaminants etc), size and size distribution, surface properties and mass. NPs can cross biological barriers, such as the blood–brain barrier or placental barriers, which could have significant health implications (Gigault et al. 2018; Schwaferts et al. 2019; Fu et al. 2020).

NPs have mainly been categorized in two types, based on their origin, that is, primary NPs, which are intentionally manufactured particles with a defined size and composition, designed for specific applications, that is, medical devices, pharmaceuticals and electronics, and secondary NPs, which are formed through the degradation of larger plastic items, that is, degraded fragments of PET bottles and breakdowns of polyethylene bags into nanoscale particles (Jambeck et al. 2015; Wagner and Reemtsma 2019). There is no standardized definition for "nano-plastics" because they do not fit a single material or composition. Their variability in size, shape and chemical properties makes uniform classification challenging. Based on the literature, the scientific community classifies different plastic particle groups using the following size categories: NPs (1 nm–1 μm), MPs (1 μm–1 mm), meso-plastics (1 mm–1 cm) and macro-plastics (1–100 cm). Figure 8.1 illustrates the classification of plastic particles by size, along with size references and the definition of NPs.

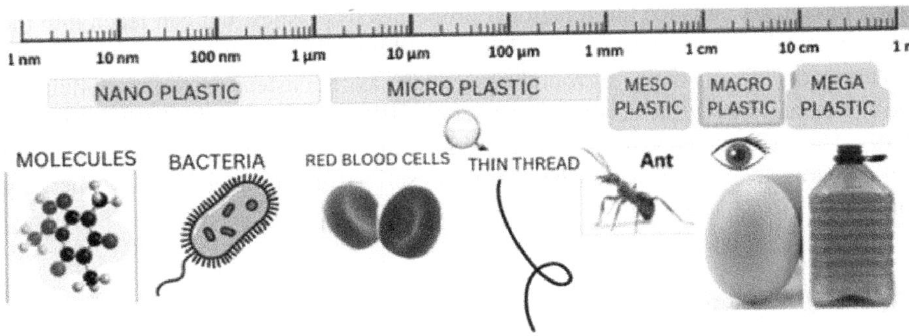

Figure 8.1. *Classifications of plastic particles by their size measurements. For a color version of this figure, see www.iste.co.uk/singh/photodegradation.zip*

8.1.3. *Environmental impact of micro- and nano-plastics*

The small size of MPs and NPs allows them to be taken up by a variety of organisms that span the ecological spectrum (i.e. starting from algae, through small invertebrates, to the entire range of large vertebrates). There has been growing alarm over the potential ecological hazards and environmental risks posed by MPs and NPs (Hanun et al. 2021; Impellitteri et al. 2023; Lamichhane et al. 2023). MPs have been detected in various environmental matrices, including water, soils and sediment, as well as in feeding organisms. The impact of MPs in regional areas, such as the Mediterranean, have been studied (Galgani et al. 2015). The pollution by MPs have been also reported for the South China Sea and continental water bodies (Adhikari et al. 2022; Rani et al. 2023; Rathore et al. 2023). The increasing accumulation of plastic in the environment poses a global threat to ecological and public health. Approximately, 400 million tons of plastic are produced worldwide every year, and due to their persistence in the environment and non-biodegradability, the plastic waste enters into the environment. Globally, more than 250,000 tons of plastics have been detected in the marine environment (Eriksen et al. 2014). Large plastic items are weathered and fragmented into small plastic debris, which are MP pollutants with a diameter less than 5 mm. The large production in recent years has caused growing environmental infection and ecological threat (Alqattaf 2020; Meneses et al. 2022; Nayanathara Thathsarani Pilapitiya and Ratnayake 2024). MPs can also adsorb pathogens and transport them across regional boundaries. Due to their large production and numerous applications, the accumulation of plastic wastes and the universal existence of MPs, they have been considered as a new type of pollutant of immediate concern and the cause of genuine environmental pollution problems in the 21st century (Arienzo et al. 2021; Huang et al. 2021; Xiang et al. 2022). When plastic waste enters into the environment, it is subjected to a range of degradation processes, including the impacts of solar radiation, mechanical forces and microbial action, all of which contribute to the fragmentation of these materials into smaller and smaller particles (Pervez et al. 2023). These small MPs in the environment were first detected by scientists in the 1970s, and since then, studies on their distribution and effects have intensified (Rai et al. 2021).

Over the last decade, there has been a significant increase in publications related to MPs and NPs. According to various sources, including academic journals and research databases, the number of publications on these topics has grown exponentially. In 2021 alone, there were over 2,000 publications on plastic pollution, a significant increase from approximately 200 papers per year at the beginning of the decade (Ivleva 2021; Allen et al. 2022). The amount of MPs in some oceanic compartments is predicted to double by 2030, emphasizing the need for continued research and understanding of plastic composition.

8.1.4. *Better alternatives to plastics*

We have learned that plastic waste is a significant issue contributing to pollution and harming wildlife. Many industries are now focusing on developing biodegradable alternatives and improving recycling methods to mitigate these issues. Long-chain polymers with high molecular weight (Mw), hydrophobicity and crystallinity exhibit significant resistance to biodegradation (Wilkes and Aristilde 2017). To overcome these barriers, microorganisms first depolymerize the large polymer chains into smaller, more manageable monomers. This depolymerization process is typically catalyzed by extracellular enzymes secreted by the microorganisms (Shah et al. 2008). This process is essential for the biodegradation of polymers, as the initial depolymerization is a crucial step in converting insoluble, high Mw polymers into forms that can be utilized by microbial metabolic pathways (Kolvenbach et al. 2014).

With the environmental concerns about the degradation and remediation of petrochemical plastics, which are notorious for their persistence and non-biodegradability, there has been a significant push toward developing biodegradable plastics such as poly lactic acid (PLA), poly ε-caprolactone (PCL), poly butylene succinate (PBS) or poly butylene succinate-co-butylene adipate (PBSA) (Urbanek et al. 2018). These new materials are designed to break down more easily in the environment, potentially reducing their ecological footprint. Despite advancements in biodegradable plastics, environmental concerns persist regarding plastic contamination in the food chain. Toussaint et al. (2009), critically analyzed data on MP contamination in various food products and edible animal species since 2010. They collected data on 201 edible animal species, including sea fishes, molluscs, crustaceans, birds, sweetwater fish, turtles, chickens, sea salt, sugar and honey, beer and even water. These species were found to be contaminated with MPs, which can enter the food chain through ingestion.

8.1.5. *Status of plastic recycling in India with other countries*

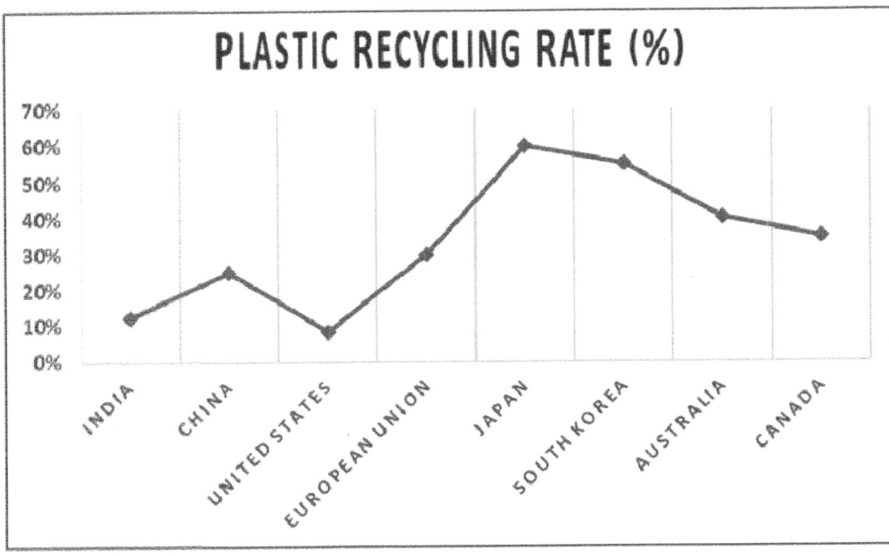

Figure 8.2. *Status of plastic recycling in India with other countries (data from different sources). For a color version of this figure, see www.iste.co.uk/singh/photodegradation.zip*

The disposal of plastics poses significant challenges regarding their ecological impact. Only 60% of the plastic produced is recycled, leaving 9,400 tonnes of plastic unmanaged in the environment, which contributes to land, air, and water pollution (https://www.motherearthliving.com/health-and-wellness/harmful-effects-of-plastic-ze0z1205zsch). Figure 8.2 depicts the status of plastic recycling in India with other countries. The graph clearly shows that India's plastic recycling rate is lower compared to other countries, with a rate of 12% in 2020. China, the European Union, Japan and South Korea have significantly higher plastic recycling rates, ranging from 25% to 60%. The United States has a lower plastic recycling rate than India at 8%. China has made significant strides in plastic recycling, with a rate of 25%, driven by government initiatives and investments in waste management infrastructure. The European Union has implemented strict regulations and targets for plastic recycling, resulting in a high recycling rate of 30%. Japan and South Korea have implemented effective waste management systems and recycling technologies, achieving high plastic recycling rates of 60% and 55%, respectively (swachhbharaturban.gov.in/writereaddata/SBM%20Plastic%20Waste%20Book.pdf; Plastic waste management (amendment), 2018; UNIDO 2018).

8.2. Graphene-based materials

GBMs have tremendous optical, electrical, mechanical and chemical properties, which make them useful to make thin films and thick sheets useful for different purposes. These materials can be used in various sectors such as nuclear physics, microelectronics, biomedicine, engineering and agriculture (Torrisi et al. 2022). Due to their awesome properties, it is possible to realize gas sensors, dosimeters for ionizing radiation, light detectors, electronic devices, membranes for gases and desalination, biocompatible surfaces, medical protheses and thin coating films to preserve underlying layers (Aliofkhazraei et al. 2016; Yu et al. 2020).

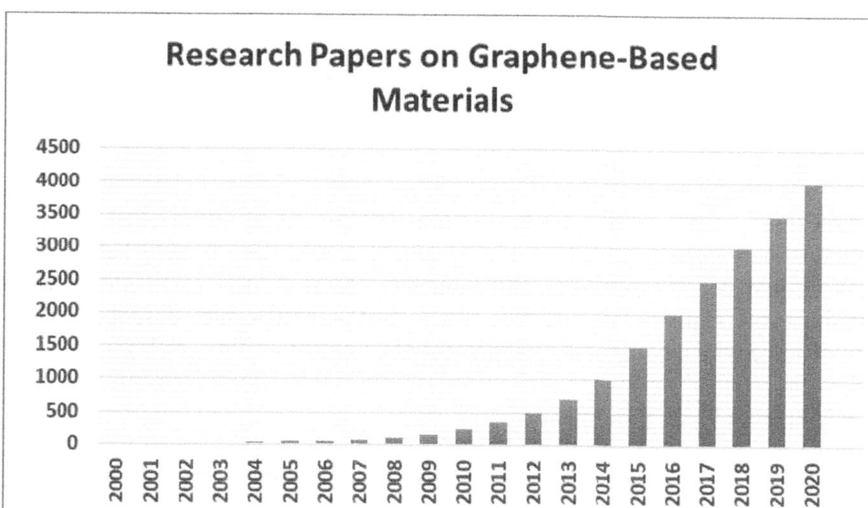

Figure 8.3. *Number of research papers published on graphene-based materials over the past 20 years (2000–2020). For a color version of this figure, see www.iste.co.uk/singh/photodegradation.zip*

Figure 8.3 depicts a graph showing the number of research papers published on GBMs over the past 20 years (2000–2020), based on data from Web of Science. Overall, the graph demonstrates the growth and diversification of research on GBMs over the past 20 years, highlighting the material's potential for various applications and its continued importance in the scientific community.

8.3. Structure and characteristics of graphene-based materials

Carbon is a fundamental element in the chemistry of life, making up the backbone of most biological molecules. Its compounds are considered more biologically and environmentally friendly compared to many inorganic materials. Graphene, a two-dimensional material composed of carbon atoms, has been widely studied for its exceptional electrical, thermal and mechanical properties. Graphene was discovered by Andre Geim and Kostya Novoselov (https://www.graphene.manchester.ac.uk/learn/discovery-of-graphene/) using the simple scotch tape method. After the discovery, it became famous as a wonderful material due to its various properties, such as large specific surface area, high thermal conductivity, high Young modulus, remarkable strength and elevated electron mobility (2.5 × 105 cm^2/Vs) (Sur 2012; Cheng et al. 2017). GBMs possess all the inherent properties of graphene, along with enhanced reactivity and a wide range of functional groups. This enables them to establish various interactions with ions, molecules and nanoparticles, resulting in sustainable, scalable and effective pollution treatment technologies (https://www.magmaThe fascinating world of carbon-based nanomaterials). The specifications of some of these materials are given as follows:

– Graphite

Graphite is a naturally occurring mineral composed of stacked layers of hexagonal carbon rings, also known as graphene layers. Each layer is held together by weak van der Waals forces, allowing them to slide past each other, making graphite a soft, slippery material. The structure of graphite can be described as a 3D crystal lattice with a hexagonal arrangement of carbon atoms in each layer.

– Graphene

Graphene is a single layer of sp^2 hybridized carbon atoms arranged in a hexagonal honeycomb crystal lattice (Allen et al. 2010; Zhu et al. 2010; Sur 2012). Carbon allotropes such as fullerenes, carbon nanotubes and graphite all derive from graphene as their fundamental building block. It can be synthesised by two methods, which are shown in Figure 8.4. Various techniques are used to analyze graphene's structure, properties and quality, ensuring its suitability for specific applications like microscopy (TEM and SEM), spectroscopy (Raman spectroscopy and X-ray photoelectron spectroscopy (XPS)), electrical characterization (electrical conductivity, carrier mobility and other electrical properties) and thermal analysis (differential scanning calorimetry (DSC) and thermal gravimetric analysis (TGA)).

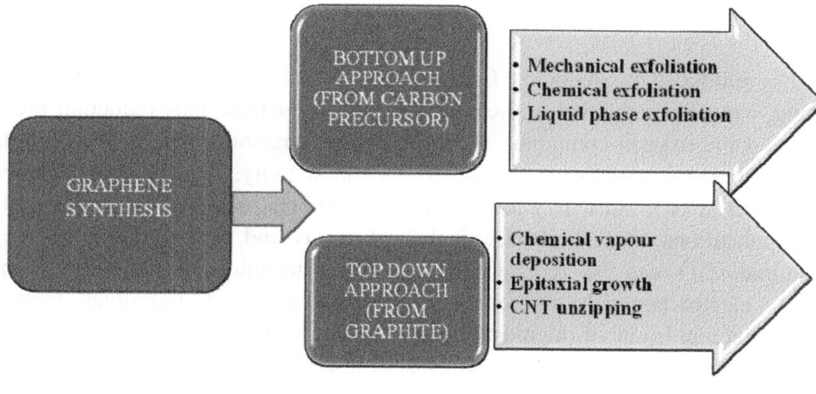

Figure 8.4. *Synthesis of graphene. For a color version of this figure, see www.iste.co.uk/singh/photodegradation.zip*

– Graphite oxide

Graphite oxide is a derivative of graphite, obtained by treating graphite with strong oxidizing agents, such as nitric acid or potassium permanganate. This process introduces oxygen-containing functional groups, such as hydroxyl (–OH), carboxyl (–COOH) and epoxy (–O–) groups, onto the graphite surface. The oxidation process disrupts the planar structure of graphite, leading to a more disordered and wrinkled morphology. Graphite oxide is commonly used as a precursor for synthesizing GO (Grajek et al. 2020). The process of converting graphite oxide into GO involves exfoliating the graphite oxide into single or few-layered sheets by mechanical stirring or sonication (Li et al. 2015; Mas'udah et al. 2016).

– Graphene oxide

GO is a single layer of graphite oxide, typically produced by exfoliating graphite oxide in water or other solvents. GO retains the oxygen-containing functional groups introduced during the oxidation process, which makes it hydrophilic and dispersible in water. The structure of GO is characterized by a 2D lattice of carbon atoms, with a mixture of sp^2 and sp^3 hybridized carbon atoms, and a high density of defects and vacancies. rGO is also synthesized with the help of GO (Yu et al. 2020). The properties of GO make it highly suitable for use in water purification during wastewater treatment. It reduces the need for researchers to undergo the often complex and time-consuming synthesis of GO in the laboratory. This availability allows scientists and engineers to focus more on applications and experimental research rather than materials preparation (Ikram et al. 2020).

– Reduced graphene oxide

rGO is produced by reducing GO through various methods, including thermal annealing, chemical reduction and electrochemical reduction. This reduction process eliminates some oxygen-containing functional groups, restoring the sp^2 hybridization of carbon atoms and enhancing the material's conductivity. The structure of rGO is similar to that of GO, but with a lower density of defects and vacancies. rGO is often used in applications where high electrical conductivity and high mechanical strength are required. GO can be reduced using various techniques, including thermal, chemical, electrochemical, microwave, photo-assisted and microbial methods. Thermal reduction is one of the most effective methods to produce rGO from GO, and it can yield sheets with properties that closely resemble pristine graphene (Agarwal and Zetterlund 2021). Preparation of rGO in various forms such as hydrogel, aerogel, films, foam, monoliths and beads significantly broadens its application potential, particularly as an adsorbent. Each form brings unique structural properties that make them highly efficient in adsorption processes, largely due to their porous nature and large surface area (Zhang et al. 2019; Zheng et al. 2021a, 2021b).

– Graphene nanosheets

Graphene nanosheets are thin, two-dimensional layers of graphene, usually produced through mechanical exfoliation of graphite or chemical vapor deposition growth. Graphene nanosheets have a hexagonal lattice structure, similar to graphite, but with a much smaller thickness (typically <10 nm). Graphene nanosheets can be synthesized by GOs, rGOs and exfoliated GOs (EGOs). They possess outstanding mechanical, thermal and electrical properties, making them ideal for a variety of applications, including electronics, energy storage and composites (Krishnamoorthy et al. 2013).

– Graphene oxide-based composite materials

Various techniques are used to produce GO-based nanocomposites, each offering distinct advantages depending on the desired application. There are various methods of preparation of these nanocomposites such as in situ polymerization, electrochemical deposition, vacuum impregnation, sol gel technique, ball milling, hydrothermal, microwave-assisted and solution mixing methods. In GO-based nanocomposites, both functional and substrate components play a crucial role. The functional component typically includes GO, which provides the necessary mechanical reinforcement, electrical conductivity, or other desired properties, while the substrate acts as a matrix that supports the GO and enables its immobilization within the structure (Norhayati et al. 2016). To further enhance its performance and tailor its characteristics for specific applications, researchers have developed various graphene-based nanocomposites by incorporating different materials into the graphene matrix. These potent graphene materials nanocomposites with numerous applications are given in Table 8.1.

Types of nanocomposites	Name	Examples	Applications
Graphene-polymer nanocomposites	Graphene-reinforced polymers (GRPs)	Graphene-polyethylene (GPE), graphene-polypropylene (GPP), graphene-polystyrene (GPS)	Aerospace, automotive and energy storage
Graphene-metal nanocomposites	Graphene-metal hybrid nanocomposites (GMHNs)	Graphene-gold (G-Au), graphene-silver (G-Ag), graphene-copper (G-Cu)	Electronics, energy storage, catalysis
Graphene-ceramic nanocomposites	Graphene-ceramic hybrid nanocomposites (GCHNs)	Graphene-silicon carbide (G-SiC), graphene-alumina (G-Al_2O_3), graphene-zirconia (G-ZrO_2)	Aerospace, biomedical, energy storage
Graphene-carbon nanotube (CNT) nanocomposites	Graphene-CNT hybrid nanocomposites (GCHNs)	Graphene-single-walled CNT (G-SWCNT), graphene-multi-walled CNT (G-MWCNT)	Energy storage, biomedical devices
Graphene-nanoparticle nanocomposites	Graphene-nano-particle hybrid nanocomposites (GNHNs)	Graphene-quantum dot (G-QD), graphene-nano-crystal (G-NC), graphene-metal oxide (G-MO)	Optoelectronics, energy storage and biomedical imaging
Graphene-biomaterial nanocomposites	Graphene-biomaterial hybrid nanocomposites (GBHNs)	Graphene-protein (G-protein), graphene-peptide (G-peptide), graphene-polysaccharide (G-polysaccharide)	Biomedical devices, tissue engineering, drug delivery
Graphene-graphene oxide (GO) nanocomposites	Graphene-GO hybrid nanocomposites (GGHNs)	Graphene-GO (G-GO), reduced graphene oxide (rGO)-GO (rGO-GO)	Energy storage, biomedical devices
Graphene-nanofiber nanocomposites	Graphene-nanofiber hybrid nanocomposites (GNHNs)	Graphene-cellulose nanofiber (G-CNF), graphene-polymer nanofiber (G-PNF)	Aerospace, automotive and energy storage
Graphene-quantum dot nanocomposites	Graphene-quantum dot hybrid nanocomposites (GQHNs)	Graphene-cadmium selenide (G-CdSe), graphene-cadmium telluride (G-CdTe)	Optoelectronics, energy storage and biomedical imaging
Graphene-metal organic framework (MOF) nanocomposites	Graphene-MOF hybrid nanocomposites (GMHNs)	Graphene-zeolitic imidazolate framework (G-ZIF), graphene-metal-organic framework (G-MOF)	

Table 8.1. *Different graphene materials nanocomposites, examples and application (Maruthamani et al. 2015; Sharif et al. 2017; Mao et al. 2018; Banu et al. 2019; Khan et al. 2019; Zhang et al. 2019; Al Kausor and Chakrabortty 2020; Botsa et al. 2020; Chang et al. 2020; Sree et al. 2020; Liu et al. 2022; Usharani et al. 2022; Ding et al. 2023; Kausar et al. 2023; Shahraki et al. 2023; Xia et al. 2023; Dar and Sahu 2024)*

The structure of graphene, GO and rGO is given in Figure 8.5.

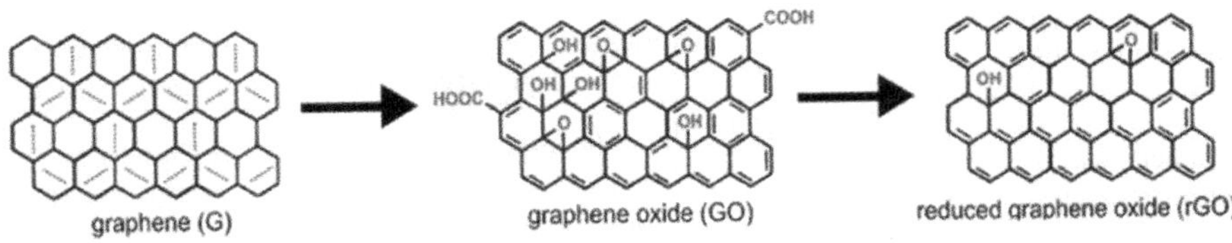

Figure 8.5. *Structure of graphene, graphene oxide (GO) and reduced graphene oxide (rGO) (https://nanografi.com/blog/what-is-the-difference-between-graphene-oxide-and-reduced-graphene-oxide/). For a color version of this figure, see www.iste.co.uk/singh/photodegradation.zip*

8.4. Photodegradation and graphene-based materials

Photocatalysis, as an advanced oxidation process (AOP), holds significant promise for addressing environmental challenges, particularly in the degradation of different pollutants. This eco-friendly technology can initiate chemical transformations in pollutants through a light-activated catalyst. Recently, the development of photocatalytic technologies for environmental remediation has rapidly accelerated, driven by the urgent need for sustainable energy solutions and cleaner environments. The photocatalytic method is a very significant approach for the degradation of dyes, pesticides, oil, other organic pollutants and pharmaceutical compounds

In this context, GBMs have shown great potential when integrated into photocatalytic systems. These materials can act as catalysts, adsorbents and photocatalysts to break down plastic molecules into smaller fragments, making them more susceptible to biodegradation. The synergy between photocatalysts and GBMs improves both the efficiency and versatility of the process, largely because graphene's exceptional conductivity enhances charge separation and carrier mobility, minimizing recombination of electron–hole pairs. This makes hybrid systems more effective under visible light. Efficient photocatalysis requires rapid charge separation and transport to the catalyst surface, where redox reactions occur. Graphene's high conductivity and ability to act as an electron acceptor improve the movement of electrons, reducing recombination and enhancing the reaction rate (Ahmed et al. 2010; Nickheslat et al. 2013; Chowdhury et al. 2017). GBM systems allow for the use of visible light, enhancing energy efficiency and environmental friendliness in the process. GBM systems allow for the use of visible light, enhancing energy efficiency and environmental friendliness in the process. GBM systems allow for the use of visible light, enhancing energy efficiency and environmental friendliness in the process.

8.5. Application of GMBs in removal/degradation/remediation of different pollutants

This chapter is mainly focused on the role of GBMs in MPs and NPs. But they play a very crucial role in the removal as well as degradation of various pollutants other than MPs and NPs. Table 8.1 shows the numerous examples and applications of GBMs in various fields. GO and rGO are two types of graphene-based components that have a variety of applications, such as catalysis, supercapacitors, photoelectric materials and adsorbents. The expanded honeycomb network of graphene serves as the fundamental building block for other important allotropes, such as fullerenes, carbon nanotubes and graphite (Tabatabai-Yazdi et al. 2021). The GBM is an emerging field and requires much needed attention from the scientific community. Bruno et al. (2022) reported the use of GBMs for water remediation and studied the pollutant sorption, photodegradation and filtration. As we know, the use of dyes has hazardous effects on human health and the environment. Parvez et al. (2024) also studied the different GBMs and synthesized a novel polyethylene glycol-functionalized graphene oxide/chitosan composite (PEG-GO/CS) to remove dyes like methyl orange and other similar organic dyes from contaminated water. In 2021, Zheng and Andou (2022) discovered the various methods for the remediation of bisphenol A (BPA), plastic additive using GMBs. GBM-based sensors are increasingly used in environmental monitoring due to their effectiveness in detecting pollutants. These sensors can detect hazardous metals such as lead (Pb), mercury (Hg) and cadmium (Cd). GBM-based sensors enhance the detection limits of these metals through their high sensitivity, making them reliable substitutes for traditional methods (Si et al. 2018; Liu et al. 2019; Bhardiya et al. 2021). The role of synthetic pesticides and their environmental hazards are well known, and GBMs are also used for detecting insecticides, fungicides and herbicides, which are often present in agricultural runoff and contaminate water bodies. The combination of graphene's high surface-to-volume ratio and functionalization strategies enhances the detection of these toxic compounds (Rajaji et al. 2019; Pérez-Fernández et al. 2020). Their versatility makes GBMs highly promising for developing portable, cost-effective and accurate sensors for environmental pollutant monitoring. Likewise, rGO and graphene-based nanocomposites have also been seen as an effective photocatalyst for the degradation of organic dyes and other organic pollutants.

8.6. Photodegradation of micro- and nano-plastics by graphene-based materials

Indeed, the use of GBMs for the photodegradation of MPs and NPs is a relatively new and emerging research area. Lots of extensive efforts have been made to efficiently remove macro-plastics, but it still remains a difficult task. The main focus of this chapter is to describe the different GBMs used in the photodegradation of MPs and NPs. When we survey the literature available from the internet through various searching platforms like web of science, we found limited literature compared to other applications like pollutant detection or catalysis. GO has been found to be effective in degrading MPs, such as polyethylene and polypropylene under UV light irradiation. Graphene-based membranes and graphene nanosheets have been developed for the removal of MPs and NPs from water (Shahraki et al. 2023). These membranes can be designed to have high selectivity and permeability, allowing for the efficient separation of plastics from water.

Graphene's high surface area and conductivity enable efficient charge transfer and ROS generation, enhancing the degradation process (Ramalingam et al. 2022). In the presence of light, the photocatalyst (e.g. TiO_2) generates electron–hole pairs. The electrons are then transferred to the RGO, which acts as an electron acceptor, preventing recombination and allowing the holes to participate in the degradation reaction. Jeong et al. (2023) fabricated Fe_3O_4/laser-induced graphene by irradiating a laser on the surface of a polymer prepared by homogeneously mixing iron oxide, lignin powder and polydimethylsiloxane, to efficiently remove household-derived MPs. Similarly, Uogintė et al. (2023) screened the ability of GO-based metal oxide nanomaterials to degrade MP particles from polyethylene in aqueous solutions and they substantiated the potential value of GO-based metal oxide nanomaterials for the removal of polyethylene MP particles with an environmentally friendly and highly efficient process.

Tan et al. (2023) synthesized GO-ZnO and studied its role for the photodegradation of low-density polyethylene (LDPE) MPs under UV light irradiation. The GO-ZnO photocatalyst was successfully synthesized through a simple hydrothermal process, which is a preferred method due to its efficiency in creating GO-based nanomaterials at high temperatures and vapor pressures. This study utilized the Box-Behnken Design to examine the interactions between pH, temperature and catalyst loading on the mass loss of LDPE. Ultimately, a reduced quadratic model was developed to predict MP mass loss under the photodegradation of GO-ZnO. This model can be used to forecast the optimal conditions for MP degradation, offering important insights for future research and applications in environmental remediation.

Apart from this, TiO_2/rGO nanocomposites have been shown to effectively degrade dyes and plastics in aqueous solutions (Tan et al. 2023). Like Tan et al. (2023), Kocijan et al. (2021) have also reported the improved photodegradation efficiency for plastics by silver-doped titanium dioxide nanosheets (Ag-TNs-RGO). Graphene can be combined with various polymer matrices to create nanocomposites that exhibit enhanced photodegradation properties (Tabatabai-Yazdi et al. 2021).

There are different types of degradation methods of plastics like primary (re-extrusion), secondary (mechanical recycling), tertiary (chemical recycling) and quaternary (oxidation to produce energy). Among all of them, chemical recycling is most important because it decomposes plastics into smaller molecular weight polymers, which can be used as raw materials for producing new products or plastics (Ibrahim et al. 2021).

8.7. Challenges and future perspectives

However, there are many challenges associated with the use of GBMs for MP and NP degradation, like limited understanding of the degradation mechanisms and pathways that need the research on further optimization of material properties and reaction conditions. Future research directions in this area should focus on elucidating the degradation mechanisms and pathways of GBMs and scaling up the production of GBMs for large-scale applications (Eda et al. 2008; Bratovcic 2019). India is gradually increasing its research in GBMs, particularly through collaborations between academic institutions like the Indian Institute of Technology (IITs) and national research bodies such as the Council of Scientific & Industrial Research (CSIR) (Becerril et al. 2008).

The production of GBMs is still in its early stages, but it is expected to become more widespread once the challenges of commercialization are addressed. As plastic production continues to increase, it is essential to develop countermeasures to mitigate the environmental impact of plastic waste, particularly MPs in marine environments. GBMs have the potential to play a crucial role in the remediation of marine debris, including MPs. This review provides a comprehensive overview of the current state of GBMs in relation to the photodegradation of MPs and NPs, highlighting their potential as a solution to this pressing environmental issue.

8.8. Environmental fate of graphene-based materials

The environmental fate of GBMs is a critical aspect of their ecotoxicity assessment. GBMs can enter the environment through various pathways, including industrial effluents, wastewater treatment plants and accidental releases during transportation or storage. Once released, GBMs can undergo various transformations, such as aggregation, sedimentation and degradation, which can affect their bioavailability and toxicity (Somani et al. 2006; Amobonye et al. 2021; Assad et al. 2024).

Several factors can influence the ecotoxicity and environmental fate of GBMs, such as:

1) Size and shape: smaller GBMs tend to be more toxic because of higher surface area and reactivity.

2) Surface functionalization: chemical modifications on the surface of GBMs can alter their interactions with biological molecules and affect their toxicity.

3) Environmental conditions: pH, temperature and ionic strength can influence the aggregation and sedimentation of GBMs, affecting their bioavailability and toxicity.

4) Biological factors: the presence of bacteria, fungi and algae can influence the degradation and transformation of GBMs.

8.9. Conclusion

Globally, GBMs are the focus of research aimed at mitigating MP and NP pollution. While international efforts are more advanced, India is steadily advancing in this domain through focused research and development initiatives. However, challenges related to scalability, environmental impact and economic feasibility must be addressed to make these technologies viable for large-scale applications. Further research is required to explore their potential for large-scale applications and develop more effective GBMs for photodegradation. Given the indispensable nature of plastics in modern society, their production is expected to continue increasing, raising concerns about their environmental impact, particularly in the form of MPs and NPs. These persistent pollutants have become a major environmental challenge, necessitating effective countermeasures. GBMs could provide viable solutions for the detection, degradation and eventual removal of MPs in various environmental settings. Although this review may not have covered every aspect of GBMs regarding their role in the detection and photodegradation of MPs and NPs, it is intended to serve as a comprehensive overview of this emerging field. Hopefully, this write-up will inspire further research and exploration into the application of GBMs in environmental remediation, ultimately contributing to the development of innovative solutions for the pressing issue of plastic pollution.

8.10. References

Adhikari, S., Kelkar, V., Kumar, R., Halden, R.U. (2022). Methods and challenges in the detection of microplastics and nanoplastics: A mini-review. *Polymer International*, 71(5), 543–551. doi: 10.1002/pi.6348.

Agarwal, V. and Zetterlund, P.B. (2021). Strategies for reduction of graphene oxide – A comprehensive review. *Chemical Engineering Journal*, 405, 127018. doi: 10.1016/j.cej.2020.127018.

Ahmed, S., Rasul, M.G., Martens, W.N. (2010). Heterogeneous photocatalytic degradation of phenols in wastewater: A review on current status and developments. *Desalination*, 261, 3–18. doi: 10.1016/j.desal.2010.04.062.

Al Kausor, M. and Chakrabortty, D. (2020). Facile fabrication of N-TiO_2/Ag_3PO_4@ GO nanocomposite toward photodegradation of organic dye under visible light. *Inorganic Chemistry Communications*, 116, 107907.

Aliofkhazraei, M., Ali, N., Milne, W.I., Ozkan, C.S., Mitura, S., Gervasoni, J.L. (eds) (2016). *Graphene Science Handbook, Electrical and Optical Properties*. CRC Press, Taylor & Francis Group, Boca Raton, FL.

Allen, M.J., Tung, V.C., Kaner, R.B. (2010). Honeycomb carbon: A review of graphene. *Chemical Reviews*, 110, 132–145.

Allen, S., Allen, D., Karbalaei, S., Maselli, V., Walker, T.R. (2022). Micro(nano)plastics sources, fate, and effects: What we know after ten years of research. *Journal of Hazardous Materials Advances*, 6, 100057. doi: 10.1016/j.hazadv.2022.100057.

Alqattaf, A. (2020). Plastic waste management: Global facts, challenges and solutions. In *Second International Sustainability and Resilience Conference: Technology and Innovation in Building Designs*. IEEE Publications.

Amobonye, A., Bhagwat, P., Raveendran, S., Singh, S., Pillai, S. (2021). Environmental impacts of microplastics and nanoplastics: A current overview. *Frontiers in Microbiology*, 12, 768297. doi: 10.3389/fmicb.2021.768297.

Arienzo, M., Ferrara, L., Trifuoggi, M. (2021). Research progress in transfer, accumulation and effects of microplastics in the oceans. *Journal of Marine Science and Engineering*, 9(4), 433.

Assad, H., Lone, I.A., Kumar, A., Kumar, A. (2024). Unveiling the contemporary progress of graphene-based nanomaterials with a particular focus on the removal of contaminants from water: A comprehensive review. *Frontiers in Chemistry*, 12, 1347129. doi: 10.3389/fchem.2024.1347129.

Bajt, O. (2021). From plastics to microplastics and organisms. *FEBS Open Bio*, 11, 954–966. doi: 10.1002/2211-5463.13120.

Banu, R., Salvi, N., Ameta, C., Ameta, R., Punjabi, P.B. (2019). Visible light driven photocatalytic degradation of brilliant green dye using graphene oxide/copper oxide binary composite. *Indian Journal of Chemistry*, 58B, 157–166.

Becerril, H.A., Mao, J., Liu, Z., Stoltenberg, R.M., Bao, Z., Chen, Y. (2008). Evaluation of solution-processed reduced graphene oxide films as transparent conductors. *ACS Nano*, 2, 463–470.

Bhardiya, S.R., Asati, A., Sheshma, H. (2021). A novel bio conjugated reduced graphene oxide-based nanocomposite for sensitive electrochemical detection of cadmium in water. *Sensors and Actuators. Part B*, 328, 129019. doi: 10.1016/j.snb.129019.

Botsa, S.M., Naidu, G.P., Ravichandra, M., Rani, S.J., Anjaneyulu, R.B., Ramana, C.V. (2020). Flower like SnO_2-Fe_2O_3-rGO ternary composite as highly efficient visible light induced photocatalyst for the degradation of organic pollutants from contaminated water. *Journal of Materials Research and Technology*, 9(6), 12461–12472.

Bratovcic, A. (2019). Degradation of micro- and nano-plastics by photocatalytic methods. *Journal of Nanoscience and Nanotechnology*, 3(304), 1–9. doi: 10.18875/2577-7920.3.304.

Bratovcic, A., Odobašić, A., Ćatić, S., Šestan, I. (2015). Application of polymer nanocomposite materials in food packaging. *Croatian Journal of Food Science and Technology*, 7, 86–94. doi: 10.17508/CJFST.2015.7.2.06.

Braun, U., Altmann, K., Bannick, C.G., Becker, R., Bitter, H., Bochow, M., Dierkes, G., Enders, K., Eslahian, K.A., Fischer, D. (2020). Status report: Analysis of microplastics sampling, preparation and detection methods within the scope of the BMBF research focus plastics in the environment: Sources, sinks. *Solutions*, 1–66.

Bruno, A., Agostinetto, G., Fumagalli, S., Ghisleni, G., Sandionigi, A. (2022). It's a long way to the tap: Microbiome and DNA-based omics at the core of drinking water quality. *International Journal of Environmental Research and Public Health*, 19(13), 7940.

Chang, S., Zhang, Q., Lu, Y., Wu, S., Wang, W. (2020). High-efficiency and selective adsorption of organic pollutants by magnetic $CoFe_2O_4$/graphene oxide adsorbents: Experimental and molecular dynamics simulation study. *Separation and Purification Technology*, 238, 116400.

Cheng, Y., Zhou, S., Hu, P. (2017). Enhanced mechanical, thermal, and electric properties of graphene aerogels via supercritical ethanol drying and high-temperature thermal reduction. *Scientific Report*, 7, 1–11. doi: 10.1038/s41598-017-01601-x.

Chowdhury, P., Nag, S., Ray, A.K. (2017). Degradation of phenolic compounds through UV and visible-light-driven photocatalysis: Technical and economic aspects. *Intech*. doi: 10.5772/66134.

Dar, M.S. and Sahu, N.K. (2024). Graphene quantum dot-crafted nanocomposites: Shaping the future landscape of biomedical advances. *Discover Nano*, 19(1), 79. doi: 10.1186/s11671-024-04028-2.

Ding, L., Zhang, S., Tao, C., Liao, J., Zhang, Y., Zhang, L. (2023). Efficient uranium (VI) adsorption platform based on graphene oxide-supported TixAl1-xOy bimetallic oxide. *Applied Surface Science*, 615, 156337.

Eda, G., Fanchini, G., Chhowalla, M. (2008). Large-area ultrathin films of reduced graphene oxide as a transparent and flexible electronic material. *Nature Nanotechnology*, 3, 270–274.

Eriksen, M., Lebreton, L.C.M., Carson, H.S., Thiel, M., Moore, C.J., Borerro, J.C., Galgani, F., Ryan, P.G., Reisser, J. (2014). Plastic pollution in the world's oceans: More than 5 trillion plastic pieces weighing over 250,000 tons afloat at sea. *PLOS ONE*, 9(12), e111913. doi: 10.1371/journal.pone.0111913.

Fu, W., Min, J., Jiang, W., Li, Y., Zhang, W. (2020). Separation, characterization and identification of microplastics and nanoplastics in the environment. *Science of the Total Environment*, 721, 137561. doi: 10.1016/j.scitotenv.2020.137561.

Galgani, F., Hanke, G., Maes, T. (2015). Global distribution, composition and abundance of marine litter. In *Marine Anthropogenic Litter*, Bergmann, M., Gutow, L., Klages, M. (eds). Springer, Cham. doi: 10.1007/978-3-319-16510-3_2.

Gigault, J., Halle, A.T., Baudrimont, M., Pascal, P.Y., Gauffre, F., Phi, T.L., El Hadri, H., Grassl, B., Reynaud, S. (2018). Current opinion: What is a nanoplastic? *Environmental Pollution*, 235, 1030–1034. doi: 10.1016/j.envpol.2018.01.024.

Hanun, J.N., Hassan, F., Jiang, J.J. (2021). Occurrence, fate, and sorption behaviour of contaminants of emerging concern to microplastics: Influence of the weathering/aging process. *Journal of Environmental Chemical Engineering*, 9(5), 106290. doi: 10.1016/j.jece.2021.106290.

Hartmann, N.B., Hüffer, T., Thompson, R.C., Hassellöv, M., Verschoor, A., Daugaard, A.E., Rist, S., Karlsson, T., Brennholt, N., Cole, M. et al. (2019). Are we speaking the same language? Recommendations for a definition and categorization framework for plastic debris. *Environmental Science and Technology*, 53(3), 1039–1047. doi: 10.1021/acs.est.8b05297.

Huang, W., Song, B., Liang, J., Niu, Q., Zeng, G., Shen, M., Deng, J., Luo, Y., Wen, X., Zhang, Y. (2021). Microplastics and associated contaminants in the aquatic environment: A review on their ecotoxicological effects, trophic transfer, and potential impacts to human health. *Journal of Hazardous Materials*, 405, 124187.

Ibrahim, A., Klopocinska, A., Horvat, K., Hamid, Z. (2021). Graphene-based nanocomposites: Synthesis, mechanical properties, and characterizations. *Polymers*, 13(17), 2869. doi: 10.3390/polym13172869.

Ikram, R., Jan, B.M., Ahmad, W. (2020). An overview of industrial scalable production of graphene oxide and analytical approaches for synthesis and characterization. *Journal of Materials Research and Technology*, 9, 11587–11610. doi: 10.1016/j.jmrt.2020.08.050.

Impellitteri, F., Multisanti, C.R., Rusanova, P., Piccione, G., Falco, F., Faggio, C. (2023). Exploring the impact of contaminants of emerging concern on fish and invertebrates' physiology in the Mediterranean Sea. *Biology*, 12(6), 767. doi: 10.3390/biology12060767.

ISO (2020). ISO/TT 21960. Plastics–Environmental Aspects–State of Knowledge and Methodologies. Standard, ISO, Geneva.

Ivleva, N.P. (2021). Chemical analysis of microplastics and nanoplastics: Challenges, advanced methods, and perspectives. *Chemical Reviews*, 121(19), 11886–11936. doi: 10.1021/acs.chemrev.1c00178.

Jambeck, J.R., Geyer, R., Wilcox, C., Siegler, T.R., Perryman, M., Andrady, A., Narayan, R., Law, K.L. (2015). Plastic waste inputs from land into the ocean. *Science*, 347, 768–771. doi: 10.1126/science.1260352.

Jeong, S.Y., Sugita, N., Shin, B.S. (2023). Fe_3O_4/laser-induced graphene as an adsorbent for microplastics emitted from household wastewater. *International Journal of Precision Engineering and Manufacturing-Green Technology*, 10, 807–818. doi: 10.1007/s40684-022-00464-6.

Kausar, A., Ahmad, I., Zhao, T., Aldaghri, O., Ibnaouf, K.H., Eisa, M.H. (2023). Graphene nanocomposites as innovative materials for energy storage and conversion–Design and headways. *International Journal of Molecular Sciences*, 24, 11593. doi: 10.3390/ijms241411593.

Khan, M.Y., Ahmad, M., Sadaf, S., Iqbal, S., Nawaz, F., Iqbal, J. (2019). Visible light active indigo dye/graphene/WO_3 nanocomposites with excellent photocatalytic activity. *Journal of Materials Research and Technology*, 8(3), 3261–3269. doi: 10.1016/j.jmrt.2019.05.015.

Kocijan, M., Ćurković, L., Ljubas, D., Mužina, K., Bačić, I., Radošević, T., Podlogar, M., Bdikin, I., Otero-Irurueta, G., Hortigüela, M.J. et al. (2021). Graphene-based TiO_2 nanocomposite for photocatalytic degradation of dyes in aqueous solution under solar-like radiation. *Applied Sciences*, 11, 3966. doi: 10.3390/app11093966.

Koelmans, A.A., Mohamed Nor, N.H., Hermsen, E., Kooi, M., Mintenig, S.M., De France, J. (2019). Microplastics in freshwaters and drinking water: Critical review and assessment of data quality. *Water Research*, 155, 410–422. doi: 10.1016/j.watres.2019.02.054.

Kolvenbach, B.A., Helbling, D.E., Kohler, H.P.E., Corvini, P.F.X. (2014). Emerging chemicals and the evolution of biodegradation capacities and pathways in bacteria. *Current Opinion in Biotechnology*, 27, 8–14.

Krishnamoorthy, K., Kim, G.S., Kim, S.J. (2013). Graphene nanosheets: Ultrasound assisted synthesis and characterization. *Ultrasonics Sonochemistry*, 20(2), 644–649. doi: 10.1016/j.ultsonch.2012.09.007.

Lamichhane, G., Acharya, A., Marahatha, R., Modi, B., Paudel, R., Adhikari, A., Raut, B.K., Aryal, S., Parajuli, N. (2023). Microplastics in environment: Global concern, challenges, and controlling measures. *International Journal of Environmental Science and Technology*, 20(4), 4673–4694. doi: 10.1007/s13762-022-04261-1.

Li, F., Jiang, X., Zhao, J., Zhang, S. (2015). Graphene oxide: A promising nanomaterial for energy and environmental applications. *Nano Energy*, 16, 488–515.

Liu, C., Chen, X., Zong, B., Mao, S. (2019). Recent advances in sensitive and rapid mercury determination with graphene-based sensors. *Journal of Materials Chemistry A*, 7, 6616–6630. doi: 10.1039/C9TA01009B.

Liu, S., Lin, R., Pu, C., Huang, J., Zhang, J., Hou, H. (2022). Nanocomposite biomaterials for tissue engineering and regenerative medicine applications. *IntechOpen*. doi: 10.5772/intechopen.102417.

Mao, D., Zhu, L., Fei, X., Zheng, X., Fang, D. (2018). Fabrication of indium sulfide/graphene oxide photodegradation of rhodamine B-containing wastewater under visible light irradiation. *Materials Research Express*, 5(6), 065906.

Maruthamani, D., Divakar, D., Kumaravel, M. (2015). Enhanced photocatalytic activity of TiO_2 by reduced graphene oxide in mineralization of rhodamine B dye. *Journal of Industrial and Engineering Chemistry*, 30, 33–43.

Mas'udah, K.W., Nugraha, I.M.A., Abidin, S., Mufid, A., Astuti, F., Darminto, D. (2016). Solution of reduced graphene oxide synthesized from coconut shells and its optical properties. *AIP Conference Proceedings*, 1725(1).

Meneses, R.A.M., Cabrera-Papamija, G., Machuca-Martínez, F., Rodríguez, L.A., Diosa, J.E., Mosquera-Vargas, E. (2022). Plastic recycling and their use as raw material for the synthesis of carbonaceous materials. *Heliyon*, 8(3), e09028. doi: 10.1016/j.heliyon.2022.e09028.

Nayanathara Thathsarani Pilapitiya, P.G.C. and Ratnayake, A.S. (2024). The world of plastic waste: A review. *Cleaner Materials*, 11. doi: 10.1016/j.clema.2024.100220.

Nguyen, B., Claveau-Mallet, D., Hernandez, L.M., Xu, E.G., Farner, J.M., Tufenkji, N. (2019). Separation and analysis of microplastics and nanoplastics in complex environmental samples. *Accounts of Chemical Research*, 52, 858–866. doi: 10.1021/acs.accounts.8b00602.

Nickheslat, A., Amin, M.M., Izanloo, H. (2013). Phenol photocatalytic degradation by advanced oxidation process under ultraviolet radiation using titanium dioxide. *Journal of Environmental and Public Health*, 1–9. doi: 10.1155/2013/815310.

Norhayati, H., Zuhailimuna, M., Mohd Zobir, H., Illyas, M.I., Azmi, M., Azlan, K., Suriani, A.B., Mazidah, M., Adila Mohamad, J. (2016). A brief review on recent graphene oxide-based material nanocomposites: Synthesis and applications. *Journal of Materials and Environmental Science*, 7(9), 3225–3243.

Parvez, S., Ullah, H., Faruk, O., Simon, E., Czédli, H. (2024). Role of microplastics in global warming and climate change: A review. *Water Air & Soil Pollution*, 235, 201.

Pérez-Fernández, B., Costa-García, A., de la Muñiz, A. (2020). Electrochemical (bio)sensors for pesticides detection using screen printed electrodes. *Biosensors*, 10, 32.

Pervez, M.N., Jahid, M.A., Mishu, M.M.R., Talukder, M.E., Buonerba, A., Jiang, T., Liang, Y., Tang, S., Zhao, Y., Dotto, G.L., Cai, Y., Naddeo, V. (2023). Tuning the surface functionality of polyethylene glycol-modified graphene oxide/chitosan composite for efficient removal of dye. *Scientific Reports*, 13, 13460. doi: 10.1038/s41598-023-40701-9.

Pico, Y., Alfarhan, A., Barcelo, D. (2019). Nano- and microplastic analysis: Focus on their occurrence in freshwater ecosystems and remediation technologies. *TrAC Trends in Analytical Chemistry*, 113, 409–425. doi: 10.1016/j.trac.2018.08.022.

Rai, P.K., Lee, J., Brown, R.J.C., Kim, K.H. (2021). Micro- and nano-plastic pollution: Behaviour, microbial ecology, and remediation technologies. *Journal of Cleaner Production*, 291(7), 125240. doi: 10.1016/j.jclepro.2020.125240.

Rajaji, U., Murugan, K., Chen, S.M. (2019). Graphene oxide encapsulated 3D porous chalcopyrite (CuFeS2) nanocomposite as an emerging electrocatalyst for agro-hazardous (methyl paraoxon) detection in vegetables. *Composites Part B Engineering*, 160, 268–276. doi: 10.1016/j.compo sitesb.2018.10.042.

Ramalingam, G., Perumal, N., Priya, A.K., Rajendran, S. (2022). A review of graphene-based semiconductors for photocatalytic degradation of pollutants in wastewater. *Chemosphere*, 300, 134391.

Rani, M., Ducoli, S., Depero, L.E., Prica, M., Tubić, A., Ademovic, Z., Morrison, L., Federici, S. (2023). A complete guide to extraction methods of microplastics from complex environmental matrices. *Molecules*, 28(15). doi: 10.3390/molecules28155710.

Rathore, C., Saha, M., Gupta, P., Kumar, M., Naik, A. de Boer, J. (2023). Standardization of micro-FTIR methods and applicability for the detection and identification of microplastics in environmental matrices. *Science of the Total Environment*, 888, 164157. doi: 10.1016/j.scitotenv.2023.164157.

Schwaferts, C., Niessner, R., Elsner, M., Ivleva, N.P. (2019). Methods for the analysis of submicrometer and nanoplastic particles in the environment. *TrAC Trends in Analytical Chemistry*, 112, 52–65. doi: 10.1016/j.trac.2018.12.014.

Shah, A.A., Hasan, F., Hameed, A., Ahmed, S. (2008). Biological degradation of plastics: A comprehensive review. *Biotechnology Advances*, 26, 246–265.

Shahraki, H.S., Bushra, R., Shakeel, N., Ahmad, A., Quratulen, M., Ahmad, M., Ritzoulis, C. (2023). Papaya peel waste carbon dots/reduced graphene oxide nanocomposite: From photocatalytic decomposition of methylene blue to antimicrobial activity. *Journal of Bioresources and Bioproducts*, 8(2), 162–175.

Sharif, F., Gagnon, L.R., Mulmi, S., Roberts, E.P.L. (2017). Electrochemical regeneration of a reduced graphene oxide/magnetite composite adsorbent loaded with methylene blue. *Water Research*, 114, 237–245.

Shi, H., Frias, J., Sayed, A.E.-D., De-la-Torre, G., Choo, J., Uddin, S., Rajaram, R., Chavanich, S., Naji, A., Fernandez Severini, M., Ibrahim, Y., Su, L. (2023). Small plastic fragments: A bridge between large plastic debris and micro- & nano-plastics. *TrAC Trends in Analytical Chemistry*, 168, 117308. doi: 10.1016/j.trac.2023.117308.

Si, Y., Liu, J., Chen, Y. (2018). rGO/AuNPs/tetraphenyl porphyrin nanoconjugate-based electrochemical sensor for highly sensitive detection of cadmium ions. *Analytical Methods*, 10, 3631–3636. doi: 10.1039/C8AY0.

Somani, P.R., Somani, S.P., Umeno, M. (2006). Planer nano-graphenes from camphor by CVD. *Chemical Physics Letters*, 430, 56–59.

Sorasan, C., Edo, C., González-Pleiter, M., Fernández-Piñas, F., Leganés, F., Rodríguez, A., Rosal, R. (2022). Ageing and fragmentation of marine microplastics. *Science of the Total Environment*, 827, 154438. doi: 10.1016/j.scitotenv.2022.154438.

Sree, G.S., Botsa, S.M., Reddy, B.J.M., Ranjitha, K.V.B. (2020). Enhanced UV–visible triggered photocatalytic degradation of brilliant green by reduced graphene oxide based NiO and CuO ternary nanocomposite and their antimicrobial activity. *Arabian Journal of Chemistry*, 13(4), 5137–5150.

Sur, U.K. (2012). Graphene: A rising star on the horizon of materials science. *International Journal of Electrochemistry*, 1–12. doi: 10.1155/2012/237689.

Tabatabai-Yazdi, F.S., Ebrahimian Pirbazari, A., Esmaeili Khalil Saraei, F., Gilani, N. (2021). Construction of graphene based photocatalysts for photocatalytic degradation of organic pollutant and modeling using artificial intelligence techniques. *Physica. Part B*, 608, 412869. doi: 10.1016/j.physb.2021.412869.

Tan, S.Y., Chong, W.C., Sethupathi, S., Pang, Y.L., Sim, L.C., Mahmoudi, E. (2023). Optimisation of aqueous phase low density polyethylene degradation by graphene oxide-zinc oxide photocatalysts. *Chemical Engineering Research and Design*, 190, 550–565.

Thompson, R.C., Olsen, Y., Mitchell, R.P., Davis, A., Rowland, S.J., John, A.W.G., McGonigle, D., Russell, A.E. (2004). Lost at sea: Where is all the plastic? *Science*, 304, 838–838. doi: 10.1126/science.1094559.

Torrisi, L., Silipigni, L., Torrisi, A. (2022). Argon diffusion in graphene oxide and reduced graphene oxide foils. *Vacuum*, 200. doi: 10.1016/j.vacuum.2022.110993.

Toussaint, B., Raffael, B., Angers-Loustau, A., Gilliland, D., Kestens, V., Petrillo, M., Rio-Echevarria, I.M., Van den Eede, G. (2019). Review of micro- and nanoplastic contamination in the food chain. *Food Additives and Contaminants: Part A*, 36, 639–673. doi: 10.1080/19440049.2019.1583381.

UNIDO (2018). Recycling of plastics in Indian perspective. UNIDO.

Uogintė, I., Pleskytė, S., Skapas, M., Stanionytė, S., Lujanienė, G. (2023). Degradation and optimization of microplastic in aqueous solutions with graphene oxide-based nanomaterials. *International Journal of Environmental Science and Technology*, 20, 9693–9706. doi: 10.1007/s13762-022-04657-z.

Urbanek, A.K., Rymowicz, W., Mirończuk, A.M. (2018). Degradation of plastics and plastic-degrading bacteria in cold marine habitats. *Applied Microbiology and Biotechnology*, 102, 7669–7678.

Usharani, B., Murugadoss, G., Rajesh Kumar, M., Gouse Peera, S., Manivannan, V. (2022). Reduced graphene oxide–metal oxide nanocomposites (ZrO_2 and Y_2O_3): Fabrication and characterization for the photocatalytic degradation of picric acid. *Catalysts*, 12(10), 1249.

Wagner, S. and Reemtsma, T. (2019). Things we know and don't know about nanoplastic in the environment. *Nature Nanotechnology*, 14(4), 300–301. doi: 10.1038/s41565-019-0424-z.

Wilkes, R.A. and Aristilde, L. (2017). Degradation and metabolism of synthetic plastics and associated products by Pseudomonas sp.: Capabilities and challenges. *Journal of Applied Microbiology*, 123, 582–593.

Xia, H., Ren, Q., Lv, J., Wang, Y., Feng, Z., Li, Y., Wang, C., Liu, Y., Wang, Y. (2023). Hydrothermal fabrication of phytic acid decorated chitosan-graphene oxide composites for efficient and selective adsorption of uranium (VI). *Journal of Environmental Chemical Engineering*, 11, 110760.

Xiang, Y., Jiang, L., Zhou, Y., Luo, Z., Zhi, D., Yang, J., Lam, S.S. (2022). Microplastics and environmental pollutants: Key interaction and toxicology in aquatic and soil environments. *Journal of Hazardous Materials*, 422, 126843.

Yu, W., Sisi, L., Haiyan, Y., Jie, L. (2020). Progress in the functional modification of graphene/graphene oxide: A review. *RSC Advances*, 10, 15328–15345.

Zhang, W., Fu, J., Wang, Y., Zhang, X., Li, J. (2019). Enhanced visible-light photocatalytic activity of ZnS/BiOBr/graphene oxide ternary composite. *Journal of Physics and Chemistry of Solids*, 127, 19–27.

Zheng, A.L.T. and Andou, Y. (2022). Detection and remediation of bisphenol A (BPA) using graphene-based materials: Mini-review. *International Journal of Environmental Science and Technology*, 19, 6869–6888. doi: 10.1007/s13762-021-03512-x.

Zheng, A.L.T., Boonyuen, S., Ohno, T., Andou, Y. (2021a). Hydrothermally reduced graphene hydrogel intercalated with divalent ions for dye adsorption studies. *Processes*, 9, 169. doi: 10.3390/pr9010169.

Zheng, A.L.T., Boonyuen, S., Ohno, T., Andou, Y. (2021b). Accessing effects of aliphatic dicarboxylic acid towards the physical and chemical changes in low temperature hydrothermally reduced graphene hydrogel. *Journal of Porous Materials*, 28, 1291–1300. doi: 10.1007/s10934-021-01072-6.

Zhu, Y., Murali, S., Cai, W. (2010). Graphene and graphene oxide: Synthesis, properties, and applications. *Advanced Materials*, 22, 3906–3924. doi: 10.1002/adma.201001068.

9

2D Nanomaterials for Photocatalytic Degradation of Micro- and Nano-Plastics

Plastic has a complex relationship with society nowadays. While the socioeconomic benefits of plastic are clear, it has also come to play a prominent part in a culture of waste and throwaway living that poses a substantial threat to health and the natural environment. Waste plastic disposal is becoming more and more of a social issue, but it can be solved with advancements in plastic management science and technology. Photocatalytic waste plastic conversion is gaining popularity these days. Ultimately, the development of ecologically friendly plastic conversion technology is critical for transitioning away from the current plastic use environment and toward a more sustainable future. Compared to existing processes, photocatalysis may be a more environmentally benign method of recycling waste micro- and nano-plastics (NPs), since it uses sunlight as an energy source at room temperature and pressure. Furthermore, waste plastics can be recycled to increase their original worth using photocatalytic processes. Among the several methods for increasing the efficiency of the photocatalytic process, nanomaterials have proven to be crucial in reducing charge recombination. Due to these developments in material-based investigations using basic photocatalysts, attempts have recently been made to incorporate nanomaterials/nanostructures into photocatalytic plastic conversion. Due to their qualities and uses in electronics, sensors, solar cells, lithium batteries and photocatalysis, two-dimensional (2D) nanomaterials such as graphene, graphitic carbon nitride, hexagonal boron nitride (h-BN), transition metal dichalcogenides (TMDs), etc., have drawn interest. In the field of photocatalysis, graphene and graphene derivatives, nanomaterials have also made practical contributions to the study of photocatalysis at very low energy input. Since the previous decade, 2D nanomaterials have received extensive research in the area of photocatalytic degradation of plastic materials; therefore, in this chapter, we sought to highlight the straightforward, practical, affordable and environmentally friendly method for producing carbon-based graphene and derivatives nanomaterials properties, influencing factor and mechanism for photocatalytic degradation of plastics.

Chapter written by Thakur Prasad YADAV and Kalpana AWASTHI.

9.1. Introduction

New scientific discoveries and breakthroughs in a variety of fields have been made possible by the ongoing developments in science and technology. The emergence of two-dimensional (2D) nanomaterials has revolutionized the landscape of materials science, introducing a paradigm shift in our understanding of material properties and their applications. The 2D materials are crystalline and have one (or a few) layer atoms, with stronger interatomic interactions than those along the stacking direction. These materials are unique among nanomaterials and perform important roles in a variety of fields. The discovery of carbon nanomaterial (graphene, a single sheet of carbon atoms organized in a 2D honeycomb lattice) flickered interest in the isolation and research of 2D materials.

It has attracted attention because of its distinct electrical, mechanical, chemical, optical and biological characteristics. These remarkable characteristics enable graphene to have applications in electronic devices, composite materials, energy storage, sensing and drug delivery, etc. (Novoselov et al. 2004; Tiwari et al. 2016; Yadav and Awasthi 2022). The quick and effective advancement of graphene has piqued many researchers' interests in other 2D materials. Beyond graphene, a wide variety of 2D materials with distinct physicochemical properties have been found, such as silicene, phosphorene, h-BN, transition metal carbides and carbonitrides (MXenes), tungsten disulphide (WS_2) and transition metal dichalcogenides (TMDs) (Shanmugam et al. 2022). In metals ($NbSe_2$ exhibits a superconductivity effect at low temperature), semimetals ($TiSe_2$), semiconductors (MoS_2), insulators (HfS_2) and other materials, TMDs, for instance, represent a variety of structures matching the oxidation state of their atoms and coordination number (Moore et al. 2013; Andriotis and Menon 2014; Ganatra and Zhang 2014; Huang et al. 2014; Kanazawa et al. 2016). While 2D boron nitride (BN) nanosheets and graphene share the same honeycomb structure, they differ greatly from graphene in terms of band gap and conductivity. Due to their reduced dimensionality, these materials display special and amazing properties that make them suitable for a wide range of applications in fields such as electronics, sensors, thermoelectric, photoelectric, optoelectronics, superconductors, energy storage and topological insulator devices (Watanabe et al. 2004; Stoeckel et al. 2019; Fan et al. 2021; Pak et al. 2021).

2D nanoparticles often have more distinct chemical and physical properties than macroscopic ones made of the same substance. Polyvinyl chloride (PVC), polyethylene terephthalate (PET), polystyrene (PS) and polyethylene (PE) were all found to be rather stable in nature. A recent study discovered that 2D materials might degrade marine micro-plastics (MPs) without harming the surrounding bacteria. Low-density polyethylene (LDPE) film was successfully destroyed in water consuming visible light-activated heterogeneous zinc oxide photocatalysts, as well as PE plastic with a photocatalyst composed of polypyrrole/TiO_2.

Our understanding of nanoplastics (NPs) and MPs in the environment is limited, with very little scientific research available. As a result, MPs and NPs are seen as a developing source of major worry. Lower molecular weight intermediates produced by plastic degradation can be used as raw materials in organic synthesis and as petrochemical or plastic products in the chemical industry.

This chapter provides an overview of recent advances and degrading technologies for MPs and NPs using 2D materials, as well as a description of global plastic production and its impact on human health.

9.2. 2D materials

In 2004, scientists used adhesive tape to isolate a single-layer sheet of graphene from graphite, and graphene demonstrated improved electrical characteristics. The mass manufacture of 2D materials has gained attention as a result of this discovery. In the past 10 years, exfoliated van der Waals materials have been used by scientists to generate single and few-layer 2D materials (both chemically and mechanically). The 2D nanomaterials can be classified into the graphene family, chalcogenides, transition metal carbides and carbonitrides (MXene), phosphorene and oxide/hydroxide according to their components and structures (Figure 9.1). In this section, we summarized the structure of 2D materials.

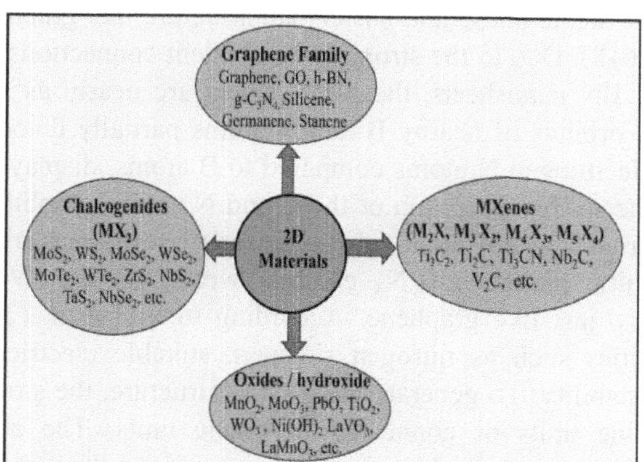

Figure 9.1. *Classification of 2D materials*

9.2.1. *Graphene family*

Graphene is a 2D carbon allotrope composed of a single sheet of carbon atoms organized in a hexagonal lattice (C-C bond length: 0.142 nm) (Figure 9.2).

Each carbon atom in graphene is attached by covalent bond to three other carbon atoms, forming strong sigma (σ) bonds which give planar structure. In addition to the sigma bonds, there are also delocalized π (pi) bonds above and below the plane of the carbon atoms. These π-bonds provide the high electrical conductivity to graphene. The derivatives of graphene contain diverse hybridized carbon atoms or heterogeneous elements, and these are graphene oxide (GO), fluorographene and chlorographene; the saturated carbon atoms (sp^3 hybridization) bind with non-carbon elements and form a cyclic pattern. The GO is formed by oxidation of graphite and contains oxygen functional groups. The hydroxyl (C–OH) and epoxide (C–O–C) groups were present on the basal plane of GO and carbonyl (C=O) and carboxyl (COOH) groups were attached to sheet edges. These groups raise sp^3 defect sites to the nanosheets, distorting the intrinsic conjugated pi-system which lowers the conductivity and strength of GO. In graphene structure, the lattice of benzene rings joined with acetylene bonds. Like graphene, two or more elements can replace the original carbon atoms to generate complicated layered structures such as BN, boron carbon nitrogen (BCN) and graphitic carbon nitride (g-C_3N_4). BN nanosheets are made up of boron and nitrogen atoms organized in a hexagonal lattice (a = 0.25 nm; c = 0.666 nm, interlayer spacing = 0.33 nm), comparable to graphene. Unlike graphene, which is made up of carbon atoms, BN nanosheets are made up of boron and nitrogen atoms in alternating layers. This gives BN nanosheets unique properties that are distinct from graphene. The BN exists in four crystalline forms, that is, hexagonal BN (h-BN), cubic (c-BN), rhombohedral (r-BN) and wurtzite (w-BN), in which h-BN is the stable phase and has crystal structure like graphite (Pakdel et al. 2014; Yu et al. 2018). Due to the strong B-N covalent connections that form the h-BN layers inside BN nanosheets, the h-BN layers are nearly as hard to break as graphene. The p_z orbitals of nearby B and N atoms partially delocalize due to the more restricted electrons in N atoms compared to B atoms, displaying an insulating feature. Due to the sp^2 hybridization of the C and N atoms, graphitic carbon nitride (g-C_3N_4) is a nitrogen hetero atom substituted graphite framework made up of p-conjugated graphitic planes. g-C_3N_4 exhibits weak van der Waals interactions between its layers, just like graphene. According to Liu et al. (2016), it exhibits exceptional qualities such as nitrogen richness, suitable electrical structure and unusual thermal stability. To generate the g-C_3N_4 structure, the structures are based on either s-triazine units or connected heptazine units. The atoms of silicon, germanium and tin are organized in a single layer to form the xenes, which include silicene, germanene and stanine (Jose and Datta 2014; Acun et al. 2015). This structure is comparable to that of graphene. Silica has a bent structure, in contrast to graphene's flat hexagonal shape. Compared to graphene, it is far more reactive because of its inherent spontaneity in forming sp^3 hybridization.

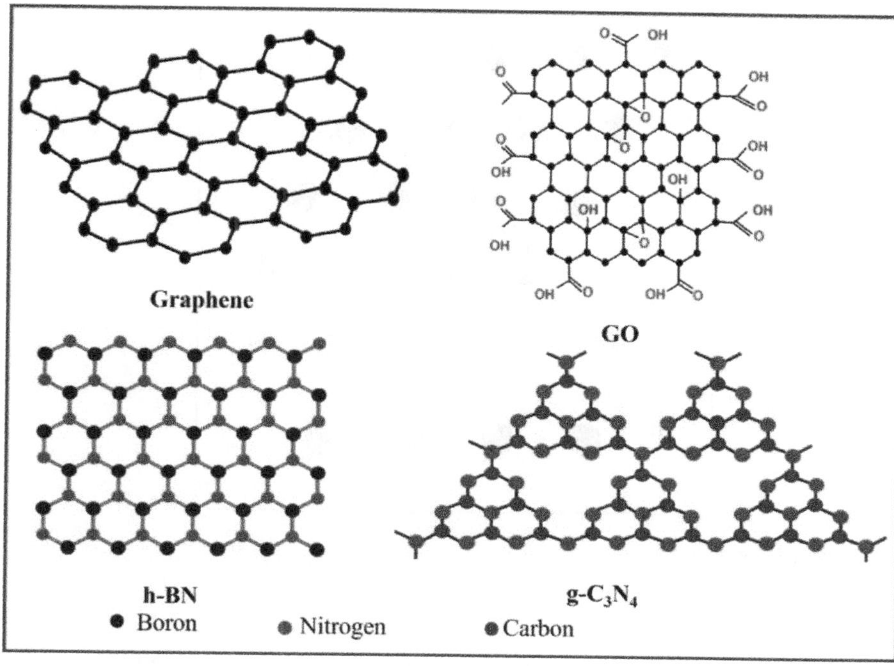

Figure 9.2. *Structure of graphene and graphene-like materials (GO, h-BN and g-C_3N_4). For a color version of this figure, see www.iste.co.uk/singh/photodegradation.zip*

9.2.2. Transition metal dichalcogenides and MXenes

Recent years have seen a study of 2D TMDs, one type of chalcogenide. The layered structure of the TMDs is made up of transition metal atoms positioned between chalcogen atoms. Its structural formula is MX_2, where X stands for the chalcogen atoms (tellurium (Te), sulfur (S) or selenium (Se)) and M is a transition element (molybdenum (Mo), tungsten (W), vanadium (V), etc.). Molybdenum disulfide (MoS_2), diselenide ($MoSe_2$) and tungsten disulfide (WS_2) are the most prevalent TMDs with hexagonal structures (Lv et al. 2015; Da Silva et al. 2019; Pi et al. 2019). The palladium diselenide ($PdSe_2$) is a TMD with a pentagonal structure (Oyedele et al. 2017; Pi et al. 2019). Figure 9.3 shows the structure of MoS_2. It is possible to control these materials' electrical characteristics by adjusting the layer thickness from bulk to monolayer.

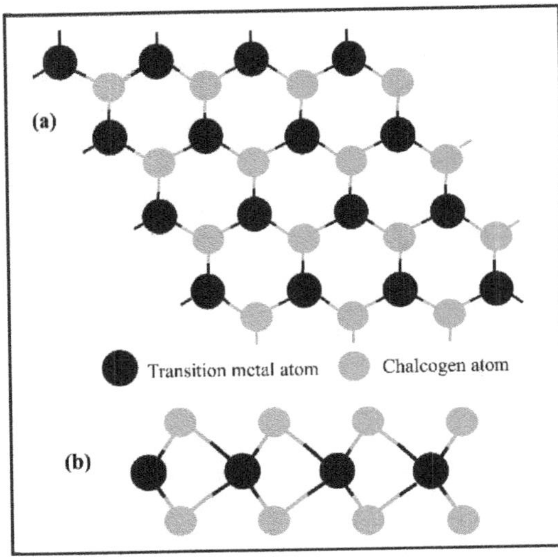

Figure 9.3. *(a) Top and (b) side views of monolayer TMDs. For a color version of this figure, see www.iste.co.uk/singh/photodegradation.zip*

According to Naguib et al. (2014) and Dahlqviset al. (2010), the MXene family of 2D transition metal carbides, nitrides and carbonitrides has the formula M_{n+1} A X_n, where M is a transition metal (such as Ti, V, Cr, Mn, Zr, Nb, Mo or Ta), A is a group 13 or 14 elements (such as Al, Ga, In, Si, Ge, As, Sn or Pb) and X can be either carbon or nitrogen, with n = 1,2,3. Only their parent compounds, stacked ternary transition metal carbides and carbonitrides, commonly referred to as MAX phases, can be converted into MXenes. The M planes are closely packed with the X planes in an octahedral prism, forming the layered hexagonal lattice that houses the MAX phases. The M-X bond has strong covalent/metallic/ionic properties, while the bond between M and A is metallic in nature. The M-A bonds are slightly weaker than the M-X bonds; therefore powerful acids or molten salts can selectively etch the A atoms from the MAX phases without disturbing the M-X bonds (Naguib et al. 2014). The 2D material created by selectively etching A layers from its MAX phase has the composition $M_{n+1}X_nT_x$, where T_x refers to the surface-terminated functional groups, such as hydroxyl (OH), oxygen (O), or fluorine (F), connected to the M atoms following the etching technique. Naguib et al. (2011) initially created MXenes (Ti_3C_2) by etching aluminum atoms off the MAX complex Ti_3AlC_2.

9.2.3. *Phosphorene*

Phosphorene is a 2D phosphorus material composed of a single layer of purposefully manufactured black phosphorus, the most stable allotrope of phosphorus. Black phosphorus has a puckered hexagonal crystal structure in which each phosphorus atom bonds to three other phosphorus atoms in plane (Bridgman 1914; Kou et al. 2015). Phosphorene has a non-zero bandgap and layers are held together by van der Waals interactions. It also possesses a finite bandgap, which may be modified by varying the number of layers or applying strain. Figure 9.4 depicts the crystal structure of phosphorene.

9.2.4. *Oxides and hydroxide materials*

Layered oxides are a type of substance with the chemical formula A_nMO_2, where A is an alkali metal (e.g. lithium or potassium), M is a transition metal, O is oxygen and $0.5 \leq n \leq 1$ (Delmas et al. 1980). These include metal dioxides, trioxides, perovskites and niobates. Layered oxides are typically made up of MO planes separated by A planes. The ions between the layers can balance their surface charges, creating a stable structure.

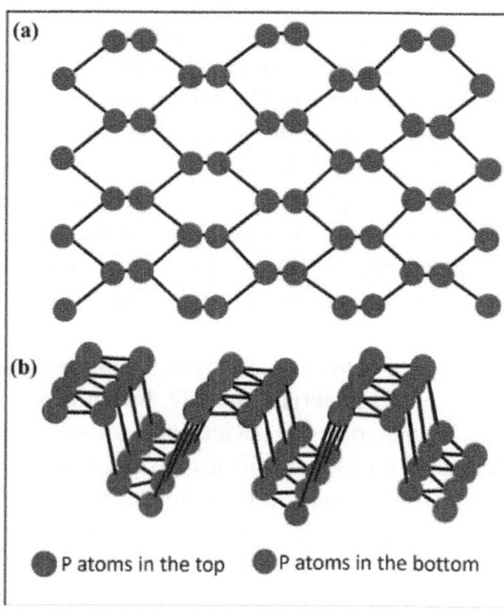

Figure 9.4. *(a) Top and (b) side views of 2D few-layer phosphorene. For a color version of this figure, see www.iste.co.uk/singh/photodegradation.zip*

9.3. Synthesis of 2D materials

The 2D materials were developed using a number of procedures, which are classified as top-down and bottom-up approaches. Top-down synthesis creates single-few-layer 2D materials by dissociating bulk thin crystals using mechanical, liquid-phase, ultrasonic, electrochemical, ion-change and lithium-intercalated exfoliation or chemical reduction techniques. To generate nanostructures, bottom-up synthesis manipulates atoms, molecules or clusters one at a time. 2D materials are mass-produced via chemical vapor deposition (CVD), pulsed laser deposition, wet chemical methods and microwave-assisted procedures. Figure 9.5 exhibits the schematic diagram of 2D material synthesis approaches. In 2004, Andre Geim's group isolated the single layer of graphite (Novoselov et al. 2004) using the mechanical exfoliation method. This methodology was also used to prepare inorganic materials. The scotch tape method can be used to manually exfoliate layered TMDs from the bulk crystals, which involves peeling off layers from a bulk crystal using adhesive tape. Mechanical exfoliation was used to create the single-layer and multilayer MoS_2 and WSe_2 nanosheets (Li et al. 2014).

The free-standing h-BN nanosheets were fabricated on SiO_2 using the mechanical exfoliation method (Pacilé et al. 2008). Massive crystalline formations are broken down into single or multi-layered samples for usage in micro-sized optical and electrical devices by the micromechanical process. The need for a big bulk material to make nanosheets with an appropriate size and low yield is a significant difficulty with this process. The liquid exfoliation process was used to prepare the 2D nanomaterials because the scotch-tape approach cannot be scaled up. This technique involves dispersing bulk crystals in a solution and using mechanical or ultrasonic forces to separate the layers. The resulting colloidal solution contains dispersed mono- or few-layer flakes 2D flakes that can be deposited onto substrates. The good solvent reduced the re-aggregation of the exfoliated layers due to adsorption of solvent molecules on the surface of the layer. N-methyl-pyrrolidone (NMP) and dimethylformamide (DMF) chemicals were used as effective solvents. The exfoliation of bulk crystal depends on ultrasonic duration and intensity. Another method for the synthesis of 2D materials is CVD, which is a suitable technique for growth of high-quality 2D materials for device applications. The uniform large area graphene film was deposited on cooper foils using the CVD method (Li et al. 2009). After that, nanosheets of numerous 2D materials, such as MoS_2, WS_2, h-BN transition metal carbides and nitrides, were prepared (Lee et al. 2012; Gao et al. 2015; Zhu et al. 2016)

The synthesis of MXenes typically involves the selective etching of A layers from layered MAX phases. The hydrofluoric acid, ammonium hydrogen bifluoride and ammonium bifluoride can be used as etching agents for the MAX layer (Karlsson et al. 2015). Dimethylsulfoxide, hydrazine, isopropylamine, nbutylamine and tetramethylammonium were used as interacting molecules in loosely packed $M_{n+1}X_n$ layers. The mild sonication process of mixture produced the MXene layers.

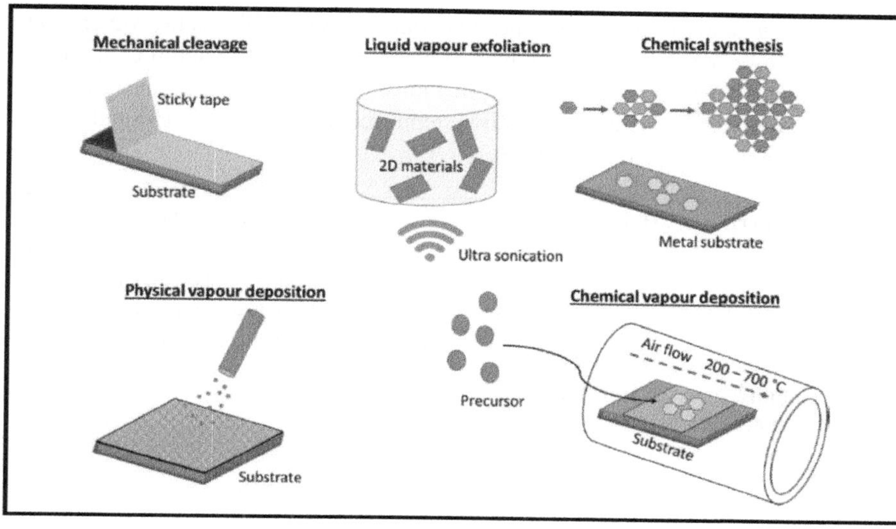

Figure 9.5. *Different synthesis approaches for 2D materials (Shanmugam et al. 2022). For a color version of this figure, see www.iste.co.uk/singh/photodegradation.zip*

9.4. Properties and applications of 2D materials

The 2D nanomaterials have gained attention due to their attractive properties, including electrical, mechanical and optical. The 2D carbon nanomaterial, that is, graphene exhibits zero bandgap and has a large surface area, extraordinary electrical conductivity, highest tensile strength and modulus (Allen et al. 2010; Soldano et al. 2010). Due to these properties, graphene is used in different fields including electronics, sensing technology, energy storage, water filtration, polymer composites and others (Zang et al. 2014; Tiwari et al. 2016; Shanmugam et al. 2022). Graphene's enormous surface area and electrical conductivity make it a potential material for energy storage devices like supercapacitors and batteries. Unlike graphene, 2D materials, such as TMDs, have a variable bandgap, making them useful for electronics and optoelectronics applications. The properties and applications of different 2D materials are listed in Table 9.1.

Other 2D materials, like TMDs and phosphorene, shows the unique optical properties (Li and Zhu 2015). They can be used in optoelectronics for applications such as light-emitting diodes (LEDs), photodetectors and sensors. Due to its atomically flat surface and lack of dangling bonds, h-BN nanosheets can serve as a dielectric substrate for other 2D materials in device manufacturing. For example, h-BN nanosheets were utilized as a substrate for graphene transistors instead of SiO_2, increasing device carrier mobility. The g-C_3N_4 represents the semiconductor behavior, photocatalytic properties and potential for use in solar cells. MXenes are either metallic or have a narrow bandgap (Khazaei et al. 2013; Naguib et al. 2014). MXenes have a wide range of applications, including electronics, sensors, supercapacitors, batteries and conductive reinforcement additives (Naguib et al. 2014; Wang et al. 2015; Xu et al. 2016).

Properties of 2D material	Applications
Graphene (Soldano et al. 2010) Zero band gap Electrical conductivity: ~$10^4\ \Omega^{-1}\ cm^{-1}$ Thermal conductivity: 3 × 10^3 W/m K Electron mobility: 2.5 × 10^5 cm^2/Vs (Novoselov et al. 2005a) Young's modulus: 1.0 TPa (Lee et al. 2008) Strength:130 GPa Surface area ~2,630 m^2/g	Field-effect transistors (FETs), transparent electrodes, batteries, solar cells, water filtration and catalysis energy storage devices such as supercapacitors, gas, chemical and bio sensors
h-BN Band gap: 4–5.9 eV Young's modulus: 856 GPa Thermal conductivity: >600 W/m K	Substrate for graphene transistor, polymer nanocomposites, building block or van der Waal heterostructures
Phosphorene Bandgap: 0.3–2.0 eV Carrier mobility: 200–1,000 cm^2/Vs (Khandelwal et al. 2017)	FET, photodetectors, photovoltaic devices, gas and bio sensors, superconductors, photocatalyst and electrocatalyst
MXenes Metallic conductivity of $Ti_3C_2T_x$: 2,400 S/cm Thermal conductivity of $Ti_3C_2T_x$: 2.84 W/mK	Sensors, electrochemical energy storage devices, that is, batteries, supercapacitors, fuel cells, electrocatalysis of oxygen reduction reaction, electromagnetic interference shielding

Table 9.1. *Some properties of 2D nanomaterials and their application in different sectors*

9.5. Application of 2D materials in photocatalytic degradation

People have known for years how important the problem of industrial chemical waste-related water pollution is, and one common remedy to these issues is solar radiation, which causes water molecules to break down along with pollutants. There are certain challenges when using traditional materials in a catalytic process, like high cost and trouble regulating reaction efficiency (Kumar Maddheshiya et al. 2023). 2D layered materials have a huge specific surface area with only one or a few nanometer-thick layers. This exposes more atomic and functional surface structures, resulting in more active sites and faster reactions (Yadav et al. 2018, 2020a). According to Huang et al. (2021), introducing impurities, vacancies, dislocations and functional groups into 2D materials can alter their catalytic characteristics, transforming them into unique catalysts.

The materials' properties can be enhanced by forming composites with other nanomaterials and narrowing band gaps to boost photocatalyst efficiency. The architectures and characteristics of various 2D nanocomposites are also thoroughly investigated both theoretically and practically (Zhao et al. 2021). Catalytic activity in 2D materials is also influenced by reaction temperature, reaction duration and material shape. Different reaction conditions also influence the catalytic element's performance (Qu et al. 2018a; Santos et al. 2022). However, 2D materials' catalytic activity can be divided into five categories: size effect, layer effect, edge and defect effect, curvature effect, dopant effect and functional group impact. The photocatalytic separation of water-soluble hazardous chemical waste with 2D materials is also successful at producing hydrogen. Graphene can effectively separate photogenerated charges as a conductive channel, improving the photocatalytic activity of the g-C_3N_4 catalyst. Consequently, it has been discovered that when graphene is used as the foundation, the optical characteristics and structure of the g-C_3N_4 catalyst are influenced by the graphene's structure (Yadav et al. 2020a). Li et al. (2019) explains photocatalytic degradation as the process by which organic pollutants and free radicals react to transform pollutants into harmless by-products. By using radiation, the reaction mechanism creates extremely active free radicals and photocatalysts. The success of the new material's development led to many examples, including the creation, a larger catalytic surface, an enhanced charge separation rate and a higher light absorption range. Additionally, they used spectrum analysis to highlight several significant aspects influencing the catalytic performance. 2D materials, such as graphene, can absorb contaminants and act as a semiconductor photocatalyst, degrading dye-containing wastewater. This type of recombination 2D material has a wide range of potential applications and research value in the environmental field because it can be used for both environmental detection and the adsorption and catalytic degradation of contaminants such as heavy metals, dyes and persistent organic compounds. Singh et al. (2018) speculate that the composite material is an excellent candidate for novel catalysts, since it inhibits ZnO photolithography. They show that more MoS_2 semiconductor nanocomposites may be able to display this dual effect, which could lead to the creation of MoS_2 hybrid photocatalysts with exceptionally high efficiencies. The g-C_3N_4 is a typical 2D metal-free conjugated polymer semiconductor material with a configurable electronic structure, cheap cost, little toxicity and, most importantly, an acceptable band edge. Its other distinguishing feature is its tiny band gap (2.7 eV).

These properties account for its widespread use in photocatalysis to generate H_2. The study discovered that g-C_3N_4 has disadvantages that greatly hampered its effectiveness in photocatalysis, such as weak electric conductivity, limited charge flexibility and poor visible light consumption (Qu et al. 2017).

9.6. Micro- and nano-plastics

Plastics are composed of natural materials that have undergone various chemical and physical reactions. The primary processes employed are polymerization and polycondensation, which fundamentally turn the basic constituents into polymer chains (Li et al. 2021). Despite the numerous benefits of frequent use, plastic materials' limited biodegradability, incorrect use and inadequate disposal contribute to environmental damage (Akdogan and Guven 2019). The release of plastic derivatives into the environment causes small plastic pieces to amass in both terrestrial and aquatic environments, hastening physical, chemical and biological degradation processes (Alexiadou et al. 2019). The classification of plastics is crucial for understanding emerging pollution problems. These characteristics include diversity, persistence, global concerns, combined pollution and hazards to creatures and human health. According to Zhu et al. (2023), plastic fragments are classified as NPs, which have diameters ranging from 1 nm to 1 mm, MPs (<5 mm), and macro- and meso-plastics (>5 mm). MPs are categorized as primary or secondary according to how they are released from the body. Primary MPs are intentionally produced in that form and are then released into the environment by a variety of non-biological matrixes as small-sized particles. Some non-biological matrixes that release primary MPs into the environment include tire abrasion during driving, synthetic clothing laundering and cosmetic and body care products. MPs are formed from the abiotic and biotic decomposition of big polymers following exposure to the environment (Ruth et al. 2023). Secondary MPs can be found in domestic waste, industrial manufacturing and debris from used automobile tires, among other places. It is predicted that MPs account for 70–80% of MPs released into the environment, with just 15–31% being primary MPs (Karkanorachaki et al. 2023). MPs are also classified based on physical properties such as density (light/heavy), flexibility (hard/soft) and shape (fragments, pellets, filaments and granules). According to Xu et al. (2023), the most common plastic polymers used to make MPs include nylon (PA), cellulose acetate (CA), PE, polypropylene (PP), PS, PVC and thermoplastic polyester (PET). MP content and structure are inextricably linked to the source material. MPs are easily dispersed and spread by wind and water currents in the environment due to their tiny size and low density. These properties, along with their resistance to biological degradation, make MPs especially bio-accumulative and resistant to environmental disintegration. Among other physical and chemical characteristics, MPs' size, shape, porosity, surface area, morphology, solubility and capacity to interact with other contaminants all affect how quickly they degrade (Liu et al. 2023). To enhance their manufacture, plastics are very frequently made with various chemical additions that may be harmful to human health. Plasticizers, acid scavengers, pigments, antioxidants, flame retardants, stabilizers for heat and light, lubricants, antistatic agents and health stabilizers are just a few of the many hazardous substances present in MPs. MPs may have furthermore absorbed hydrophobic or hydrophilic organic pollutants from the environment in addition to these chemical alterations because of their vast surface areas and susceptibility to these contaminants (Xia et al. 2023). For instance, MPs taken from the environment have been found to contain persistent organic pollutants such as phthalates, polychlorinated biphenyls and aromatic hydrocarbons (Ivanic et al. 2023). Indeed, MPs are frequently discovered to be associated with hazardous substances that serve as a conduct for their environmental dissemination. Although MPs carry harmful substances into ecosystems, they are also a mixture of dangerous substances that are purposefully added to polymers during production to improve their characteristics and extend their lifespan (Anastopoulos et al. 2022). Regarding risk assessment, an evaluation based on commercially available model particles is necessary to ascertain whether the size-specific toxicity paradigms established for the other engineered nanomaterials also apply to MPs/NPs, given the paucity of information in the literature regarding the amounts of MPs and NPs dispersed into the environment (Zheng et al. 2023). NPs are made up of a variety of components that vary in terms of their chemical makeup as well as the size, shape and other features of the particles. Currently, it is unknown what causes the toxicity of nanoparticles, the polymeric compounds or the particle itself (Gong et al. 2023). Analyzing MPs or NPs in separated samples from biotic and abiotic environmental matrices is becoming more and more common. Therefore, it is imperative to find trustworthy techniques for differentiating between these micro- and nano-sized particles. We go over the benefits and drawbacks of the identification techniques that are currently and potentially in use for MP analysis in this study. We go over the best approaches now available, ranging from chemical to physical, as well as the difficulties in improving the current approaches and creating new ones. Microscopical techniques such as dissect, polarized, fluorescence, scanning electron microscopy (SEM) and atomic force microscopy (AFM) are the most frequent approaches to identify MPs and NPs. However, when analyzing small particles, these approaches can yield partial data. Currently, this limit is circumvented by combining newly developed alternative techniques with chemical analysis, or spectroscopy. For instance, holographic imaging in a microscope setup captures direct images of MPs in unfiltered water, allowing for the differentiation of various plastic types, sizes and shapes as well as the separation of MPs from diatoms (Yu et al. 2024). The existing issues with the identification of MPs or NPs may be resolved by the development of new analytical equipment used in conjunction with one another or with traditional and novel microscopy (Eriksen et al. 2023).

9.7. Micro- and nano-plastics identification

Two main processes are typically involved in the identification of MPs and NPs: (i) sample preparation and sampling, and (ii) testing and the interpretation of the results. The purpose of sample preparation is to separate MPs and NPs from their complex environmental background by means of filtration, concentration, digestion/washing and density separation/floating. Sadly, there is not a set procedure or standard for preparing samples, especially when it comes to NPs. The sample's history and the ensuing analytical method have a significant influence on the method choice. The methods of analysis have been discussed in this chapter.

9.7.1. *Microscopy: stereo microscopy and dissecting microscopy*

The stereomicroscope allows for three-dimensional analysis by looking at the sample from two slightly different angles to obtain the two images required for stereoscopic vision (Etabonate et al. 2023). According to Thaiba et al. (2023), objects can therefore be primarily observed using reflected light at low magnification, usually between 8 and 50 times. In contrast to optical microscopes, which use a light beam that travels over the material from below, stereoscopic microscopes use illumination from above (Delgado et al. 2021). Certain models of stereomicroscopes have two lights. It can be used to see whole fresh microorganisms, cells, fungus and permanent slides because of its higher magnification compared to an optical microscope. Although the stereomicroscope's lower magnification may seem like a disadvantage, it actually has advantages since it makes it easier to examine objects that are visible to the unaided eye. To identify MPs with dimensions between hundreds and microns, the stereomicroscope is a practical and often utilized tool (Kalaronis et al. 2022). According to several studies, in the case of transparent particles, the proportion of plastic-like particles detected by stereomicroscopy and then classified by other techniques is between 20 and 70% of all particles observed (Kalaronis et al. 2022). MPs are identified using a stereomicroscope based on their outward appearance. This is the first fast-screening technique that enables quick identification of the particles' size, shape and color; additional techniques will be used to better define them. Thus, it is necessary to combine stereomicroscopy with additional methods like spectroscopy.

9.7.2. *Fluorescence microscopy*

The fluorescence microscope gathers fluorescent emission from samples that are activated by a certain wavelength, in contrast to the optical microscope, which depends on the contrast of the image created by light reflection on the sample (Lichtman and Conchello 2005). The component known as a fluorophore is present in the material under examination, which is primarily made up of living creatures such as cells, bacteria or tissues. It may have been added from the outside or already function as a fluorophore for proteins or neurotransmitters in the sample. The laser light strikes and stimulates this material, which then releases fluorescence light. Using conventional methods, the laser typically continually emits in the visible (blue) and UV wavelengths, passes through a tiny hole (pinhole) and concentrates on a specific area of the sample that must be studied. Each fluorophore molecule that is struck by a photon is stimulated by the short wavelength and releases fluorescence into the focus and its surrounds. Due to the need for the incident photon to provide more energy, the exciter laser's wavelength needs to be around 50–200 nm shorter than the light released by the fluorophore. A fluorophore, for instance, can emit light at 450 nm, or in the blue-green light spectrum, and absorb light at 360 nm, or in the UV range. In a confocal laser scanning microscope, or point, the light beam illuminates only a section of the sample. Consequently, the detector does not record the entire image; instead, it records the intensity of light coming from a single preparation point. A point can be used to drag along a plane or even inside a volume to replicate the sample image in sections or three dimensions. The excitation system has a lot of shortcomings. At first, the focus the concentrated region of laser intensity as well as the cone that results from focusing above and below the focal plane, excite the fluorophore. This results in undesired fluorescence even close to the focus, which the pinhole partially eliminates. Additionally, the addition of background noise lowers the signal-to-noise ratio and, thus, the resolution. Particularly constrained are the three-dimensional and in-depth scans of the samples. Moreover, while examining organic material, the UV laser may be hazardous and poisonous to the sample. MPs/NPs and biological samples have both been studied using fluorescence microscopy. To assess the impact of MP size on survival, development and fecundity, three distinct sizes of microbeads were chosen (Karakolis et al. 2019). As MPs naturally emit fluorescence, fluorescence microscopy is an excellent technique to identify MPs, especially for white and transparent polymers. This approach lowers the size limit of the found MPs and lowers the MP detection failure rate when combined with imaging. The detection of MPs in different matrices can also be based on the measurement of fluorescent spheres using microscopy techniques. Another limitation on MP vision is the use of chemical additives during manufacture, which can also alter the fluorescence properties.

9.7.3. *Transmission electron microscopy*

Transmission electron microscopy (TEM) is the most often used technique in electron microscopy for characterizing nanomaterials, delivering chemical information and pictures of nanomaterials at atomic-level spatial resolution (Yadav et al. 2007, 2013; Hars et al. 2016). The electron beam that transports the incident light becomes elastically or inelastically scattered electrons when it interacts with a thin foil material. TEM has not actually discovered any NPs in soft matrices yet, and detection methods are still being developed (Agarwal et al. 2017). Heavy-metal stains are required to see NPs and MPs under a TEM because of their amorphous shape, which makes them non-electron dense. The weak elastic interactions of elements with electrons in polymers are usually the cause of elements exhibiting low contrast in TEM. For this reason, it is still usual practice to apply heavy element stains in TEM to view the microstructure of organic specimens (Yadav et al. 2014; Mishra et al. 2017). It has been demonstrated that the stains themselves alter the polymer's chemical structure. Numerous physical processes, including plasmon excitation, phonon excitation and the ionization of core shell electrons, cause the electrons to be inelastically scattered. A spectroscopic method known as electron energy loss spectroscopy can be used to investigate these inelastically scattered electrons. A TEM can be made fluorescence-integrated, which offers fluorescent data that are superimposed in color on TEM pictures and can be used to characterize chemicals. TEM is frequently used to investigate how MPs affect model systems. Another study examined the possible effects of MPs (PP, PE, PET and PVC) on microalgae, including growth suppression and structural changes to the cells, using TEM. The need to characterize MPs according to their dimensions, thickness, surface characteristics and other physicochemical attributes makes the use of additional microscopic methods, like SEM, imperative.

9.7.4. *Scanning electron microscopy*

By acquiring high-resolution pictures of MPs' surface states, SEM is a microscopic method that can reveal details about the morphological surface structure of MPs. Since it has EDS detectors, it can also provide information about the chemical composition of materials. When the first electrons penetrate a solid object, they produce distinct scattering processes that can be both elastic and inelastic. These processes are subsequently collected by different sensor systems to provide an image. The secondary electrons in particular create a comprehensive image that aids in understanding object morphology, while the back-scattered electrons flowing through the specimens provide an increasing intensity that provides information on the material's topography and contrast, dependent on atomic number. More back-scattered electrons are released by elements with higher atomic numbers than by those with lower atomic numbers. This phenomenon can be used to detect changes in sample composition. EDS is not a distinct technique from SEM; rather, it is an additional detector that gives the observer qualitative and quantitative information on the elemental analysis of the sample. It consists of an electron beam generated by an electron microscopes cathode. X-rays are among the many reactions that the primary electron beam induces when it hits the sample's surface. These characterize the sample's constituent elements and their spatial distribution. Most people consider SEM-EDS to be an expensive method. The SEM is a flexible tool that may also be used to characterize NPs because it contains a sample holder for grid analysis. Without the high energy and voltage needs of a traditional electron microscope, small particles can be seen.

9.7.5. *Atomic force microscopy*

The necessity to measure non-conductive materials without employing surface metallization led to the creation of AFM, which, while sharing some general characteristics with prior microscopies, works on an entirely new principle. AFM works by connecting a small tip of highly conductive material to the end of a rod or micro-lever (cantilever), which presses the tip against the sample during measurement. A piezoelectric mechanism causes the cantilever to flow across the surface of a sample as it moves along the three Cartesian axes. The tip can be kept under "constant force" or "constant height" conditions using a control system to collect data on the strength of the interaction between the tip and the sample surface, as well as variations in the sample's height. An optical device that captures the microscopic motions of the bar supporting the tip also detects cantilever oscillations. The actual image of the sample is reconstructed by analyzing data on (i) the relative movement mode between the sample and tip, as well as (ii) the precise results on the sample-tip distance. AFMs can operate in the following modes, depending on how the tip interacts with the sample surface.

Contact mode: when the lever's tip comes into actual contact with the surface under examination, it bends due to van der Waals repulsive forces and electrostatic interactions with an average value of 10^{-9} N. To avoid applying excessive pressure to the sample and modifying its surface, the lever must have a small elastic constant. To prevent the tip from breaking, the feedback loop can either maintain or change the lever's height. This technique is used to evaluate hard materials that are unaffected by contact with the tip. The engagement phase, or the probe's approach to the surface, is the most important aspect of the examination; a mistake here could break the tip or damage the sample surface.

Non-contact mode: the tip is positioned a few nanometers away from the sample and driven into resonance to generate a continuous vibration. van der Waals contact forces, which depend on the tip's distance from the sample, cause variations in oscillation frequency, which are used to reconstruct the surface image. This form of examination is employed when working with materials that are extremely soft and may be destroyed by contact with the tip.

Tapping mode: maximum resolutions are possible with this technology, but large elastic constant levers are needed. When the oscillation amplitude reaches its maximum, the lever tip briefly contacts the sample surface because it is vibrating at frequencies that are near resonance. Generally, the sample absorbs the force provided by the tip without any damage, unless it encounters extremely soft materials. To create the image, variations in the oscillation's amplitude are noted.

Images can be created in AFM by applying varying forces between the sample and the tip. van der Waals forces, as well as electrostatic, magnetic and capillary forces, are responsible for the images in the "no contact mode". The forces of ion repulsion are greater in the "contact mode". Furthermore, for a full and in-depth analysis of the sample, the friction force acting between the tip and the surface is critical. Friction, often known as "lateral force" or "lateral deflection", not only reflects the sample's properties but also how the tip and surface interact. Numerous nanoscale materials, including soil particles, polymeric membranes, artificial nanoparticles and other nanostructures, have been comprehensively characterized using AFM. AFM analysis, as opposed to SEM analysis, enables the performance of further characterizations such as conductivity, magnetism, stiffness and hydrophobicity. AFM is therefore a useful tool for characterizing MPs or NPs, but it must be used in conjunction with other methods. The coupling with other analytical procedures will be explained in more detail in the paragraphs that follow.

9.7.6. *FTIR spectroscopy*

An absorption spectroscopic technique known as infrared (IR) or Fourier-transform infrared (FTIR) spectroscopy is commonly used in the field of material characterization to investigate chemical bonding. A molecule changes from its basic vibrational state to an excited vibrational state when an infrared photon is absorbed by it. The transmittance is found on the ordinate and the incident photon's wavenumber is found on the abscissa of a typical infrared spectrum. Two forms of vibrations are possible: bending, which is the distortion of the bond angle, and stretching, which is the stretching of the chemical bond. FTIR spectroscopy is performed using an interferometer that allows for the scanning of all frequencies contained in the source's IR radiation. A moving mirror changes the optical path by moving and interfering with the light reflected from a stationary mirror, allowing for scanning. This generates an interferogram that shows the intensity representation in the time domain. The Fourier transform converts the interferogram into a spectrum with peaks representing specific chemical bonds or molecular vibrations. There are four methods utilized in FTIR: attenuated total reflection, transmission, reflectance and true specular reflectance/reflection–adsorption. One of the key advantages of FTIR is its high energy availability, which creates a substantially better signal-to-noise ratio than traditional IR spectroscopy and ensures improved performance (Park and Kim 2022). Moreover, there is a notable reduction in the analytical times. Every investigation involves the excitation of MPs samples, which result in certain vibrations that can be detected and used to create a spectrum with a fingerprint range. By contrasting it with the recognized reference spectra, we can determine the composition of the material from this spectrum. The method is helpful for characterizing complicated samples and enables the collection of infrared signals with a high degree of spatial resolution. In fact, the characterization of unidentified samples may reveal some of the method's shortcomings.

9.7.7. *Raman spectroscopy*

Raman spectroscopy is a method based on the interaction of radiation and matter. With this method, light is re-emitted at wavelengths specific to atomic groups by interacting with the vibrational movements of molecules via laser radiation. The molecules in the sample scatter photons from incident radiation inelastically, resulting in a Raman spectrum. When the sample is exposed to a monochromatic laser beam of frequency v, a portion of the radiation is elastically diffused at the same initial frequency v. This produces photons with identical energy. Due to inelastic diffusion, the dispersed radiation spectra will also show multiple lines with frequencies other than the Rayleigh line. Raman spectroscopy is another useful chemical analysis tool for finding MPs in different environmental matrices. According to Raman spectroscopy data, some of the microfilaments found in the intestinal contents could have come from fishing waste (Lv et al. 2020). Raman micro-spectroscopy, which analyzes particles down to the μm range, was used for detection and quantification. While MPs up to 1 μm can be analyzed using micro-Raman spectroscopy (μ-Raman spectroscopy), conventional Raman typically detects MPs larger than 10 μm (Sobhani et al. 2020). The visual selection of the precise region of the sample to be studied is made possible by an integration of an optical microscope and a Raman spectroscope. Due to this, it can be applied in a wide range of scientific domains, including the environmental, forensic, materials, biology, medical, geology, pharmaceutical and art restoration areas (Orlando et al. 2021). It enables non-destructive research of materials in any aggregate condition with minimal sample preparation by applying lasers of varying strengths that can impact the material and generate a Raman spectrum corresponding to the substance under study. Furthermore, the thickness of the samples had no impact on the outcome. Analyses can be done on samples in gas, solution, film, surface, solid or single crystal form.

9.7.8. *Thermal analysis*

Plastic polymers vary in their temperature stability in addition to their IR absorption and Raman scattering characteristics (Bitter and Lackner 2021). The identification of MPs via thermo-analytical methods makes use of modifications in the chemical and physical characteristics of polymers (Sorolla-Rosario et al. 2022). The process is predicted by identifying the polymer based on the products of its degradation. The characterization of low-solubility MPs and additives that are difficult to dissolve, extract or hydrolyze depends critically on the development of heat techniques. Thermogravimetric thermogravimetry (TGA), differential thermal analysis (DTA), differential scanning calorimetry (DSC) and pyrolysis–gas chromatography–mass spectrometry (py-GC-MS) techniques are examples of thermal analysis techniques. DSC, a thermal analysis technique, can be used to determine the temperature and heat flux related to sample transitions, melting enthalpies, glass transitions and polymeric material crystallization dynamics. The basic idea behind this technology is to extract information from a material by controlling the heating or chilling. Specifically, the DSC analyzes the variance in heat transfer between the sample being tested and a reference sample, both of which are kept at a varied temperature determined by a previous program. According to Mukhopadhyay et al. (2002), DSC is a helpful approach for studying polymeric materials, but it needs reference materials. In fact, its principal application is in the detection of primary MPs with well-established properties, such as PE microbeads. However, because polycarbonate (PC), PS, polymethyl methacrylate (PMMA) and acrylonitrile butadiene (ABS) melt at different temperatures, the utility of DSC for the study of these plastics is limited. Another old method, thermogravimetry (TGA), uses weight loss from a sample at a specified temperature to offer quantitative thermal analysis. However, TGA does not indicate the composition of the sample's elements. This approach allows for the creation of graphs depicting the link between mass and temperature. This method of thermal analysis is frequently applied to polymeric materials. It is well known that the experimental settings under which the heating is carried out can have a substantial impact on the polymer's thermal degradation mechanism. Consequently, as much control as possible over the experiment's operating parameters such as the sample's size and shape, its rate of heating and the kind of atmosphere it is heated in necessary to ensure the reproducibility of polymer thermogravimetry results (Mansa and Zou 2021). While enthalpy changes are often the first sign of polymeric material deterioration, a mass loss is recorded in TGA (Sorolla-Rosario et al. 2022). DSC measurements can be used to determine enthalpy changes, which are not detectable in TGA. Thus, in MPs analysis, a mixture of two approaches is recommended (Abbasi et al. 2023). Many MPs are identifiable, but the overlapping phase transition signals of other polymers make detection impossible. It is a destructive procedure, which makes it impossible to examine MPs using the other ways that follow. Due to this, using analytical approaches to identify MPs is detrimental, while they can be helpful for the first screening of bulk samples that will subsequently be subjected to spectroscopic analysis.

9.7.9. *New approaches and new identification strategies*

Expensive equipment and a lengthy analysis period are necessary for the analytical methods, particularly when many particles need to be examined. This applies to the quantification of controlled laboratory experimental samples as well as environmental samples with complex matrices. As a result, it is imperative to create new, alternative methods for the rapid and easy identification of MPs and NPs so that they may be studied in the lab and observed in the field to assess their toxicity, accumulation, aging and other characteristics. To solve these issues, a staining procedure can offer a fresh take or a supplement (Tarafdar et al. 2022). This technique, which is still among the most sophisticated currently in use, is adequate to identify a polymeric particle on its own without the need for additional spectroscopic analysis (Lv et al. 2021).

9.7.10. *Impact of micro- and nano-plastics on human health*

Global plastic consumption is increasing this year, as seen by the numbers for plastic manufacturing that indicate there are currently over 170 trillion plastic particles floating in the world's oceans (Eriksen et al. 2023). Since plastics are so common in the environment, they are also a popular subject in academia when it comes to human health. Every aspect of daily life, including household goods, technology and medical ailments and therapies, is significantly impacted by plastics. Most used plastics that consumers throw away after only one use end up in landfills, the ocean and other bodies of water, which is a serious environmental problem. Numerous research has examined the degradation of plastics in the environment, their source, abundance, transit, fate and potential risks to human health, wildlife and the ecosystem. The substantial impacts that environmental plastic pollution has on the growth, survival and development of numerous living species, including humans, are prompting the scientific community to create new monitoring and decontamination methods. Furthermore, incorrect rubbish disposal of the created waste happens. Plastic is a ubiquitous material that gradually breaks down into tiny particles, which has raised concerns about the possible impacts of MPs and NPs on the environment and human health (Feng et al. 2023; Ali et al. 2024). Plastic trash decomposes into enormous volumes after being disposed of and is subjected to biological, chemical, and environmental factors, leading to the implementation of hazardous classifications of the harmful plastics (Kroon et al. 2016). According to Halden (2010), a single MP particle can decompose into billions of other NP particles, indicating that NP pollution will likely be a global issue. Since NPs can pass through biological membranes, they are probably more harmful than MPs. Despite this, not enough research has been done on the possible impacts of NP exposure on human health. Many of these plastics are thrown away every day, and as they degrade from micro- to nano-sizes, concerns about their toxicity to both humans and the environment have grown. Although the impacts of MP and NPs on the environment have been the subject of numerous previous studies, nothing is known about how they affect the human body at the molecular or subcellular levels (Rochman et al. 2013). Specifically, not enough research has been done on the possibility that systemic exposure may result from the way that NPs pass through the skin, lungs and stomach (Bhat et al. 2023). This chapter delves further into the behavior and breakdown of MPs and NPs in the environment, their levels of toxicity and pollution and the potential health effects on people.

9.8. Photocatalytic degradation of micro- and nano-plastic

MPs and NPs can be extracted from liquids using a variety of technologies, which can be broadly classified into physical, chemical and biological approaches. Typically, physical procedures include screening, adsorption and filtering. The disc filter is one type of industrial liquid filter used in wastewater treatment. In order to remove MP particles from wastewater as it travels through the filter, it is made up of several layers of overlapping filter screen plates and flange rings. Lower molecular weight intermediates are created when plastics are broken down by photocatalysis. These intermediates can then be utilized as a raw material in organic synthesis or as a petrochemical product's manufacturing input by the chemical industry. Both bare and composite photocatalysts are major categories of photocatalysts, and they have distinct degradation mechanisms. Since photocatalysis often exhibits far higher degradation efficiency than biodegradation and is carried out in softer settings than chemical degradation (catalysis), it is a technique of great interest. Two main subgroups of this technology are heterogeneous and homogeneous photocatalytic systems. Due to their high reaction rate, quick processing times and ease of use, homogeneous systems are widely used; nevertheless, they have some significant disadvantages as well, including expensive costs, difficult extraction, convoluted recycling processes and reactor corrosion. Furthermore, because heterogeneous photocatalysts are easier to recover and have higher stability, using process-engineering tactics in photocatalytic processes promotes heterogeneous photocatalysis.

9.9. Photocatalytic degradation of micro- and nano-plastic through 2D materials

The extensive and inevitable use of plastics in our daily lives and industries has resulted in a tremendous amount of plastic rubbish in urban water systems (such as stormwaters, drinking waters, wastewaters, sludges and receiving waters). As a result, MP/NP pollution has become a major environmental problem due to the growing fear of negative effects on living things. It is therefore imperative to segregate and degrade MPs/NPs in metropolitan waters. Various techniques have been used up to this point to isolate and break down MPs/NPs in urban waterways. The primary goal of the separation process is to remove microscopic MPs/NPs particles from the solution; in most cases, MPs/NPs characteristics do not change throughout this process. To remove MPs/NPs from waterways, a variety of advanced solid–liquid separation techniques have been used, including adsorption, coagulation/flocculation, flotation, filtering and magnetic separation. The physicochemical features of MPs/NPs (e.g. hydrophilicity, size, shape and magnetic properties) influence the choice of separation techniques, and the functional materials (e.g. adsorbents) and technical factors have a major influence on the separation performance. For instance, the magnetic separation method makes it simple to extract plastics that have been magnetized by nano-Fe_3O_4. Integrated approaches, which incorporate the advantages of many procedures, are frequently employed to improve separation efficiency in addition to the usage of a single process. In many water treatment facilities (WTPs), the coagulation–flocculation–sedimentation–filtration process is commonly used for MP/NP separation. Apart from the physical separation, there has been an increasing interest in the degradation of MPs/NPs. Through a variety of processes, MPs and NPs are broken down into smaller, less dangerous molecules during the degradation phase. Biological, photocatalytic, electrochemical and photodegradation processes have all demonstrated significant promise in the last 10 years for the removal or mineralization of MPs and NPs from urban waterways. Optimizing parameters (like temperature and duration) and using effective functional materials (like catalysts) are essential for achieving a high rate of degradation. According to reports, employing the C, N-TiO_2 catalyst, low temperatures and low pH values can result in a high MP photocatalytic degradation efficiency of 70% in less than 50 h (Ariza-Tarazona et al. 2020). Furthermore, a variety of recently developed integrated techniques, such as the AOP-hydrolysis and the electro-Fenton-like method enable good degrading performance. Recent intriguing research indicates that MPs and NPs may be transformed into value-added nanomaterials and fuels as opposed to degrading (Chen et al. 2022). All things considered, effective MPs/NPs separation and deterioration in urban waterways work together to provide our society with clean water. However, there is currently no thorough analysis of the separation and degradation of MP/NP in urban waterways in the literature. 2D materials can be used to enhance the effects of essential functional materials/organisms and operational parameters on degrading performance. Many of the pure 2D materials have good photocatalytic activity, but their energy conversion efficiencies are still too low to be used in commercial electrochemical devices. Thus, interface engineering and materials engineering, that is, band gap adjustment in particular, could increase conversion efficiency. Different techniques, including thickness reduction, surface functionalization and dopant inclusion, can be used to modify the band gap of 2D materials. Furthermore, artificial 2D/2D nanostructures for higher activity than their bulk counterparts could be constructed by forming heterostructures of different 2D materials. For the photocatalytic degradation of NPs/MPs, nanosheets of 2D semiconductor materials are often of great importance (Shanker et al. 2021). The first topic we covered in this chapter is the various materials engineering techniques used in the case of 2D materials to achieve better photocatalytic NPs/MPs degrading activities.

9.10. Summary and conclusion

The extraordinary qualities of 2D materials make them superior to commercial bulk materials and offer a wide range of applications. In this chapter, we mostly examined the structure, characteristics, synthesis methods and uses of various 2D materials. The majority of individuals believe that 2D materials have high catalytic potential. The application of 2D materials in photocatalyst research is not limited to the many effective 2D materials or 2D materials-based catalysts that have already been mentioned; there is still plenty of potential for research and development into unique hybrid 2D material components. More background study and theoretical and scientific examination of the 2D material structure, in our opinion, will lead to more appropriate production, stability and efficiency of 2D photocatalysts. As MP/NP contamination poses a risk to all relevant living creatures, it has received priority attention in metropolitan waters. We also give a summary of current developments in the use of 2D materials for MP/NP separation and degradation in this chapter.

9.11. Acknowledgments

The authors express their sincere gratitude for the conversations they had with each of their partners, who have made significant contributions to the collaborative research project that has been covered in part in this chapter. TPY acknowledges receipt of funding for the Collaborative Research Scheme (CRS) (CRS/2023-24/01/1057) from the University Grants Commission–Department of Atomic Energy (UGC-DAE) Consortium for Scientific Research, Indore and the Council of Science and Technology, Uttar Pradesh. K.A. expresses gratitude to the Uttar Pradesh Higher Education Department for financial support under the research development scheme.

9.12. References

Abbasi, S., Razeghi, N., Yousefi, M.R., Podkościelna, B., Oleszczuk, P. (2023). Microplastics identification in water by TGA-DSC method: Maharloo Lake. *Environmental Science and Pollution Research International*, 30, 67008–67018.

Acun, A., Zhang, L., Bampoulis, P., Farmanbar, M., van Houselt, A., Rudenko, A.N. (2015). Germanene: The germanium analogue of graphene. *Journal of Physics. Condensed Matter*, 27, 443002.

Agarwal, H., Yadav, T.P., Srivastava, O.N., Shaz, M.A. (2017). Dielectric response and alternating current conductivity in (Co, Ni) Al_2O_4 nano-spinel. *Ceramics International*, 43, 16986–16992.

Akdogan, Z. and Guven, B. (2019). Microplastics in the environment: A critical review of current understanding and identification of future research needs. *Environmental Pollution*, 254, 113011.

Alexiadou, P., Foskolos, I., Frantzis, A. (2019). Ingestion of macroplastics by odontocetes of the Greek Seas, Eastern Mediterranean: Often deadly! *Marine Pollution Bulletin*, 146, 67–75.

Ali, N., Katsouli, J., Marczylo, E.L., Gant, T.W., Wright, S., Bernardino de la Serna, J.B. (2024). The potential impacts of micro-and-nano plastics on various organ systems in humans. *EBiomedicine*, 99, 104901.

Allen, M.J., Tung, V.C., Kaner, R.B. (2010). Honeycomb carbon: A review of graphene. *Chemical Reviews*, 110, 132–145.

Anastopoulos, I., Pashalidis, I., Kayan, B., Kalderis, D. (2022). Microplastics as carriers of hydrophilic pollutants in an aqueous environment. *Journal of Molecular Liquids*, 350, 118182.

Andriotis, A.N. and Menon, M. (2014). Tunable magnetic properties of transition metal doped MoS_2. *Physical Review. Part B*, 90, 125304.

Ariza-Tarazona, M.C., Villarreal-Chiu, J.F., Hernández-López, J.M., Rivera De la Rosa, J., Barbieri, V., Siligardi, C., Cedillo-González, E.I. (2020). Microplastic pollution reduction by a carbon and nitrogen-doped TiO_2: Effect of pH and temperature in the photocatalytic degradation process. *Journal of Hazardous Materials*, 395, 122632.

Bhat, M.A., Gedik, K., Gaga, E.O. (2023). Atmospheric micro (nano) plastics: Future growing concerns for human health. *Air Quality, Atmosphere, & Health*, 16, 233–262.

Bitter, H. and Lackner, S. (2021). Fast and easy quantification of semi-crystalline microplastics in exemplary environmental matrices by differential scanning calorimetry (DSC). *Chemical Engineering Journal*, 423, 129941.

Bridgman, P.W. (1914). Two new modifications of phosphorus. *Journal of the American Chemical Society*, 36, 1344–1363.

Chen, Z., Liu, X., Wei, W., Chen, H., Ni, B.J. (2022). Removal of microplastics and nanoplastics from urban waters: Separation and degradation. *Water Research*, 221, 118820.

Da Silva A.C.H., Caturello, N.A.M.S., Besse, R., Lima, M.P., Da Silva, J.L.F. (2019). Edge, size, and shape effects on WS_2, WSe_2, and WTe_2 nanoflake stability: Design principles from an ab initio investigation. *Physiological Chemistry and Physics*, 21, 23076–23084.

Dahlqvist, M., Alling, B., Rosén, J. (2010). Stability trends of MAX phases from first principles. *Physical Review. Part B*, 81, 220102.

Delgado, G., Sullivan, J., Esteban, G.L., Wang, P., Arar, Z., Li, Z., Watson, T., Sarp, S. (2021). From sampling to analysis: A critical review of techniques used in the detection of micro-and nanoplastics in aquatic environments. *ACS ES&T Water*, 1, 748–764.

Delmas, C., Fouassier, C., Hagenmuller, P. (1980). Structural classification and properties of the layered oxides. *Physica B+C*, 99, 81–85.

Eriksen, M., Cowger, W., Erdle, L.M., Coffin, S., Villarrubia-Gómez, P., Moore, C.J., Carpenter, E.J., Day, R.H., Thiel, M., Wilcox, C. (2023). A growing plastic smog, now estimated to be over 170 trillion plastic particles afloat in the world's oceans–Urgent solutions required. *PLOS ONE*, 18, e0281596.

Estahbanati, M.K., Rostami, S., Ghasemian, M., Kiendrebeogo, M., Drogui, P., Tyagi, R.D. (2023). Quantitative and qualitative identification, characterization, and analysis of microplastics and nanoplastics in water. In *Current Developments in Biotechnology and Bioengineering*, Mannina, G., Pandey, A., Sirohi, R. (eds). Elsevier, Amsterdam.

Fan, F.R., Wang, R., Zhang, H., Wu, W. (2021). Emerging beyond-graphene elemental 2D materials for energy and catalysis applications. *Chemical Society Reviews*, 50, 10983–11031.

Feng, Y., Tu, C., Li, R., Wu, D., Yang, J., Xia, Y., Peijnenburg, W.J.G.M., Luo, Y. (2023). A systematic review of the impacts of exposure to micro- and nano-plastics on human tissue accumulation and health. *Eco-Environment and Health*, 2, 195–207.

Ganatra, R. and Zhang, Q. (2014). Few-layer MoS_2: A promising layered semiconductor. *ACS Nano*, 8, 4074–4099.

Gao, Y., Liu, Z., Sun, D.M., Huang, L., Ma, L.P., Yin, L.C., Ma, T., Zhang, Z., Ma, X.L., Peng, L.M. et al. (2015). Large-area synthesis of high-quality and uniform monolayer WS_2 on reusable Au foils. *Nature Communications*, 6, 8569.

Gao, H., Suh, J., Cao, M.C., Joe, A.Y., Mujid, F., Lee, K.H., Xie, S., Poddar, P., Lee, J.U., Kang, K. et al. (2020). Tuning electrical conductance of MoS_2 monolayers through substitutional doping. *Nano Letters*, 20, 4095–4101.

Gong, H., Li, R., Li, F., Guo, X., Xu, L., Gan, L., Yan, M., Wang, J. (2023). Toxicity of nanoplastics to aquatic organisms: Genotoxicity, cytotoxicity, individual level and beyond individual level. *Journal of Hazardous Materials*, 443, 130266.

Halden, R.U. (2010). Plastics and health risks. *Annual Review of Public Health*, 31, 179–194.

Hars, S.S., Sharma, H.R., Smerdon, J.A., Yadav, T.P., Al-Mahboob, A., Ledieu, J., Fournée, V., Tamura, R., McGrath, R. (2016). Surface structure of the Ag-In(rare earth) complex intermetallics. *Physical Review. Part B*, 93, 205428.

Huang, Y.H., Peng, C.C., Chen, R.S., Huang, Y.S., Ho, C.H. (2014). Transport properties in semiconducting NbS_2 nanoflakes. *Applied Physics Letters*, 105, 093106.

Huang, Y., Zhu, G., Zou, K., Tian, F., Yadav, T.P., Xu, H., Yang, G., Li, H., Qu, L. (2021). Highly efficient removal of organic pollutants from wastewater using a recyclable graphene oxide membrane intercalated with g-C_3N_4@ TiO_2-nanowires. *Journal of Molecular Liquids*, 337, 116461.

Ivanic, F.M., Guggenberger, G., Woche, S.K., Bachmann, J., Hoppe, M., Carstens, J.F. (2023). Soil organic matter facilitates the transport of microplastic by reducing surface hydrophobicity. *Colloids and Surfaces A*, 676, 132255.

Jose, D. and Datta, A. (2014). Structures and chemical properties of silicene: Unlike graphene. *Accounts of Chemical Research*, 47, 593–602.

Kalaronis, D., Ainali, N.M., Evgenidou, E., Kyzas, G.Z., Yang, X., Bikiaris, D.N., Lambropoulou, D.A. (2022). Microscopic techniques as tools for the determination of microplastics and nanoplastics in the aquatic environment: A short review. *Green Analy. Chem.*, 3, 100036.

Kanazawa, T., Amemiya, T., Ishikawa, A., Upadhyaya, V., Tsuruta, K., Tanaka, T., Miyamoto, Y. (2016). Few-layer HfS_2 transistors. *Scientific Reports*, 6, 22277.

Karakolis, E.G., Nguyen, B., You, J.B., Rochman, C.M., Sinton, D. (2019). Fluorescent dyes for visualizing microplastic particles and fibers in laboratory-based studies. *Environmental Science and Technology Letters*, 6, 334–340.

Karkanorachaki, K., Syranidou, E., Maravelaki, P.N., Kalogerakis, N. (2023). Intertwined synergistic abiotic and biotic degradation of polypropylene pellets in marine mesocosms. *Journal of Hazardous Materials*, 457, 131710.

Karlsson, L.H., Birch, J., Halim, J., Barsoum, M.W., Persson, P.O.Å. (2015). Atomically resolved structural and chemical investigation of single MXene sheets. *Nano Letters*, 15, 4955–4960.

Khandelwal, A., Mani, K., Karigerasi, M.H., Lahiri, I. (2017). Phosphorene – The two dimensional black phosphorous: Properties, synthesis and applications. *Materials Science and Engineering*, 221, 17–34.

Khazaei, M., Arai, M., Sasaki, T., Chung, C.Y., Venkataramanan, N.S., Estili, M., Sakka, Y., Kawazoe, Y. (2013). Novel electronic and magnetic properties of two-dimensional transition metal carbides and nitrides. *Advanced Functional Materials*, 23, 2185–2192.

Kou, L., Chen, C., Smith, S.C. (2015). Phosphorene: Fabrication, properties, and applications. *Journal of Physical Chemistry Letters*, 6, 2794–2805.

Kroon, R., Mengistie, D.A., Kiefer, D., Hynynen, J., Ryan, J.D., Yu, L., Müller, C. (2016). Thermoelectric plastics: From design to synthesis, processing and structure-property relationships. *Chemical Society Reviews*, 45, 6147–6164.

Kumar Maddheshiya, A., Yadav, N., Yadav, S., Singh Yadav, P., Prasad Yadav, T. (2023). Facial synthesis of the trimetallic Au-Pt-Pd nanofluids and performance for methylene blue degradation. *Materials Letters*, 349, 134735.

Lai, B., Singh, S.C., Bindra, J.K., Saraj, C.S., Shukla, A., Yadav, T.P., Wu, W., McGill, S.A., Dalal, N.S., Srivastava, A. et al. (2019). Hydrogen evolution reaction from bare and surface-functionalized few-layered MoS_2 nanosheets in acidic and alkaline electrolytes. *Materials Today. Chemistry*, 14, 100207.

Lee, W.H. and Park, Y.D. (2018). Tuning electrical properties of 2D materials by self-assembled monolayers. *Advanced Materials Interfaces*, 5, 1700316.

Lee, C., Wei, X., Kysar, J.W., Hone, J. (2008). Measurement of the elastic properties and intrinsic strength of monolayer graphene. *Science*, 321, 385–388.

Lee, Y.H., Zhang, X.Q., Zhang, W., Chang, M.T., Lin, C.T., Chang, K.D., Yu, Y.C., Wang, J.T.W., Chang, C.S., Li, L.J. et al. (2012). Synthesis of large-area MoS_2 atomic layers with chemical vapor deposition. *Advanced Materials*, 24, 2320–2325.

Li, X. and Zhu, H. (2015). Two-dimensional MoS_2: Properties, preparation, and applications. *Journal of Materiomics*, 1, 33–44.

Li, X., Cai, W., An, J., Kim, S., Nah, J., Yang, D., Piner, R., Velamakanni, A., Jung, I., Tutuc, E. et al. (2009). Large-area synthesis of high-quality and uniform graphene films on copper foils. *Science*, 324, 1312–1314.

Li, H., Wu, J., Yin, Z., Zhang, H. (2014). Preparation and applications of mechanically exfoliated single-layer and multilayer MoS_2 and WSe_2 nanosheets. *Accounts of Chemical Research*, 47, 1067–1075.

Li, P., Wang, X., Su, M., Zou, X., Duan, L., Zhang, H. (2021). Characteristics of plastic pollution in the environment: A review. *Bulletin of Environmental Contamination and Toxicology*, 107, 577–584.

Lichtman, J.W. and Conchello, J.A. (2005). Fluorescence microscopy. *Nature Methods*, 2, 910–919.

Liu, J., Wang, H., Antonietti, M. (2016). Graphitic carbon nitride 'reloaded': Emerging applications beyond (photo)catalysis. *Chemical Society Reviews*, 45, 2308–2326.

Liu, R., Wang, Y., Yang, Y., Shen, L., Zhang, B., Dong, Z., Gao, C., Xing, B. (2023). New insights into adsorption mechanism of pristine and weathered polyamide microplastics towards hydrophilic organic compounds. *Environmental Pollution*, 317, 120818.

Lv, R., Robinson, J.A., Schaak, R.E., Sun, D., Sun, Y., Mallouk, T.E., Terrones, M. (2015). Transition metal dichalcogenides and beyond: Synthesis, properties, and applications of single- and few-layer nanosheets. *Accounts of Chemical Research*, 48, 56–64.

Lv, L., He, L., Jiang, S., Chen, J., Zhou, C., Qu, J., Lu, Y., Hong, P., Sun, S., Li, C. (2020). In situ surface-enhanced Raman spectroscopy for detecting microplastics and nanoplastics in aquatic environments. *Science of the Total Environment*, 728, 138449.

Lv, L., Yan, X., Feng, L., Jiang, S., Lu, Z., Xie, H., Sun, S., Chen, J., Li, C. (2021). Challenge for the detection of microplastics in the environment. *Water Environment Research*, 93, 5–15.

Mansa, R. and Zou, S. (2021). Thermogravimetric analysis of microplastics: A mini review. *Environmental Advances*, 5, 100117.

Mishra, S.S., Mukhopadhyay, S., Yadav, T.P., Yadav, R.M., Radhakrishnan, S., Vajtai, R., Ajayan, P.M., Mukhopadhyay, N.K., Singh, H.K., Srivastava, O.N. (2017). Effect of Fe substitution by Co on off-stoichiometric Ni-Fe-Co-Mn-Sn Heusler alloy ribbons. *Materials Research Express*, 4, 086507.

Moore, D.B., Beekman, M., Disch, S., Zschack, P., Häusler, I., Neumann, W., Johnson, D.C. (2013). Synthesis, structure, and properties of turbostratically disordered $(PbSe)_{1.18}(TiSe_2)_2$. *Chemistry of Materials*, 25, 2404–2409.

Mukhopadhyay, N.K., Yadav, T.P., Srivastava, O.N. (2002). An investigation on the transformation of the icosahedral phase in the Al-Fe-Cu system during mechanical milling and subsequent annealing. *Philosophical Magazine A*, 82, 2979–2993.

Naguib, M., Kurtoglu, M., Presser, V., Lu, J., Niu, J., Heon, M., Hultman, L., Gogotsi, Y., Barsoum, M.W. (2011). Two-dimensional nanocrystals produced by exfoliation of Ti_3AlC_2. *Advanced Materials*, 23, 4248–4253.

Naguib, M., Mochalin, V.N., Barsoum, M.W., Gogotsi, Y. (2014). MXenes: A new family of two dimensional materials. *Advanced Materials*, 26, 992–1005.

Novoselov, K.S., Geim, A.K., Morozov, S.V., Jiang, D., Zhang, Y., Dubonos, S.V., Grigorieva, I.V., Firsov, A.A. (2004). Electric field effect in atomically thin carbon films. *Science*, 306, 666–669.

Novoselov, K.S., Geim, A.K., Morozov, S.V., Jiang, D., Katsnelson, M.I., Grigorieva, I.V., Dubonos, S.V., Firsov, A.A. (2005a). Two-dimensional gas of massless dirac fermions in graphene. *Nature*, 438, 197–200.

Novoselov, K.S., Jiang, D., Schedin, F., Booth, T.J., Khotkevich, V.V., Morozov, S.V., Geim, A.K. (2005b). Two-dimensional atomic crystals. *Proceedings of the National Academy of Sciences of the United States of America*, 102, 10451–10453.

Orlando, A., Franceschini, F., Muscas, C., Pidkova, S., Bartoli, M., Rovere, M., Tagliaferro, A. (2021). A comprehensive review on Raman spectroscopy applications. *Chemosensors*, 9, 262.

Oyedele, A.D., Yang, S., Liang, L., Puretzky, A.A., Wang, K., Zhang, J., Yu, P., Pudasaini, P.R., Ghosh, A.W., Liu, Z. et al. (2017). $PdSe_2$: Pentagonal two-dimensional layers with high air stability for electronics. *Journal of the American Chemical Society*, 139, 14090–14097.

Pacilé, D., Meyer, J.C., Girit, Ç.Ö., Zettl, A. (2008). The two-dimensional phase of boron nitride: Few-atomic-layer sheets and suspended membranes. *Applied Physics Letters*, 92, 133107.

Pak, S., Jang, S., Kim, T., Lim, J., Hwang, J.S., Cho, Y., Chang, H., Jang, A.R., Park, K.H., Hong, J. et al. (2021). Electrode-induced self-healed monolayer MoS_2 for high performance transistors and phototransistors. *Advanced Materials*, 33, e2102091.

Pakdel, A., Bando, Y., Golberg, D. (2014). Nano boron nitride flatland. *Chemical Society Reviews*, 43, 934–959.

Park, S.Y. and Kim, C.G. (2022). A comparative study on the distribution behaviour of microplastics through FT-IR analysis on different land uses in agricultural soils. *Environmental Research*, 215, 114404.

Pi, L., Li, L., Liu, K., Zhang, Q., Li, H., Zhai, T. (2019). Recent progress on 2D noble-transition-metal dichalcogenides. *Advanced Functional Materials*, 29, 1904932.

Qu, L., Wang, N., Hui, X., Weipeng, W., Liu, Y., Kuo, L., Yadav, T.P., Wu, J., Joyner, J., Song, Y. et al. (2017). Gold nanoparticles and g-C_3N_4-intercalated graphene oxide membrane for recyclable surface enhanced Raman scattering advanced functional materials. *Advanced Functional Materials*, 27, 1701714.

Qu, L., Zhu, G., Ji, J., Yadav, T.P., Chen, Y., Yang, G., Xu, H., Li, H. (2018a). Recyclable visible light-driven O-g-C_3N_4/graphene oxide/N-carbon nanotube membrane for efficient removal of organic pollutants. *ACS App. Mat. Int.*, 10, 42427–42435.

Qu, L.L., Wang, N., Zhu, G., Yadav, T.P., Shuai, X., Bao, D., Yang, G., Li, D., Li, H. (2018b). Facile fabrication of ternary TiO_2-gold nanoparticle-graphene oxide nanocomposites for recyclable surface enhanced Raman scattering. *Talanta*, 186, 265–271.

Rochman, C.M., Browne, M.A., Halpern, B.S., Hentschel, B.T., Hoh, E., Karapanagioti, H.K., Rios-Mendoza, L.M., Takada, H., Teh, S., Thompson, R.C. (2013). Policy: Classify plastic waste as hazardous. *Nature*, 494, 169–171.

Ruth, A., Samavi, M., Rakshit, S.K. (2023). Biological degradation of microplastics and nanoplastics in water and wastewater. *Current Dev. Biotech. Bioengg.*, 293–314.

Santos, B.G., Gonçalves, J.M., Rocha, D.P., Higino, G.S., Yadav, T.P., Pedrotti, J.J., Ajayan, P.M., Angnes, L. (2022). Electrochemical sensor for isoniazid detection by using a WS_2/CNTs nanocomposite. *Sensors and Actuators Reports*, 4, 100073.

Shanker, G.S., Biswas, A., Ogale, S. (2021). 2D materials and their heterostructures for photocatalytic water splitting and conversion of CO_2 to value chemicals and fuels. *Journal of Physics: Energy*, 3, 022003.

Shanmugam, V., Mensah, R.A., Babu, K., Gawusu, S., Chanda, A., Tu, Y., Neisiany, R.E., Försth, M., Sas, G., Das, O. (2022). A review of the synthesis, properties, and applications of 2D materials. *Particle & Particle Systems Characterization*, 39, 2200031.

Singh, A.K., Pal, P., Gupta, V., Yadav, T.P., Gupta, V., Singh, S.P. (2018). Green synthesis, characterization and antimicrobial activity of zinc oxide quantum dots using *Eclipta alba*. *Materials Chemistry and Physics*, 203, 40–48.

Sobhani, Z., Zhang, X., Gibson, C., Naidu, R., Megharaj, M., Fang, C. (2020). Identification and visualisation of microplastics/nanoplastics by Raman imaging (i): Down to 100 nm. *Water Research*, 174, 115658.

Soldano, C., Mahmood, A., Dujardin, E. (2010). Production, properties and potential of graphene. *Carbon*, 48, 2127–2150.

Sorolla-Rosario, D., Llorca-Porcel, J., Pérez-Martínez, M., Lozano-Castelló, D., Bueno-López, A. (2022). Study of microplastics with semicrystalline and amorphous structure identification by TGA and DSC. *Journal of Environmental Chemical Engineering*, 10, 106886.

Stoeckel, M.A., Gobbi, M., Leydecker, T., Wang, Y., Eredia, M., Bonacchi, S., Verucchi, R., Timpel, M., Nardi, M.V., Orgiu, E. et al. (2019). Boosting and balancing electron and hole mobility in single- and bilayer WSe_2 devices via tailored molecular functionalization. *ACS Nano*, 13, 11613–11622.

Tarafdar, A., Choi, S.H., Kwon, J.H. (2022). Differential staining lowers the false positive detection in a novel volumetric measurement technique of microplastics. *Journal of Hazardous Materials*, 432, 128755.

Thaiba, B.M., Sedai, T., Bastakoti, S., Karki, A., Anuradha, K.C., Khadka, G., Acharya, S., Kandel, B., Giri, B., Neupane, B.B. (2023). A review on analytical performance of micro-and nanoplastics analysis methods. *Arabian Journal of Chemistry*, 16, 104686.

Tiwari, S.K., Kumar, V., Huczko, A., Oraon, R., Adhikari, A.D., Nayak, G.C. (2016). Magical allotropes of carbon: Prospects and applications. *Critical Reviews in Solid State and Materials Sciences*, 41, 257–317.

Wang, X., Kajiyama, S., Iinuma, H., Hosono, E., Oro, S., Moriguchi, I., Okubo, M., Yamada, A. (2015). Pseudocapacitance of MXene nanosheets for high-power sodium-ion hybrid capacitors. *Nature Communications*, 6, 6544.

Watanabe, K., Taniguchi, T., Kanda, H. (2004). Direct-bandgap properties and evidence for ultraviolet lasing of hexagonal boron nitride single crystal. *Nature Materials*, 3, 404–409.

Xia, Y., Niu, S., Yu, J. (2023). Microplastics as vectors of organic pollutants in aquatic environment: A review on mechanisms, numerical models, and influencing factors. *Science of the Total Environment*, 887, 164008.

Xu, B., Zhu, M., Zhang, W., Zhen, X., Pei, Z., Xue, Q., Zhi, C., Shi, P. (2016). Ultrathin MXene micropattern-based field-effect transistor for probing neural activity. *Advanced Materials*, 28, 3333–3339.

Xu, L., Li, K., Zhang, M., Guo, J., Jia, W., Bai, X., Tian, X., Huang, Y. (2023). Plastic substrate and residual time of microplastics in the urban river shape the composition and structure of bacterial communities in plastisphere. *Journal of Environmental Management*, 345, 118710.

Yadav, T.P. and Awasthi, K. (2022). Carbon nanomaterials: Fullerene to graphene. *Transactions of the Indian National Academy of Engineering*, 7, 715–737.

Yadav, T.P., Mukhopadhyay, N.K., Tiwari, R.S., Srivastava, O.N. (2007). On the evolution of quasicrystalline and crystalline phases in rapidly quenched Al–Co–Cu–Ni alloy. *Materials Science and Engineering*, 449–451, 1052–1056.

Yadav, T.P., Singh, D., Shahi, R.R., Shaz, M.A., Tiwari, R.S., Srivastava, O.N. (2013). Synthesis of quasicrystalline film of Al–Ga–Pd–Mn alloy. *Thin Solid Films*, 534, 265–269.

Yadav, T.P., Mishra, S.S., Pandey, S.K., Singh, D., Lowe, M., Tamura, R., Mukhopadhyay, N.K., Srivastava, O.N., McGrath, R., Sharma, H.R. (2014). Leaching of Al-based polygrain quasicrystalline and related crystalline surfaces. *Acta Physica Polonica. Part A*, 126, 629–632.

Yadav, T.P., Woellner, C.F., Sinha, S.K., Sharifi, T., Apte, A., Mukhopadhyay, N.K., Srivastava, O.N., Vajtai, R., Galvao, D.S., Tiwary, C.S. et al. (2018). Liquid exfoliation of icosahedral quasicrystals. *Advanced Functional Materials*, 28, 8.

Yadav, T.P., Palma, J.L., Singh, D.P., Michea, S. (2020a). Facile synthesis and magnetic behaviour of 1D g-C_3N_4. *Journal of Solid State Chemistry*, 290, 121539.

Yadav, T.P., Woellner, C.F., Sharifi, T., Sinha, S.K., Qu, L.L., Apte, A., Mukhopadhyay, N.K., Srivastava, O.N., Vajtai, R., Galvão, D.S. et al. (2020b). Extraction of two-dimensional aluminum alloys from decagonal quasicrystals. *ACS Nano*, 14, 7435–7443.

Yu, S., Wang, X., Pang, H., Zhang, R., Song, W., Fu, D., Hayat, T., Wang, X. (2018). Boron nitride-based materials for the removal of pollutants from aqueous solutions: A review. *Chemical Engineering Journal*, 333, 343–360.

Yu, Z., Xu, X., Guo, L., Jin, R., Lu, Y. (2024). Uptake and transport of micro/nanoplastics in terrestrial plants: Detection, mechanisms, and influencing factors. *Science of the Total Environment*, 907, 168155.

Zang, X., Chen, Q., Li, P., He, Y., Li, X., Zhu, M., Li, X., Wang, K., Zhong, M., Wu, D. et al. (2014). Highly flexible and adaptable, all-solid-state supercapacitors based on graphene woven fabric film electrodes. *Small*, 10, 2583–2588.

Zhao, G., Hao, S.H., Guo, J.H., Xing, Y.P., Zhang, L., Xu, X.J. (2021). Design of p-n homojunctions in metal-free carbon nitride photocatalyst for overall water splitting. *Chinese Journal of Catalysis*, 42, 501–509.

Zheng, R., Li, Q., Li, P., Li, L., Liu, J. (2023). Total organic carbon content as an index to estimate the sorption capacity of micro- and nano-plastics for hydrophobic organic contaminants. *Chemosphere*, 313, 137374.

Zhu, W., Gao, X., Li, Q., Li, H., Chao, Y., Li, M., Mahurin, S.M., Li, H., Zhu, H., Dai, S. (2016). Controlled gas exfoliation of boron nitride into few-layered nanosheets. *Angewandte Chemie*, 55, 10766–10770.

Zhu, X., Wang, C., Duan, X., Liang, B., Genbo Xu, E., Huang, Z. (2023). Micro- and nanoplastics: A new cardiovascular risk factor? *Environment International*, 171, 107662.

10

Hybrid 2D-Smart Materials in Photocatalytic Degradation of Micro- and Nano-Plastics

This chapter explores the potential of hybrid two-dimensional (2D)-smart materials in addressing the pressing environmental issue of micro- and nano-plastic pollution. These materials, combining the unique properties of 2D structures with other functional components, demonstrate enhanced photocatalytic efficiency in degrading persistent plastic pollutants. Through the generation of reactive oxygen species (ROS) under light irradiation, hybrid 2D materials such as graphene, metal oxides and metal sulfides exhibit remarkable abilities to break down plastics into harmless by-products. This chapter explores various synthesis methods, the role of material morphology and the underlying degradation mechanisms. It also presents recent advancements in the design of hybrid photocatalysts tailored for environmental applications. The chapter contributes to the growing body of knowledge on sustainable materials and their role in tackling global plastic waste.

10.1. Introduction

The pervasive presence of micro- and nano-plastics in the environment has emerged as a critical global challenge, causing significant concern among scientists, policymakers and the public. Micro-plastics, typically defined as plastic particles less than 5 mm in size, and nano-plastics, which are even smaller, are ubiquitous contaminants found in water bodies, soils and even in the air (Liu et al. 2022). These tiny plastic fragments originate from the degradation of larger plastic items, as well as from primary sources such as cosmetic products, industrial abrasives and synthetic textiles. The persistence of these plastics in the environment,

Chapter written by Niranjan PATRA, Gudiguntla RAVI, Muddada Jaya SURYA and Akil AHMAD.

coupled with their potential to adsorb harmful chemicals and enter the food chain, poses severe risks to ecosystems and human health (Thompson et al. 2009). Addressing the issue of micro- and nano-plastic pollution requires innovative and effective strategies for their removal and degradation. Traditional methods, such as filtration and chemical treatments, have limitations, including inefficiency at nanoscale levels and the potential generation of secondary pollutants (He et al. 2023). In recent years, the development of advanced materials for environmental remediation has gained momentum, with a particular focus on photocatalysis as a promising approach to degrade these persistent pollutants (Xu et al. 2013; Zheng et al. 2016).

Photocatalysis, a process driven by light energy to accelerate chemical reactions, offers a sustainable and energy-efficient solution for breaking down micro- and nano-plastics into harmless by-products. This process typically involves the generation of ROS through the excitation of a photocatalyst under light irradiation, which then interacts with the plastic polymers, leading to their degradation (Xie et al. 2023). However, the effectiveness of photocatalysis in degrading micro- and nano-plastics is highly dependent on the nature of the photocatalyst, necessitating the development of materials with enhanced catalytic activity and stability (Lee et al. 2023).

2D materials, characterized by their ultrathin structure and high surface area, have garnered significant attention in the field of photocatalysis. Graphene, MXenes, transition metal dichalcogenides (TMDs) and other 2D materials exhibit unique electronic, optical and surface properties that make them ideal candidates for photocatalytic applications (Zhu et al. 2010; Xu et al. 2013; Anegbe et al. 2024). Their ability to absorb light, generate charge carriers and provide reactive sites for catalytic reactions makes them particularly suitable for the degradation of micro- and nano-plastics. The concept of hybrid 2D-smart materials represents an evolution in material design, where 2D materials are combined with other functional components to create synergistic effects that enhance their photocatalytic performance. These hybrids can incorporate nanoparticles, polymers or other nanostructures, each contributing to improved light absorption, charge separation and surface reactivity (Raza et al. 2022). Moreover, the "smart" aspect of these materials refers to their ability to respond to external stimuli, such as changes in pH, temperature or light, which can further optimize their performance in real-world environmental conditions.

This chapter explores into the development and application of hybrid 2D-smart materials for the photocatalytic degradation of micro- and nano-plastics. It explores the underlying mechanisms of photocatalysis, the unique properties of 2D materials that make them suitable for this purpose and the innovative strategies employed to design and synthesize hybrid materials with enhanced capabilities. Through a comprehensive review of recent advancements and case studies, this chapter aims to provide a detailed understanding of the potential of hybrid 2D-smart materials in addressing one of the most pressing environmental issues of recent times. The discussion will also highlight the challenges and future directions in this rapidly evolving field, including the need for scalable production methods, environmental and economic considerations, and the exploration of new materials and hybrid structures. By bridging the gap between fundamental research and practical applications, this chapter seeks to contribute to the ongoing efforts to develop sustainable solutions for the degradation of micro- and nano-plastics.

10.2. 2D materials: properties and functionalities

2D materials, characterized by their atomic-scale thickness and extensive lateral dimensions, have emerged as a frontier in materials science due to their unique properties and versatile functionalities. Since the discovery of graphene in 2004, a single layer of carbon atoms arranged in a honeycomb lattice, the field of 2D materials has rapidly expanded to include a variety of compounds with diverse chemical compositions and structures, such as TMDs, hexagonal boron nitride (h-BN) and phosphorene (Choi et al. 2017; Chaves et al. 2020; Dutta et al. 2024). These materials possess a range of electronic, optical, mechanical and thermal properties, which make them highly attractive for various applications, including electronics, energy storage, catalysis and environmental remediation.

10.2.1. *Electronic properties*

One of the most compelling features of 2D materials is their electronic versatility. Graphene, for example, is renowned for its exceptional electrical conductivity and high carrier mobility, which arise from its Dirac cone structure and linear energy–momentum relationship (Yue et al. 2020; Dutta et al. 2024). This makes graphene an excellent candidate for use in high-speed transistors, transparent conductive electrodes and other electronic devices. TMDs such as molybdenum disulfide (MoS_2) and WS_2 exhibit a range of electronic properties, including semiconducting behavior with an inherent bandgap, which is absent in graphene. This bandgap allows TMDs to function as transistors with high on/off ratios, making them suitable for digital electronics and optoelectronic applications (Choi et al. 2017; Nandee et al. 2022). The ability to tune the bandgap of TMDs through methods such as layer number control, strain engineering and chemical doping further enhance their potential for use in next-generation electronic devices (Nandee et al. 2022). h-BN, another prominent 2D material, is an insulator with a wide bandgap of around 6 eV, making it an excellent dielectric material and a substrate for graphene and TMDs in electronic devices. Its high thermal stability and resistance to oxidation also contribute to its utility in high-temperature applications and protective coatings (Laturia et al. 2018; Chen et al. 2021).

10.2.2. *Optical properties*

The optical properties of 2D materials are equally remarkable and highly tunable, making them valuable for photonic and optoelectronic applications. Graphene, while primarily known for its electronic properties, also exhibits notable optical properties, such as broadband absorption and high transparency. These characteristics have led to the exploration of graphene in applications like transparent conductive films, photodetectors and ultrafast lasers (Tan et al. 2020; Qiao et al. 2021). TMDs display strong photoluminescence when thinned down to a monolayer, transitioning from an indirect to a direct bandgap. This direct bandgap enables efficient light emission, which is particularly useful in light-emitting diodes (LEDs) and solar cells (Chen et al. 2020a; Kang et al. 2022a). The excitonic effects in TMDs, driven by strong Coulomb interactions due to reduced dielectric screening in two dimensions, result in the formation of tightly bound excitons, trions and biexcitons, further enhancing their optical performance (Pospischil and Mueller 2016). Black phosphorus, or phosphorene, offers a tunable bandgap depending on the number of layers, ranging from 0.3 eV in bulk to 2.0 eV in a monolayer, making it a versatile material for photodetectors that operate across a broad spectrum, including the visible and infrared regions (Lin et al. 2019; Zeng and Guo 2021). This tunability, combined with its high carrier mobility, positions phosphorene as a promising material for photodetection and optoelectronics.

10.2.3. *Mechanical properties*

The mechanical properties of 2D materials are another area of significant interest, particularly in the context of flexible and wearable electronics. Graphene is renowned for its exceptional mechanical strength, with a tensile strength of over 130 GPa and a Young's modulus of approximately 1 TPa, making it one of the strongest known materials. This strength, coupled with its flexibility and lightweight nature, makes graphene an ideal material for reinforcing composites, flexible electronics and high-strength coatings (Lei et al. 2022; Uddin et al. 2023). TMDs, while not as mechanically robust as graphene, still exhibit considerable mechanical flexibility, which is advantageous for applications requiring bendable or stretchable materials. The ability to form monolayers and multilayers of TMDs with varying mechanical properties allows for the design of heterostructures and composites tailored to specific mechanical requirements (Choudhary et al. 2018; Peng et al. 2020; Askari et al. 2023). h-BN, though more brittle than graphene, provides excellent mechanical support when used as a substrate or an encapsulating layer in

2D-material heterostructures. Its mechanical rigidity, combined with its insulating properties, contributes to the overall mechanical and electronic stability of 2D material-based devices (Zhang et al. 2017; Juma et al. 2021).

10.2.4. *Thermal properties*

The thermal properties of 2D materials play a crucial role in their performance, particularly in high-power and high-temperature applications. Graphene exhibits an exceptionally high thermal conductivity, exceeding 3,000 W/mK, which is beneficial for thermal management in electronic devices, heat dissipation materials and thermal interface materials. This high thermal conductivity is attributed to the strong sp^2 carbon–carbon bonds and the 2D-phonon transport in graphene (Fu et al. 2020; Guo et al. 2021; Huang et al. 2021). TMDs and other 2D materials generally exhibit lower thermal conductivities compared to graphene, but they still offer adequate thermal management capabilities for certain applications. For instance, MoS_2 has a thermal conductivity in the range of 30–50 W/mK, which is sufficient for use in flexible electronics and low-power devices (Chenet al. 2020b; Kalantari and Zhang 2022). h-BN also possesses high thermal conductivity, around 200–400 W/mK, along with electrical insulation properties, making it an excellent material for thermal management in electronic devices where electrical insulation is also required (Zhang et al. 2017).

10.2.5. *Chemical properties and functionalization*

The chemical properties of 2D materials, particularly their surface chemistry and reactivity, are critical to their functionality in various applications. Graphene's surface can be chemically modified through functionalization with different chemical groups, enabling the tailoring of its properties for specific applications such as sensors, catalysis and energy storage. Functionalization also improves graphene's solubility and processability, facilitating its integration into various matrices and composite materials (Georgakilas et al. 2012; Kuila et al. 2012). TMDs and other 2D materials can also be functionalized to enhance their chemical properties and compatibility with other materials. For example, the functionalization of MoS_2 with organic molecules or nanoparticles can improve its catalytic activity for hydrogen evolution reactions (HER) and other catalytic processes (Presolski and Pumera 2016). The high surface area-to-volume ratio of 2D materials further amplifies their chemical reactivity, making them effective catalysts and adsorbents for environmental applications. This high reactivity also allows for the adsorption of pollutants, such as heavy metals and organic contaminants, making 2D materials suitable for water purification and environmental remediation technologies (Saleem et al. 2020; Uddin et al. 2023).

10.2.6. *Synergistic effects in hybrid 2D materials*

Hybrid 2D materials, formed by combining different 2D materials or incorporating 2D materials with other nanostructures, offer synergistic effects that enhance their overall performance. For instance, combining graphene with TMDs can result in hybrid structures that benefit from the high electrical conductivity of graphene and the semiconducting properties of TMDs, making them suitable for applications in optoelectronics and energy conversion (Maggini and Ferreira 2021; Kumbhakar et al. 2023). The incorporation of nanoparticles, polymers or other nanomaterials into 2D material matrices can further enhance properties such as catalytic activity, mechanical strength and thermal stability. These hybrids can be engineered to exhibit specific functionalities by optimizing the composition and structure of the constituent materials (Khdary et al. 2023). The development of hybrid 2D-smart materials represents a significant advancement in the field, allowing for the design of materials with tailored properties for specific applications, such as photocatalytic degradation of micro- and nano-plastics, energy storage and biomedical devices (Porcu et al. 2022; Li et al. 2023; Maity et al. 2023; Mishra et al. 2023; Singh et al. 2024).

10.3. Hybrid 2D-smart materials: design and synthesis

Hybrid 2D-smart materials represent an innovative class of composites that integrate the unique properties of 2D materials with other nanomaterials or molecular structures. These hybrid materials are designed to harness the synergistic effects between different components, leading to enhanced performance in applications ranging from catalysis and energy storage to environmental remediation and sensing. The design and synthesis of these materials involve a careful selection of components, strategic integration methods and tailored structural configurations to achieve desired functionalities. The design of hybrid 2D-smart materials is guided by several principles aimed at optimizing their performance for specific applications. One key principle is the selection of complementary components that enhance each other's properties. For instance, combining graphene with TMDs can yield hybrids with improved electronic conductivity, catalytic activity and mechanical strength compared to their individual components (Saleem et al. 2020). Another important design consideration is the interface between different components. The interaction at the interface, whether through covalent bonding, van der Waals forces or electrostatic interactions, can significantly influence the overall properties of the hybrid material. A strong interface can facilitate efficient charge transfer, enhance mechanical stability and improve chemical durability. For example, the strong coupling between graphene and metal nanoparticles has been shown to enhance the catalytic performance of the resulting hybrid material (Jiang et al. 2022). The structural configuration of the hybrid material also plays a crucial role. This includes factors such as the number of layers in 2D materials, the dispersion of secondary components and the overall morphology of the composite. Controlling these structural aspects allows for fine-tuning the material's properties. For example, the creation of vertically aligned 2D heterostructures can provide an enhanced surface area for catalytic reactions, while layered structures can facilitate the separation and transport of charge carriers (Shanker et al. 2021; Zhang et al. 2022; Cho and Kang 2024; Hill 2024; Singh et al. 2024; Supriya et al. 2025).

10.3.1. *Synthesis techniques*

The synthesis of hybrid 2D-smart materials involves various strategies that can be broadly categorized into two approaches: bottom-up and top-down methods. Each approach offers distinct advantages and can be chosen based on the desired properties and application of the material.

10.3.1.1. *Bottom-up synthesis*

This approach involves the construction of hybrid materials from molecular or atomic precursors, allowing for precise control over composition and structure. One common bottom-up method is chemical vapor deposition (CVD), which is widely used for the growth of 2D materials such as graphene and TMDs. By introducing different precursors during the CVD process, it is possible to synthesize hybrid materials with integrated functionalities (You et al. 2018; Alam et al. 2021; Kang et al. 2022b). Solution-based synthesis is another bottom-up technique that offers flexibility in the design of hybrid materials. In this method, 2D materials are typically exfoliated or synthesized in solution, followed by the incorporation of secondary components through chemical reactions, self-assembly or physical mixing. This approach is particularly useful for creating hybrid materials with well-dispersed nanoparticles, polymers or other nanostructures within a 2D matrix (Cai et al. 2022; Durga Devi et al. 2022; González et al. 2022; Jiang et al. 2022; Khavar et al. 2022; Quiroz-Cardoso et al. 2022; Tai et al. 2022; Winayu et al. 2022).

10.3.1.2. *Top-down synthesis*

Top-down methods involve the exfoliation or reduction of bulk materials into their 2D counterparts, followed by the integration of additional components to form hybrids. Mechanical exfoliation, liquid-phase exfoliation and laser ablation are common techniques for producing 2D materials from bulk crystals. Once the 2D material is obtained, secondary components can be introduced through techniques such as physical vapor deposition, electrochemical deposition or surface functionalization (Bonaccorso et al. 2012; Zhang 2015; Zheng and Lee 2022; Baig 2023; Grubišić-Čabo et al. 2023). Top-down approaches are advantageous for producing large-scale hybrid materials with uniform properties. However, they may offer less control over the precise arrangement of components compared to bottom-up methods. The choice between these approaches depends on factors such as the scalability of production, the desired material properties and the specific application.

10.3.2. *Examples of hybrid 2D-smart materials*

Several examples of hybrid 2D-smart materials demonstrate the versatility and potential of this class of materials. One notable example is the combination of graphene with metal oxides, such as titanium dioxide (TiO_2) or zinc oxide (ZnO), to create photocatalytic materials for environmental applications. The hybridization of graphene with TiO_2 enhances the charge separation and reduces electron–hole recombination, leading to improved photocatalytic efficiency for the degradation of pollutants, including micro- and nano-plastics (Fu et al. 2012; Yaqoob et al. 2020; Akyüz 2021; Mirikaram et al. 2021; Padmanabhan et al. 2021). Another example is the integration of TMDs with noble metal nanoparticles, such as gold or platinum, to create hybrid catalysts for HER. The presence of TMDs provides active sites for catalysis, while the metal nanoparticles facilitate electron transfer and enhance catalytic activity. These hybrids have shown promise in energy conversion applications, such as water splitting and fuel cells (Li et al. 2018; Kaushik et al. 2024; Mehta et al. 2024). Hybrid 2D materials have also been explored in the field of energy storage. For instance, the combination of graphene with metal-organic frameworks (MOFs) has led to the development of high-performance supercapacitors and batteries. The porous structure of MOFs provides a high surface area for ion adsorption, while graphene enhances electrical conductivity and mechanical stability, resulting in improved energy-storage capacity and cycle life.

10.4. Mechanisms of photocatalytic degradation of micro- and nano-plastics

The global prevalence of micro- and nano-plastics in aquatic environments has raised significant environmental and health concerns, necessitating the development of effective degradation strategies. Among these, photocatalytic degradation using semiconductor materials has emerged as a promising method due to its ability to harness solar energy for the breakdown of these persistent pollutants. The photocatalytic degradation of micro- and nano-plastics involves complex mechanisms driven by the interaction of light with photocatalysts, leading to the generation of reactive species that initiate and propagate degradation processes.

Photocatalysis is a light-driven process that involves the excitation of electrons in a semiconductor material upon absorption of photons with energy equal to or greater than its bandgap. This excitation generates electron–hole pairs, which migrate to the surface of the photocatalyst. The separated electrons and holes then participate in redox reactions with adsorbed species, typically leading to the formation of ROS such as hydroxyl radicals (•OH), superoxide radicals (O2•−) and hydrogen peroxide (H_2O_2). These ROS are highly reactive and can degrade organic pollutants, including micro- and nano-plastics, into smaller, less harmful molecules (Andrady 2011; Khavar et al. 2022; Liu et al. 2022; Mishra et al. 2023) as shown in Figure 10.1.

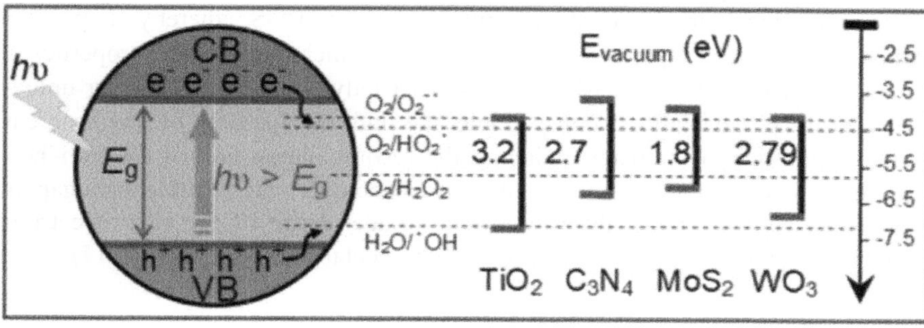

Figure 10.1. *The generation and chemical potential of highly reactive species and free radicals in semiconductors after absorbing light energy, compared to the band structures of TiO_2, g-C_3N_4, MoS_2 and WO_3 materials (reproduced with permission from Li et al. (2016)). For a color version of this figure, see www.iste.co.uk/singh/photodegradation.zip*

10.4.1. *Initiation of degradation*

The degradation of micro- and nano-plastics by photocatalysis begins with the absorption of UV or visible light by the semiconductor photocatalyst, such as titanium dioxide (TiO_2), zinc oxide (ZnO), or hybrid 2D-smart materials like graphene-based composites. Upon light absorption, electrons in the valence band (VB) of the photocatalyst are excited to the conduction band (CB), leaving behind holes in the VB. The excited electrons and holes migrate to the surface, where they interact with water and oxygen molecules adsorbed on the surface of the photocatalyst to generate ROS (He et al. 2023). The hydroxyl radical (•OH) is one of the most potent ROS involved in the degradation of micro- and nano-plastics. These radicals are capable of abstracting hydrogen atoms from the polymer chains, leading to the formation of carbon-centered radicals. These carbon-centered radicals undergo further reactions with oxygen to form peroxyl radicals, which decompose into smaller molecular fragments such as alcohols, ketones and carboxylic acids. This process, known as oxidative degradation, progressively breaks down the plastic polymers into smaller, less complex molecules that can eventually be mineralized into carbon dioxide (CO_2) and water (H_2O) (He et al. 2023).

10.4.2. *Role of photocatalyst morphology and composition*

The efficiency of photocatalytic degradation of micro- and nano-plastics is significantly influenced by the morphology and composition of the photocatalyst. Nanostructured photocatalysts with high surface area, such as nanorods, nanosheets and hierarchical structures, provide more active sites for the adsorption of pollutants and the generation of ROS. Additionally, the incorporation of co-catalysts, such as noble metals (e.g. platinum and gold) or other semiconductors, can enhance charge separation and increase the production of ROS, thereby improving degradation efficiency. Hybrid 2D-smart materials, which combine the properties of 2D materials like graphene or MoS_2 with other catalytic components, offer unique advantages for photocatalysis. For example, the high conductivity of graphene can facilitate rapid electron transfer, reducing the recombination rate of electron–hole pairs and enhancing the generation of ROS. Moreover, the tunable bandgap of 2D materials allows for the absorption of a broader range of the solar spectrum, making them suitable for visible-light-driven photocatalysis (Chen et al. 2017).

10.4.3. *Pathways of degradation*

The photocatalytic degradation of micro- and nano-plastics can proceed through various pathways depending on the nature of the polymer, the type of photocatalyst and the environmental conditions. Common degradation pathways include the following:

1) Direct photocatalytic oxidation: in this pathway, ROS generated on the photocatalyst surface directly interact with the polymer chains of the plastics. The •OH radicals abstract hydrogen atoms from the polymer, initiating chain scission and the formation of smaller molecular fragments. This process can lead to complete mineralization of the plastics, converting them into CO_2 and H_2O.

2) Sensitized photocatalysis: some photocatalysts can transfer energy to adsorbed oxygen molecules, generating singlet oxygen ($1O^2$), a highly reactive form of oxygen that can oxidize plastic polymers. Sensitized photocatalysis is particularly effective for the degradation of aromatic polymers, such as polystyrene, where singlet oxygen can react with the aromatic rings, leading to ring-opening and polymer chain cleavage.

3) Photodegradation with co-catalysts: the presence of co-catalysts or dopants in the photocatalyst can introduce new degradation pathways. For example, the doping of TiO_2 with nitrogen can enhance visible-light absorption and promote the formation of ROS through alternate reaction mechanisms. Similarly, the inclusion of transition metal oxides can facilitate multi-electron transfer processes, increasing the overall degradation rate.

4) Degradation via surface plasmon resonance: photocatalysts such as gold or silver nanoparticles exhibit surface plasmon resonance (SPR) under light irradiation, leading to the generation of hot electrons. These hot electrons can participate in redox reactions with adsorbed oxygen and water molecules, producing ROS that degrade micro- and nano-plastics.

10.4.4. *Environmental factors and degradation efficiency*

The efficiency of photocatalytic degradation is also influenced by various environmental factors, such as pH, temperature and the presence of natural organic matter (NOM). The pH of the solution can affect the surface charge of the photocatalyst and the adsorption of plastic particles, influencing the generation of ROS. For instance, acidic conditions can enhance the formation of •OH radicals, increasing the degradation rate of certain plastics. Temperature plays a dual role in photocatalytic degradation. While higher temperatures can increase the kinetic energy of molecules, leading to more frequent collisions and reactions, they can also promote the recombination of electron–hole pairs, reducing ROS generation. Thus, an optimal temperature range must be maintained to achieve efficient degradation. The presence of NOM, such as humic acids, can act as a scavenger for ROS, reducing the availability of these reactive species for plastic degradation. However, NOM can also sensitize the photocatalytic process by absorbing light and transferring energy to the photocatalyst, generating additional ROS. The net effect of NOM on degradation efficiency depends on its concentration and the type of photocatalyst used.

10.5. Degradation of micro-plastics in marine environments

One of the most pressing environmental challenges is the accumulation of micro-plastics in marine environments. In a study conducted by Ariza-Tarazona et al. (2023), the degradation of polyethylene terephthalate (PET) micro-plastics in marine environments was achieved using bio-inspired C,N-TiO_2/SiO_2 photocatalysts under visible-light photocatalysis. The results showed that the novel photocatalysts were effective in degrading PET micro-plastics in an aqueous environment with mass loss values ranging between 9.35% and 16.22% after 120 h of irradiation. The process was conducted at room temperature and evaluated at two pH levels (6 and 8), which closely reflect natural conditions in marine environments. Photocatalysis proved to be a clean and efficient approach to reducing the environmental impact of micro-plastics, as the mass loss was attributed to the breakdown of PET's molecular structure, verified by ATR-FTIR analysis in Figure 10.2. This demonstrates the potential of 2D hybrid nanomaterials like C,N-TiO_2/SiO_2 for marine pollution remediation, especially in tackling the persistence of micro-plastics in aquatic ecosystems.

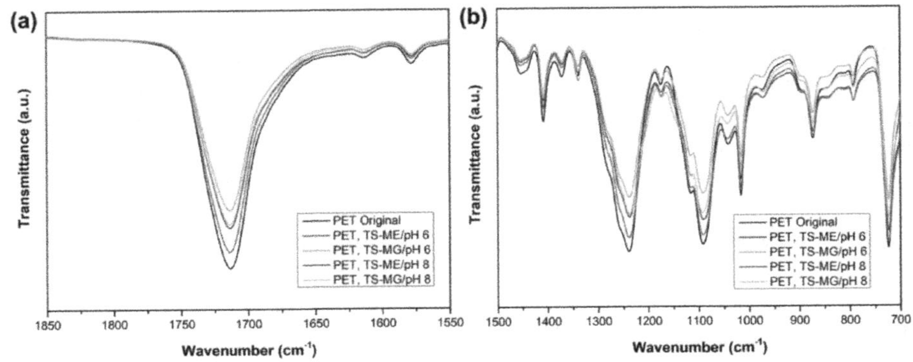

Figure 10.2. *FTIR spectra of before and after photocatalytic degradation (reproduced with permission from Ariza-Tarazona et al. (2023)). For a color version of this figure, see www.iste.co.uk/singh/photodegradation.zip*

In a study by He et al. (2023), TiO_2/ZnO tetrapod photocatalysts were employed for the degradation of various types of micro-plastics, specifically polyethylene (PE) micro-particles and polyester (PES) microfibers, under UV illumination. The findings revealed that the degradation rates were influenced by the morphology of the plastics, with PE showing complete degradation after 480 h and PES after 624 h. Environmental micro-plastics, primarily composed of polypropylene (PP), also exhibited complete mass loss after 816 h. The study further demonstrated that the addition of sodium persulfate as an electron scavenger significantly enhanced the photocatalytic activity by preventing charge carrier recombination as shown in degradation study in Figure 10.3. This method proved effective in reducing micro-plastic pollution in marine environments, showcasing the potential of TiO_2/ZnO nanomaterials for environmental remediation in aerospace-related water systems, where contamination poses a significant risk.

In a recent study on the photocatalytic degradation of micro-plastics in water, polyester micro-plastics were modified with a photocatalytic g-C_3N_4/TiO_2 composite to enhance their degradation under solar illumination. The results demonstrated that polyester micro-plastics modified with 5% g-C_3N_4/TiO_2 exhibited a remarkable degradation rate, achieving 89% mass loss after 400 h of exposure to simulated sunlight as shown in Figure 10.4. This degradation was attributed to the photocatalytic activity of TiO_2, which generates hydroxyl radicals (•OH) and superoxide radicals ($•O_2^-$), effectively breaking down the polyester's molecular chains. The g-C_3N_4 component provided additional visible light absorption and improved charge separation, further enhancing the degradation process. This study highlights the potential of photocatalytic coatings to degrade micro-plastics in aquatic environments, offering a promising solution for reducing plastic pollution in water bodies (Zhong et al. 2023).

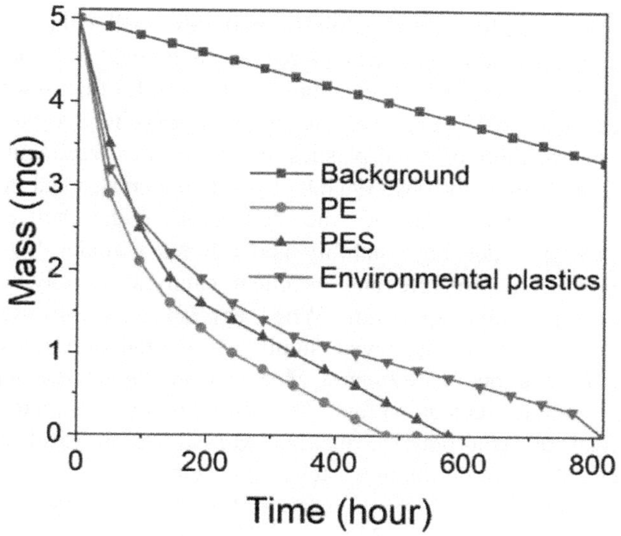

Figure 10.3. *Degradation of micro-plastics using TiO_2/ZnO photocatalyst. The experiment was conducted with 400 mg of $Na_2S_2O_8$ under 365 nm UV light at an intensity of 480 mW/cm² (reproduced with permission from He et al. (2023)). For a color version of this figure, see www.iste.co.uk/singh/photodegradation.zip*

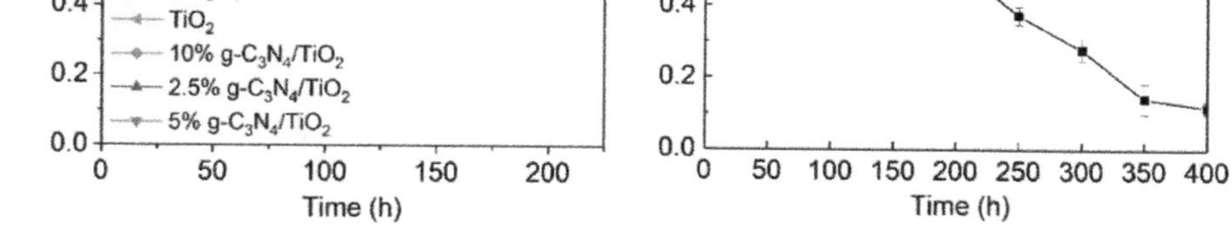

Figure 10.4. *Weight reduction of polyester fibers under simulated light exposure. (a) Comparison of polyester fibers with various catalysts. (b) Weight loss of polyester fibers containing 5% $g\text{-}C_3N_4/TiO_2$ catalyst (reproduced with permission from Zhong et al. (2023)). For a color version of this figure, see www.iste.co.uk/singh/photodegradation.zip*

10.5.1. *Photocatalytic degradation of nano-plastics in wastewater treatment*

Nano-plastics, typically defined as plastic particles smaller than 1,000 nm, are increasingly being detected in wastewater streams. Their small size and high surface area make them more reactive and potentially more harmful than micro-plastics. A study by (Chen et al. 2017) explored the use of MoS_2-TiO_2 composites for the photocatalytic degradation of nano-plastics in wastewater treatment plants. The study demonstrated that the MoS_2/TiO_2 hybrid material exhibited enhanced photocatalytic performance due to the synergetic effect between the two components. TiO_2 provides high stability and efficiency under UV light, while MoS_2, with its layered structure and excellent catalytic activity, extends light absorption into the visible spectrum. When applied to wastewater containing nano-plastics, the composite achieved significant degradation rates, with TiO_2 facilitating the production of ROS and MoS_2 enhancing charge separation and reducing electron–hole recombination. This combination resulted in a highly effective removal of nano-plastics, making MoS_2/TiO_2 an ideal candidate for addressing micro-plastic pollution in wastewater systems.

Hybrid 2D materials, such as MoS_2, graphene and g-C_3N_4, provide efficient degradation of organic pollutants through visible-light-driven photocatalytic reactions. These materials have been extensively studied for their ability to absorb solar energy, promote electron–hole separation and accelerate the transfer of charge carriers as shown in Figure 10.5. By hybridizing different 2D materials, it is possible to enhance their photocatalytic efficiency for degrading nano-plastics in aquatic environments. For instance, MoS_2 hybridized with graphene facilitates faster electron transfer, resulting in improved photodegradation rates of organic pollutants, making them ideal for application in wastewater treatment.

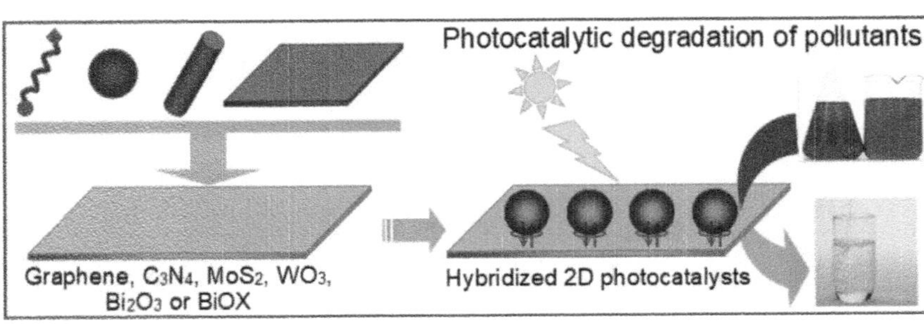

Figure 10.5. *Schematic representation of the hybridization of functional molecules and 0D, 1D, and 2D nanostructures on nanosheets, highlighting their improved photocatalytic efficiency in the degradation of organic pollutants using solar radiation (reproduced with permission from Guan et al. (2020)). For a color version of this figure, see www.iste.co.uk/singh/photodegradation.zip*

10.5.2. *Integration of photocatalytic coatings in water purification systems*

Beyond laboratory-scale studies, the integration of hybrid 2D-smart materials into practical water purification systems represents a significant application. Photocatalytic coatings based on graphene-TiO_2 composites have been developed for use in water treatment facilities, where they serve as active layers on filtration membranes or as stand-alone coatings on reactor walls. A novel graphene oxide (GO)-based nanofiltration membrane, intercalated with TiO_2 nanorods, demonstrated remarkable self-cleaning properties under visible light irradiation. The intercalation of TiO_2 nanorods not only expanded the interlayer spacing of GO, improving water flux by approximately 400%, but also provided significant photocatalytic activity, enabling the degradation of organic contaminants, including dyes such as methylene blue (MB), rhodamine B (RhB) and methyl orange (MO). The membrane achieved a rejection rate of over 99% for these dyes while maintaining high flux, even after multiple filtration cycles. This self-cleaning ability, driven by photocatalytic reactions, significantly reduces membrane fouling, making the system more efficient for long-term water treatment applications (Liu et al. 2020).

One notable application involves the use of graphene-TiO_2 coated membranes have shown remarkable improvements in water purification performance, especially in desalination. A novel thin-film nanocomposite reverse osmosis (TFN-RO) membrane was developed by incorporating reduced graphene oxide (rGO) and TiO_2 nanocomposites into its polyamide layer. This modification significantly enhanced the membranes hydrophilicity and antifouling properties, as well as its resistance to chlorine degradation. The study demonstrated that an optimized concentration of 0.02 wt.% rGO/TiO_2 resulted in a water flux of 51.3 L/m^2h and a salt rejection of 99.45%, outperforming conventional membranes. Furthermore, the modified membrane exhibited exceptional resistance to fouling, maintaining 75% of its water flux after filtration of bovine serum albumin (BSA) solution, compared to only 49% in unmodified membranes. The incorporation of rGO/TiO_2 also improved the membrane's chlorine resistance, with only a 3% drop in salt rejection after chlorination, compared to a 30% decrease in the unmodified RO membrane (Safarpour et al. 2015).

10.5.3. *Photocatalytic degradation of micro-plastics in agricultural soils*

Micro-plastics are increasingly being found in agricultural soils, where they can affect soil health and crop productivity. ZnO-graphene composites have shown significant potential in degrading micro-plastics due to their enhanced photocatalytic properties. The synergy between ZnO high redox potential and graphene excellent electron mobility enhances the separation of photo-generated charge carriers, making the composite highly effective in breaking down micro-plastics under UV and visible light irradiation. Safarpour et al. (2015) studied ZnO-graphene composites demonstrated superior degradation performance in aqueous environments, facilitating the breakdown of polymer chains through ROS like hydroxyl radicals (•OH). These radicals attack the polymer surface, causing scission and oxidation, resulting in the degradation of micro-plastics into smaller, less harmful molecules such as CO_2 and H_2O. This innovative approach highlights the composites potential for reducing micro-plastic pollution in aquatic systems. The mechanism of the degradation is shown in Figure 10.6.

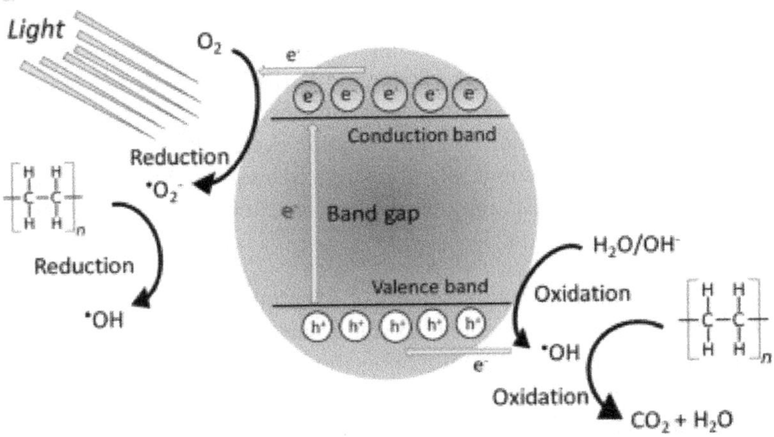

Figure 10.6. *Mechanism of photocatalytic degradation of micro-plastics (Safarpour et al. 2015). For a color version of this figure, see www.iste.co.uk/singh/photodegradation.zip*

10.6. Challenges, limitations and future scopes

The application of hybrid 2D-smart materials for the photocatalytic degradation of micro- and nano-plastics offers significant potential, but there are several challenges that hinder their widespread use. One major issue is the complexity of synthesizing these materials. Techniques such as CVD, hydrothermal methods and sol-gel processes demand precise control over reaction conditions, making large-scale production difficult. Additionally, the cost of synthesis, especially when expensive or rare precursors are involved, can be prohibitively high, posing economic challenges for extensive environmental applications. Another key challenge is the environmental stability and durability of these materials. Many 2D materials, such as TMDs and MOFs, are sensitive to environmental factors such as moisture, oxygen and UV radiation, which can degrade their structure and reduce their photocatalytic efficiency over time. This degradation becomes particularly concerning in harsh environments, such as marine settings or industrial wastewater, where materials could lose effectiveness or release harmful by-products. Improving the stability of these materials through surface modifications is an ongoing area of research.

Photocatalytic efficiency and selectivity also pose limitations. The recombination of photogenerated electron–hole pairs often reduces the availability of reactive species needed for the efficient degradation of plastics. While strategies such as doping or coupling with other materials have been explored, achieving consistently high efficiency remains difficult. Additionally, the selectivity of these materials varies depending on the type of plastic being targeted, which complicates their application across different plastic pollutants. There are also concerns regarding the environmental impact and potential toxicity of the materials themselves. The release of nanoparticles or degradation by-products from photocatalysts may pose risks to ecosystems and human health. Research into the ecotoxicity of 2D materials, such as GO and metal nanoparticles, has shown potential harmful effects on aquatic organisms and oxidative stress in cells. Comprehensive environmental impact assessments are needed to address these risks. Finally, economic and regulatory barriers play a role in limiting the adoption of hybrid 2D-smart materials. The high cost of synthesis, along with the need for specialized equipment and infrastructure, makes it difficult for industries to adopt these technologies on a large scale. Additionally, the regulatory framework for the use of nanomaterials in environmental applications is still developing, leading to uncertainty around safety standards and environmental guidelines. Addressing these challenges requires researchers to collaborate, and industry stakeholders and regulatory bodies to ensure the safe and effective use of these materials in environmental remediation.

The field of photocatalytic degradation of micro- and nano-plastics is rapidly evolving, with ongoing research focused on enhancing the efficiency and scalability of this technology. Future directions include the development of photocatalysts with broader light absorption ranges, such as those activated by near-infrared light, to utilize a greater portion of the solar spectrum. Additionally, the integration of photocatalysts into membrane systems or reactors could facilitate continuous degradation processes, making the technology more practical for large-scale applications. Advancements in material science, such as the creation of defect-engineered 2D materials or the incorporation of single-atom catalysts, are expected to lead to the next generation of photocatalysts with superior activity and selectivity for plastic degradation. Furthermore, understanding the fate of degradation by-products and ensuring that they do not pose additional environmental risks will be critical for the successful implementation of photocatalytic technologies in environmental remediation.

10.7. Conclusions

The field of hybrid 2D-smart materials presents a transformative opportunity in addressing the environmental challenges posed by micro- and nano-plastic pollution. As demonstrated throughout this chapter, the unique properties of 2D materials – such as their large surface area, enhanced charge separation and tunable electronic and optical characteristics – enable them to act as highly efficient photocatalysts. Hybrid systems, which combine 2D materials with other nanostructures, amplify these properties, resulting in superior photocatalytic performance for the degradation of persistent plastic pollutants. Photocatalytic degradation, driven by light energy, offers a sustainable approach to breaking down micro- and nano-plastics into nontoxic by-products. Through various hybrid systems, such as those combining graphene, metal oxides and TMDs, significant advancements have been made in enhancing the efficiency, stability and applicability of these photocatalysts under real-world conditions. These systems are not only effective in aquatic environments but also show potential in other settings, such as wastewater treatment and agricultural soils. Despite these promising developments, challenges remain in the large-scale production and practical deployment of hybrid 2D materials.

The high cost of synthesis, potential environmental toxicity of nanomaterials and durability under harsh conditions are issues that require further research. Nevertheless, ongoing advancements in material science, such as defect engineering and the integration of single-atom catalysts, suggest that these obstacles can be overcome, leading to more effective and scalable solutions. The hybrid 2D-smart materials offer a viable path forward in mitigating the global crisis of micro- and nano-plastic pollution. By harnessing their unique capabilities in photocatalytic degradation, these materials have the potential to significantly reduce plastic pollution, contributing to a cleaner and more sustainable environment.

10.8. References

Akyüz, D. (2021). rGO-TiO_2-CdO-ZnO-Ag photocatalyst for enhancing photocatalytic degradation of methylene blue. *Optical Materials*, 116, 111090. doi: 10.1016/j.optmat. 2021.111090.

Alam, S., Asaduzzaman Chowdhury, M., Shahid, A., Alam, R., Rahim, A. (2021). Synthesis of emerging two-dimensional (2D) materials – Advances, challenges and prospects. *FlatChem*, 30, 100305. doi: 10.1016/j.flatc.2021.100305.

Andrady, A.L. (2011). Microplastics in the marine environment. *Marine Pollution Bulletin*, 62(8), 1596–1605. doi: 10.1016/j.marpolbul.2011.05.030.

Anegbe, B., Ifijen, I.H., Maliki, M., Uwidia, I.E., Aigbodion, A.I. (2024). Graphene oxide synthesis and applications in emerging contaminant removal: A comprehensive review. *Environmental Sciences Europe*, 36, 15. doi: 10.1186/s12302-023-00814-4.

Ariza-Tarazona, M.C., Siligardi, C., Carreón-López, H.A., Valdéz-Cerda, J.E., Pozzi, P., Kaushik, G., Villarreal-Chiu, J.F., Cedillo-González, E.I. (2023). Low environmental impact remediation of microplastics: Visible-light photocatalytic degradation of PET microplastics using bio-inspired C,N-TiO_2/SiO_2 photocatalysts. *Marine Pollution Bulletin*, 193, 115206. doi: 10.1016/j.marpolbul.2023.115206.

Askari, M.B., Salarizadeh, P., Veisi, P., Samiei, E., Saeidfirozeh, H., Tourchi Moghadam, M.T., Di Bartolomeo, A. (2023). Transition-metal dichalcogenides in electrochemical batteries and solar cells. *Micromachines*, 14(3), 691. doi: 10.3390/mi14030691.

Baig, N. (2023). Two-dimensional nanomaterials: A critical review of recent progress, properties, applications, and future directions. *Composites Part A*, 165, 107362. doi: 10.1016/j.compositesa.2022.107362.

Bonaccorso, F., Lombardo, A., Hasan, T., Sun, Z., Colombo, L., Ferrari, A.C. (2012). Production and processing of graphene and 2D crystals. *Materials Today*, 15(12), 564–589. doi: 10.1016/S1369-7021(13)70014-2.

Cai, D., Tao, E., Yang, S., Ma, Z., Li, Y., Liu, L., Wang, D., Qian, J. (2022). Effect of mixed-phase TiO_2 doped with Ca^{2+} on charge transfer at the TiO_2/graphene interface. *Electrochimica Acta*, 422, 140503. doi: 10.1016/j.electacta.2022.140503.

Chaves, A., Azadani, J.G., Alsalman, H., da Costa, D.R., Frisenda, R., Chaves, A.J., Song, S.H., Kim, Y.D., He, D., Zhou, J. et al. (2020). Bandgap engineering of two-dimensional semiconductor materials. *npj 2D Materials and Applications*, 4, 29. doi: 10.1038/s41699-020-00162-4.

Chen, B., Meng, Y., Sha, J., Zhong, C., Hu, W., Zhao, N. (2017). Preparation of MoS_2/TiO_2 based nanocomposites for photocatalysis and rechargeable batteries: Progress, challenges, and perspective. *Nanoscale*, 10(1), 34–68. doi: 10.1039/c7nr07366f.

Chen, E., Xu, W., Chen, J., Warner, J.H. (2020a). 2D layered noble metal dichalcogenides (Pt, Pd, Se, S) for electronics and energy applications. *Materials Today Advances*, 7, 100076. doi: 10.1016/j.mtadv.2020.100076.

Chen, X.K., Zeng, Y.J., Chen, K.Q. (2020b). Thermal transport in two-dimensional heterostructures. *Frontiers in Materials*, 7, 1–30. doi: 10.3389/fmats.2020.578791.

Chen, V., Shin, Y.C., Mikheev, E., Lin, Q., Martis, J., Zhang, Z., Chatterjee, S., Majumdar, A., Philip Wong, H.S., Goldhaber-Gordon, D. et al. (2021). Application-driven synthesis and characterization of hexagonal boron nitride deposited on metals and carbon nanotubes. *2D Materials*, 8(4), 1–16. doi: 10.1088/2053-1583/ac10f1.

Cho, Y.S. and Kang, J. (2024). Two-dimensional materials as catalysts, interfaces, and electrodes for an efficient hydrogen evolution reaction. *Nanoscale*, 16, 3936–3950. doi: 10.1039/d4nr00147h.

Choi, W., Choudhary, N., Han, G.H., Park, J., Akinwande, D., Lee, Y.H. (2017). Recent development of two-dimensional transition metal dichalcogenides and their applications. *Materials Today*, 20(3), 116–130. doi: 10.1016/j.mattod.2016.10.002.

Choudhary, N., Islam, M.A., Kim, J.H., Ko, T.J., Schropp, A., Hurtado, L., Weitzman, D., Zhai, L., Jung, Y. (2018). Two-dimensional transition metal dichalcogenide hybrid materials for energy applications. *nanotoday*, 19, 16–40. doi: 10.1016/j.nantod.2018.02.007.

Durga Devi, A., Pushpavanam, S., Singh, N., Verma, J., Kaur, M.P., Roy, S.C. (2022). Enhanced methane yield by photoreduction of CO_2 at moderate temperature and pressure using Pt coated, graphene oxide wrapped TiO_2 nanotubes. *Results in Engineering*, 14, 100441. doi: 10.1016/j.rineng.2022.100441.

Dutta, T., Yadav, N., Wu, Y., Cheng, G.J., Liang, X., Ramakrishna, S., Sbai, A., Gupta, R., Mondal, A., Hongyu, Z., Yadav, A. (2024). Electronic properties of 2D materials and their junctions. *Nano Materials Science*, 6(1), 1–23. doi: 10.1016/j.nanoms.2023.05.003.

Fu, D., Han, G., Chang, Y., Dong, J. (2012). The synthesis and properties of ZnO-graphene nano hybrid for photodegradation of organic pollutant in water. *Materials Chemistry and Physics*, 132, 673–681. doi: 10.1016/j.matchemphys.2011.11.085.

Fu, Y., Hansson, J., Liu, Y., Chen, S., Zehri, A., Samani, M.K., Wang, N., Ni, Y., Zhang, Y. Zhang, Z.B. et al. (2020). Graphene related materials for thermal management. *2D Materials*, 7(1), 012001. doi: 10.1088/2053-1583/ab48d9.

Georgakilas, V., Otyepka, M., Bourlinos, A.B., Chandra, V., Kim, N., Kemp, K.C., Hobza, P., Zboril, R., Kim, K.S. (2012). Functionalization of graphene: Covalent and non-covalent approaches, derivatives and applications. *Chemical Reviews*, 112(11), 6156–6214. doi: 10.1021/cr3000412.

González, V.J., Vázquez, E., Villajos, B., Tolosana-Moranchel, A., Duran-Valle, C., Faraldos, M., Bahamonde, A. (2022). Eco-friendly mechanochemical synthesis of titania-graphene nanocomposites for pesticide photodegradation. *Separation and Purification Technology*, 289, 120638. doi: 10.1016/j.seppur.2022.120638.

Grubišić-Čabo, A., Michiardi, M., Sanders, C.E., Bianchi, M., Curcio, D., Phuyal, D., Berntsen, M.H., Guo, Q., Dendzik, M. (2023). In situ exfoliation method of large-area 2D materials. *Advancement of Science*, 10(22), 1–9. doi: 10.1002/advs.202301243.

Guan, G., Ye, E., You, M., Li, Z. (2020). Hybridized 2D nanomaterials toward highly efficient photocatalysis for degrading pollutants: Current status and future perspectives. *Small*, 16(19), e1907087. doi: 10.1002/smll.201907087.

Guo, X., Cheng, S., Cai, W., Zhang, Y., Zhang, X. (2021). A review of carbon-based thermal interface materials: Mechanism, thermal measurements and thermal properties. *Materials and Design*, 209, 109936. doi: 10.1016/j.matdes.2021.109936.

He, Y., Rehman, A.U., Xu, M., Not, C.A., Ng, A.M.C., Djurišić, A.B. (2023). Photocatalytic degradation of different types of microplastics by TiOx/ZnO tetrapod photocatalysts. *Heliyon*, 9(11), e22562. doi: 10.1016/j.heliyon.2023.e22562.

Hill, E.H. (2024). Layered 2D material heterostructures – A colloidal perspective. *Journal of Materials Chemistry C*, 12, 11285–11318. doi: 10.1039/D4TC01102C.

Huang, P., Li, Y., Yang, G., Li, Z.X., Li, Y.Q., Hu, N., Fu, S.Y., Novoselov, K.S. (2021). Graphene film for thermal management: A review. *Nano Materials Science*, 3, 1–16. doi: 10.1016/j.nanoms.2020.09.001.

Jiang, M., Zhang, M., Wang, L., Fei, Y., Wang, S., Núñez-Delgado, A., Bokhari, A., Race, M., Khataee, A., Jaromír Klemeš, J.J. et al. (2022). Photocatalytic degradation of xanthate in flotation plant tailings by TiO_2/graphene nanocomposites. *Chemical Engineering Journal*, 431(Part 2), 134104. doi: 10.1016/j.cej.2021.134104.

Juma, I.G., Kim, G., Jariwala, D., Behura, S.K. (2021). Direct growth of hexagonal boron nitride on non-metallic substrates and its heterostructures with graphene. *iScience*, 24(11), 103374. doi: 10.1016/j.isci.2021.103374.

Kalantari, M.H. and Zhang, X. (2022). Thermal transport in 2D materials. *Nanomaterials*, 13(1), 117. doi: 10.3390/nano13010117.

Kang, S., Eshete, Y.A., Lee, S., Won, D., Im, S., Lee, S., Cho, S., Yang, H. (2022a). Bandgap modulation in the two-dimensional core-shell-structured monolayers of WS_2. *iScience*, 25(1), 103563. doi: 10.1016/j.isci.2021.103563.

Kang, T., Tang, T.W., Pan, B., Liu, H., Zhang, K., Luo, Z. (2022b). Strategies for controlled growth of transition metal dichalcogenides by chemical vapor deposition for integrated electronics. *ACS Materials Au*, 2(6), 665–685. doi: 10.1021/acsmaterialsau.2c00029.

Kaushik, T., Ghosh, S., Dolkar, T., Biswas, R., Dutta, A. (2024). Noble metal plasmon-molecular catalyst hybrids for renewable energy relevant small molecule activation. *ACS Nanoscience Au*, 4(5), 273–289. doi: 10.1021/acsnanoscienceau.4c00009.

Khavar, A.H.C., Khazaee, Z., Mahjoub, A.R., Nejat, R. (2022). TiO_2 supported-reduced graphene oxide co-doped with gallium and sulfur as an efficient heterogeneous catalyst for the selective photochemical oxidation of alcohols; DFT and mechanism insights. *Journal of Photochemistry and Photobiology A*, 431, 114020. doi: 10.1016/j.jphotochem.2022.114020.

Khdary, N.H., Almuarqab, B.T., El Enany, G. (2023). Nanoparticle-embedded polymers and their applications: A review. *Membranes*, 13(5), 537. doi: 10.3390/membranes13050537.

Kuila, T., Bose, S., Mishra, A.K., Khanra, P., Kim, N.H., Lee, J.H. (2012). Chemical functionalization of graphene and its applications. *Progress in Materials Science*, 57, 1061–1105. doi: 10.1016/j.pmatsci.2012.03.002.

Kumbhakar, P., Jayan, J.S., Sreedevi Madhavikutty, A., Sreeram, P.R., Saritha, A., Ito, T., Tiwary, C.S. (2023). Prospective applications of two-dimensional materials beyond laboratory frontiers: A review. *iScience*, 26(5), 106671. doi: 10.1016/j.isci.2023.106671.

Laturia, A., Van de Put, M.L., Vandenberghe, W.G. (2018). Dielectric properties of hexagonal boron nitride and transition metal dichalcogenides: From monolayer to bulk. *npj 2D Materials and Applications*, 2(6). doi: 10.1038/s41699-018-0050-x.

Lee, D.E., Kim, M.K., Danish, M., Jo, W.K. (2023). State-of-the-art review on photocatalysis for efficient wastewater treatment: Attractive approach in photocatalyst design and parameters affecting the photocatalytic degradation. *Catalysis Communications*, 183, 106764. doi: 10.1016/j.catcom.2023.106764.

Lei, Y., Zhang, T., Lin, Y.C., Granzier-Nakajima, T., Bepete, G., Kowalczyk, D.A., Lin, Z., Zhou, D., Schranghamer, T.F., Dodda, A. et al. (2022). Graphene and beyond: Recent advances in two-dimensional materials synthesis, properties, and devices. *ACS Nanoscience Au*, 2(6), 450–485. doi: 10.1021/acsnanoscienceau.2c00017.

Li, X., Yu, J., Wageh, S., Al-Ghamdi, A.A., Xie, J. (2016). Graphene in photocatalysis: A review. *Small*, 12(48), 6640–6696. doi: 10.1002/smll.201600382.

Li, H., Jia, X., Zhang, Q., Wang, X. (2018). Metallic transition-metal dichalcogenide nanocatalysts for energy conversion. *Chem*, 4(7), 1510–1537. doi: 10.1016/j.chempr.2018.03.012.

Li, W., Wang, J., Xiong, Z., Li, D. (2023). Micro/nanorobots for efficient removal and degradation of micro/nanoplastics. *Cell Reports Physical Science*, 4(11), 101639. doi: 10.1016/j.xcrp.2023.101639.

Lin, S., Li, Y., Qian, J., Lau, S.P. (2019). Emerging opportunities for black phosphorus in energy applications. *Materials Today Energy*, 12, 1–25. doi: 10.1016/j.mtener.2018.12.004.

Liu, Y., Yu, Z., Peng, Y., Shao, L., Li, X., Zeng, H. (2020). A novel photocatalytic self-cleaning TiO_2 nanorods inserted graphene oxide-based nanofiltration membrane. *Chemical Physics Letters*, 749, 137424. doi: 10.1016/j.cplett.2020.137424.

Liu, L., Xu, M., Ye, Y., Zhang, B. (2022). On the degradation of (micro)plastics: Degradation methods, influencing factors, environmental impacts. *Science of the Total Environment*, 806(Part 3), 151312. doi: 10.1016/j.scitotenv.2021.151312.

Maggini, L. and Ferreira, R.R. (2021). 2D material hybrid heterostructures: Achievements and challenges towards high throughput fabrication. *Journal of Materials Chemistry C*, 9(44), 15721–15734. doi: 10.1039/D1TC04253j.

Maity, D., Sahoo, S.R., Saha, S. (2023). Synthesis and characterization of nanomaterials for electrochemical sensors. *ACS Symposium Series*, 1437, 193–222. doi: 10.1021/bk-2023-1437.ch009.

Mehta, S., Thakur, R., Rani, S., Nagaraja, B.M., Mehla, S., Kainthla, I. (2024). Recent advances in ternary transition metal dichalcogenides for electrocatalytic hydrogen evolution reaction. *International Journal of Hydrogen Energy*, 82, 1061–1080. doi: 10.1016/j.ijhydene.2024.08.051.

Mirikaram, N., Pérez-Molina, Á., Morales-Torres, S., Salemi, A., Maldonado-Hódar, F.J., Pastrana-Martínez, L.M. (2021). Photocatalytic performance of ZnO-graphene oxide composites towards the degradation of vanillic acid under solar radiation and visible-LED. *Nanomaterials*, 11(6), 1576. doi: 10.3390/nano11061576.

Mishra, K., Devi, N., Siwal, S.S., Gupta, V.K., Thakur, V.K. (2023). Hybrid semiconductor photocatalyst nanomaterials for energy and environmental applications: Fundamentals, designing, and prospects. *Advanced Sustainable Systems*, 7(8), 2300095. doi: 10.1002/adsu.202300095.

Nandee, R., Chowdhury, M.A., Shahid, A., Hossain, N., Rana, M. (2022). Band gap formation of 2D material in graphene: Future prospect and challenges. *Results in Engineering*, 15, 100474. doi: 10.1016/j.rineng.2022.100474.

Padmanabhan, N.T., Thomas, N., Louis, J., Mathew, D.T., Ganguly, P., John, H., Pillai, S.C. (2021). Graphene coupled TiO_2 photocatalysts for environmental applications: A review. *Chemosphere*, 271, 129506. doi: 10.1016/j.chemosphere.2020.129506.

Peng, Z., Chen, X., Fan, Y., Srolovitz, D.J., Lei, D. (2020). Strain engineering of 2D semiconductors and graphene: From strain fields to band-structure tuning and photonic applications. *Light, Science and Applications*, 9(190). doi: 10.1038/s41377-020-00421-5.

Porcu, S., Secci, F., Ricci, P.C. (2022). Advances in hybrid composites for photocatalytic applications: A review. *Molecules*, 27(20), 6828. doi: 10.3390/molecules27206828.

Pospischil, A. and Mueller, T. (2016). Optoelectronic devices based on atomically thin transition metal dichalcogenides. *Applied Sciences*, 6(3), 1–19. doi: 10.3390/app6030078.

Presolski, S. and Pumera, M. (2016). Covalent functionalization of MoS_2. *Materials Today*, 19(3), 140–145. doi: 10.1016/j.mattod.2015.08.019.

Qiao, H., Liu, H., Huang, Z., Hu, R., Ma, Q., Zhong, J., Qi, X. (2021). Tunable electronic and optical properties of 2D monoelemental materials beyond graphene for promising applications. *Energy and Environmental Materials*, 4(4), 522–543. doi: 10.1002/eem2.12154.

Quiroz-Cardoso, O., Suárez, V., Oros-Ruiz, S., Quintana, M., Ramírez-Rave, S., Suárez-Quezada, M., Gómez, R. (2022). Synthesis of Ni/GO-TiO_2 composites for the photocatalytic hydrogen production and CO_2 reduction to methanol. *Topics in Catalysis*, 65, 1015–1027. doi: 10.1007/s11244-022-01643-0.

Raza, A., Ikram, M., Qumar, U., Rafiq, A. (2022). Hybrid 2D nanomaterials for photocatalytic degradation of wastewater pollutants. In *Innovative Nanocomposites for the Remediation and Decontamination of Wastewater*, Kumar, A. (ed.). IGI Global, Pennsylvania. doi: 10.4018/978-1-6684-4553-2.ch005.

Safarpour, M., Khataee, A., Vatanpour, V. (2015). Thin film nanocomposite reverse osmosis membrane modified by reduced graphene oxide/TiO_2 with improved desalination performance. *Journal of Membrane Science*, 489, 43–54. doi: 10.1016/j.memsci.2015.04.010.

Saleem, Z., Pervaiz, E., Yousaf, M.U., Niazi, M.B.K. (2020). Two-dimensional materials and composites as potential water splitting photocatalysts: A review. *Catalysts*, 10(4), 464. doi: 10.3390/catal10040464.

Shanker, G.S., Biswas, A., Ogale, S. (2021). 2D materials and their heterostructures for photocatalytic water splitting and conversion of CO_2 to value chemicals and fuels. *Journal of Physics: Energy*, 3(2), 022003. doi: 10.1088/2515-7655/abdcab.

Singh, A., Dhau, J., Kumar, R., Badru, R., Singh, P., Kumar Mishra, Y.K., Kaushik, A. (2024). Tailored carbon materials (TCM) for enhancing photocatalytic degradation of polyaromatic hydrocarbons. *Progress in Materials Science*, 144, 101289. doi: 10.1016/j.pmatsci.2024.101289.

Supriya, S., Senapati, S., Naik, R. (2025). 2D chalcogenide heterostructures: A discussion on its synthesis, properties and emerging applications. *Next Materials*, 7, 100368. doi: 10.1016/j.nxmate.2024.100368.

Tai, X.H., Lai, C.W., Yang, T.C.K., Johan, M.R., Lee, K.M., Chen, C.Y., Juan, J.C. (2022). Highly effective removal of volatile organic pollutants with p-n heterojunction photoreduced graphene oxide-TiO_2 photocatalyst. *Journal of Environmental Chemical Engineering*, 10(2), 107304. doi: 10.1016/j.jece.2022.107304.

Tan, T., Jiang, X., Wang, C., Yao, B., Zhang, H. (2020). 2D material optoelectronics for information functional device applications: Status and challenges. *Advanced Science*, 7(11), 2000058. doi: 10.1002/advs.202000058.

Thompson, R.C., Moore, C.J., vom Saal, F.S.V., Swan, S.H. (2009). Plastics, the environment and human health: Current consensus and future trends. *Philosophical Transactions of the Royal Society of London. Series B, Biological Sciences*, 364(1526), 2153–2166. doi: 10.1098/rstb.2009.0053.

Uddin, M.M., Kabir, M.H., Ali, M.A., Hossain, M.M., Khandaker, M.U., Mandal, S., Arifutzzaman, A., Jana, D. (2023). Graphene-like emerging 2D materials: Recent progress, challenges and future outlook. *RSC Advances*, 13(47), 33336–33375. doi: 10.1039/d3ra04456d.

Winayu, B.N.R., Mao, W.H., Chu, H. (2022). Combination of rGO/S, N/TiO_2 for the enhancement of visible light-driven toluene photocatalytic degradation. *Sustainable Environment Research*, 32(34). doi: 10.1186/s42834-022-00143-w.

Xie, A., Jin, M., Zhu, J., Zhou, Q., Fu, L., Wu, W. (2023). Photocatalytic technologies for transformation and degradation of microplastics in the environment: Current achievements and future prospects. *Catalysts*, 13(5), 846. doi: 10.3390/catal13050846.

Xu, M., Liang, T., Shi, M., Chen, H. (2013). Graphene-like two-dimensional materials. *Chemical Reviews*, 113(5), 3766–3798. doi: 10.1021/cr300263a.

Yaqoob, A.A., Noor, N.H.B.M., Serrà, A., Ibrahim, M.N.M. (2020). Advances and challenges in developing efficient graphene oxide-based zno photocatalysts for dye photo-oxidation. *Nanomaterials*, 10(5), 932. doi: 10.3390/nano10050932.

You, J., Hossain, M.D., Luo, Z. (2018). Synthesis of 2D transition metal dichalcogenides by chemical vapor deposition with controlled layer number and morphology. *Nano Convergence*, 5(26). doi: 10.1186/s40580-018-0158-x.

Yue, S., Zhou, H., Geng, D., Sun, Z., Arita, M., Shimada, K., Cheng, P., Chen, L., Meng, S., Wu, K. et al. (2020). Experimental observation of Dirac cones in artificial graphene lattices. *Physical Review B*, 102, 201401. doi: 10.1103/PhysRevB.102.201401.

Zeng, Y. and Guo, Z. (2021). Synthesis and stabilization of black phosphorus and phosphorene: Recent progress and perspectives. *iScience*, 24(10), 103116. doi: 10.1016/j.isci.2021.103116.

Zhang, H. (2015). Ultrathin two-dimensional nanomaterials. *ACS Nano*, 9(10), 9451–9469. doi: 10.1021/acsnano.5b05040.

Zhang, K., Feng, Y., Wang, F., Yang, Z., Wang, J. (2017). Two dimensional hexagonal boron nitride (2D-hBN): Synthesis, properties and applications. *Journal of Materials Chemistry C*, 5(46), 11992–12022. doi: 10.1039/C7TC04300G.

Zhang, Z., Liu, P., Song, Y., Hou, Y., Xu, B., Liao, T., Zhang, H., Guo, J., Sun, Z. (2022). Heterostructure engineering of 2D superlattice materials for electrocatalysis. *Advancement of Science*, 9(35), 1–26. doi: 10.1002/advs.202204297.

Zheng, W. and Lee, L.Y.S. (2022). Beyond sonication: Advanced exfoliation methods for scalable production of 2D materials. *Matter*, 5(2), 515–545. doi: 10.1016/j.matt.2021.12.010.

Zheng, Q., Durkin, D.P., Elenewski, J.E., Sun, Y., Banek, N.A., Hua, L., Chen, H., Wagner, M.J., Zhang, W., Shuai, D. (2016). Visible-light-responsive graphitic carbon nitride: Rational design and photocatalytic applications for water treatment. *Environmental Science and Technology*, 50(23), 12938–12948. doi: 10.1021/acs.est.6b02579.

Zhong, Y., Zhang, B., Zhu, Z., Wang, G., Mei, X., Fang, Y., Lu, W. (2023). Photocatalytic-driven self-degradation of polyester microplastics under solar light. *Journal of Polymers and the Environment*, 31, 2415–2423. doi: 10.1007/s10924-023-02763-8.

Zhu, Y., Murali, S., Cai, W., Li, X., Suk, J.W., Potts, J.R., Ruoff, R.S. (2010). Graphene and graphene oxide: Synthesis, properties, and applications. *Advanced Materials*, 22(35), 3906–3924. doi: 10.1002/adma.201001068.

11

Design and Structural Modification of Advanced Biomaterials for Photocatalytic Degradation of Micro- and Nano-Plastics

The escalating environmental concerns associated with the persistent accumulation of micro- and nano-plastics in ecosystems demand innovative solutions for their efficient removal. This study focuses on the development of smart biomaterials tailored for photocatalytic degradation of micro- and nano-plastics. Through careful design and structural modifications, these biomaterials aim to harness the power of sunlight to initiate targeted degradation processes, contributing to the remediation of plastic pollution. The biomaterials are engineered with advanced photocatalytic properties, leveraging unique surface functionalities and catalytic centers. This research incorporates cutting-edge nanotechnology to enhance the materials' photo-responsive characteristics and increase their efficiency in capturing and degrading plastic particles. Furthermore, the smart biomaterials exhibit adaptability to environmental conditions, ensuring optimal performance across diverse ecosystems. The study employs a multidisciplinary approach, combining principles from materials science, chemistry and environmental engineering. The structural modifications are guided by a comprehensive understanding of the interaction between the biomaterials and plastic pollutants, aiming to optimize the catalytic processes for enhanced degradation rates. In addition to their photocatalytic capabilities, these biomaterials are designed with an emphasis on sustainability, biocompatibility and biodegradability. The integration of eco-friendly components ensures that the materials contribute positively to the environment throughout their lifecycle, minimizing potential ecological impact. The outcomes of this research hold promising implications for the development of effective and environmentally friendly solutions to address micro- and nano-plastic pollution. The smart biomaterials designed in this study not only exhibit high-performance photocatalytic degradation but also present a step forward in the quest for sustainable materials capable of mitigating the adverse effects of plastic contamination on our ecosystems.

Chapter written by Nisha MANDLOI, Poonam SHARMA, Aakanksha MEWAL and Ajit Kumar VARMA.

11.1. Introduction

The unrelenting rise in plastic pollution, particularly in the form of micro- and nano-plastics, poses an alarming threat to global ecosystems. As conventional plastic degradation methods fall short in addressing the scale and complexity of this issue, there is an urgent need for innovative and sustainable solutions (Alekasandrov 2024; Ratre et al. 2024). In response to this environmental crisis, the focus of this chapter is on the design and structural modification of smart biomaterials for the photocatalytic degradation of micro- and nano-plastics (Okeke et al. 2022). The convergence of materials science, nanotechnology and environmental engineering has paved the way for the development of biomaterials that not only possess advanced photocatalytic properties but are also tailored to interact synergistically with plastic pollutants. Through meticulous design and engineering, these biomaterials aim to harness solar energy efficiently, initiating targeted degradation processes that hold the promise of mitigating the adverse impacts of plastic pollution on our ecosystems (Shukla et al. 2020; Deng 2022; Kathalingam et al. 2023). This chapter explores the multidisciplinary approach adopted in the design process, emphasizing the incorporation of sustainable and eco-friendly elements, and sets the stage for unveiling the transformative potential of these smart biomaterials in combatting the escalating crisis of micro- and nano-plastic pollution.

11.1.1. *Plastic pollution: a global challenge*

Plastic pollution stands as a formidable global challenge, transcending borders and affecting ecosystems across the planet. The pervasive use of plastics in various industries, coupled with inadequate waste management practices, has led to the widespread accumulation of plastic debris in oceans, rivers and terrestrial environments. Micro- and nano-plastics, resulting from the breakdown of larger plastic items, further exacerbate the issue by infiltrating ecosystems and posing potential threats to aquatic and terrestrial life. The persistence of plastics in the environment, taking hundreds of years to degrade, intensifies the urgency of addressing this crisis. Plastic pollution not only jeopardizes biodiversity and ecosystems but also poses risks to human health as plastics enter the food chain. Global efforts are crucial to curb the production, enhance recycling infrastructure and develop innovative solutions for the effective remediation of plastic pollution, safeguarding the health of the planet and its inhabitants (Mazhandu and Muzenda 2019).

11.1.1.1. *Impact on ecosystems*

Plastic pollution exerts profound and often detrimental impacts on ecosystems worldwide. The pervasive presence of plastic debris in oceans, rivers and terrestrial environments poses a direct threat to marine and terrestrial life. Larger plastic items, such as bags and fishing gear, can entangle and suffocate marine animals, leading to injury or death. Micro- and nano-plastics, resulting from the breakdown of larger plastic items, are ingested by a variety of organisms, entering the food chain and potentially causing harm to species at different trophic levels.

The introduction of plastics into ecosystems can alter habitats and disrupt the balance of ecosystems, affecting both aquatic and terrestrial biodiversity. The persistent nature of plastics means that they can accumulate in natural habitats, causing long-term ecological damage. Moreover, the chemicals released during the degradation of plastics can leach into the environment, leading to water and soil contamination.

Plastic pollution also poses indirect threats by serving as a vector for the transport of invasive species. Floating plastic debris can act as rafts, carrying non-native species to new regions and contributing to the alteration of local ecosystems.

The overall impact on ecosystems is multifaceted, ranging from physical harm to wildlife, alteration of habitats and potential disruption to entire food webs. Addressing plastic pollution is essential not only for the preservation of biodiversity but also for the overall health and resilience of ecosystems that support life on Earth. Global efforts to mitigate plastic pollution involve reducing plastic use, improving waste management practices and developing innovative solutions to clean up and remediate affected environments (Iroegbu et al. 2021).

11.1.1.2. *Need for sustainable solutions*

The need for sustainable solutions to address plastic pollution is paramount, given the severe and far-reaching consequences of this environmental crisis. Several key reasons underscore the urgency for sustainable approaches:

1) *Environmental impact*: traditional plastic materials persist in the environment for hundreds of years, causing long-term damage to ecosystems. Sustainable solutions aim to reduce this impact by introducing materials that are biodegradable, compostable or easily recyclable, minimizing the ecological footprint of plastic waste.

2) *Biodiversity conservation*: plastic pollution poses a direct threat to wildlife through entanglement, ingestion and habitat disruption. Sustainable solutions can help protect biodiversity by minimizing harm to species, ensuring the integrity of ecosystems and preventing the introduction of harmful chemicals into food chains.

3) *Human health concerns*: plastics can release harmful chemicals, and microplastics are increasingly found in the food we consume and the water we drink. Sustainable alternatives contribute to reducing human exposure to potentially harmful substances, safeguarding public health.

4) *Resource conservation*: traditional plastic production relies heavily on finite fossil fuel resources. Sustainable solutions, such as bioplastics derived from renewable sources or recycling initiatives, help conserve resources and decrease dependency on non-renewable raw materials.

5) *Waste management challenges*: inefficient waste management systems struggle to handle the vast amounts of plastic waste generated globally. Sustainable solutions emphasize reducing overall plastic consumption, promoting reuse and improving recycling infrastructure to create a more circular economy.

6) *Climate change mitigation*: the production and incineration of plastics contribute to greenhouse gas emissions. Sustainable alternatives, coupled with improved waste management practices, can contribute to mitigating climate change by reducing carbon footprints associated with the plastic lifecycle.

7) *Public awareness and responsibility*: increasing public awareness of the environmental impact of plastic pollution has led to growing demand for sustainable alternatives. Businesses and consumers are now more inclined to choose eco-friendly options, creating a market-driven incentive for the development and adoption of sustainable solutions.

Addressing plastic pollution requires a comprehensive and collaborative effort, involving governments, industries and individuals. Sustainable solutions play a pivotal role in mitigating the environmental, social and economic challenges posed by plastic pollution and are integral to building a more resilient and responsible future (Dumbili and Henderson 2020).

11.1.2. *Photocatalytic degradation: a green approach*

Photocatalytic degradation stands out as a green and innovative approach to address environmental challenges, particularly in the context of pollutant remediation. This method harnesses the power of light, typically sunlight, to activate catalysts that facilitate the breakdown of organic and inorganic compounds. In the realm of environmental sustainability, photocatalysis offers a promising avenue for the degradation of pollutants, including organic dyes, pesticides and emerging contaminants (Zulfikar et al. 2021).

One of the key advantages of photocatalytic degradation lies in its reliance on renewable energy sources. By utilizing sunlight as the driving force, this approach aligns with green principles, reducing dependency on non-renewable energy inputs and minimizing the overall environmental footprint. Furthermore, the catalytic processes involved in photodegradation are typically non-consumptive, allowing for the catalysts to be reused, enhancing the sustainability of the method.

In the context of plastic pollution, photocatalytic degradation has emerged as a potential eco-friendly solution. Smart biomaterials designed with advanced photocatalytic properties can be tailored for the targeted breakdown of micro- and nano-plastics, contributing to the reduction of plastic waste in ecosystems. The selectivity and efficiency of photocatalytic processes make them particularly appealing for the degradation of specific pollutants without generating harmful by-products.

This green approach also holds promise in water and air purification, showcasing its versatility in addressing various environmental challenges. As the scientific community continues to explore and refine photocatalytic degradation techniques, the potential for scalable and sustainable applications across different domains becomes increasingly evident. Embracing photocatalysis as a green solution underscores the importance of leveraging natural processes to combat pollution, aligning with broader efforts toward a more environmentally conscious and sustainable future (Helaïli et al. 2017; Nair et al. 2019).

11.1.2.1. *Advantages over traditional methods*

Photocatalytic degradation, as a green approach, offers several distinct advantages over traditional methods of pollutant remediation. These advantages highlight its potential as an environmentally friendly and sustainable solution:

1) *Renewable energy utilization*: photocatalytic degradation relies on sunlight as a clean and renewable energy source. Unlike traditional methods that often require external energy inputs, the use of solar energy aligns with green principles, reducing dependence on non-renewable resources and minimizing the carbon footprint associated with the degradation process.

2) *Selective and targeted degradation*: photocatalysis enables the selective and targeted degradation of specific pollutants. The catalysts used in the process can be designed to exhibit high specificity for certain contaminants, minimizing the impact

on non-target compounds. This selectivity is advantageous in achieving more effective and environmentally friendly pollutant removal.

3) *Minimal chemical additives*: traditional methods may involve the use of chemical additives or reagents, which can contribute to secondary pollution or have environmental impacts. In photocatalytic degradation, the process often occurs in the presence of oxygen and water, and catalysts are typically used in low concentrations, reducing the need for additional chemical inputs and minimizing the risk of generating harmful by-products.

4) *Non-consumptive catalysts*: photocatalytic processes usually involve catalysts that remain unchanged during the reaction and can be reused, making the approach more sustainable. Unlike consumptive methods that may require frequent replenishment of reactants or catalysts, photocatalysis contributes to resource conservation and waste reduction.

5) *Mild operating conditions*: photocatalytic degradation typically occurs under mild operating conditions, such as ambient temperature and pressure. This is in contrast to some traditional methods that may necessitate harsh conditions, posing challenges in terms of energy consumption and potential environmental harm.

6) *Applicability to a range of pollutants*: photocatalysis has demonstrated efficacy in degrading various pollutants, including organic dyes, pesticides and emerging contaminants. Its versatility makes it a valuable tool for addressing a wide spectrum of environmental challenges, offering a holistic and adaptable approach to pollutant remediation.

7) *Potential for smart biomaterials*: incorporating photocatalysis into the design of smart biomaterials enhances its applicability to specific environmental issues, such as the degradation of micro- and nano-plastics. These biomaterials can be engineered to exhibit advanced photocatalytic properties, allowing for targeted and efficient plastic degradation.

By leveraging these advantages, photocatalytic degradation not only provides a green alternative to traditional methods but also contributes to the development of sustainable technologies for environmental remediation. Its potential to address pollution challenges while minimizing environmental impact aligns with the growing emphasis on green and eco-friendly solutions in various scientific and industrial domains (Ahmad et al. 2023; Querebillo 2023).

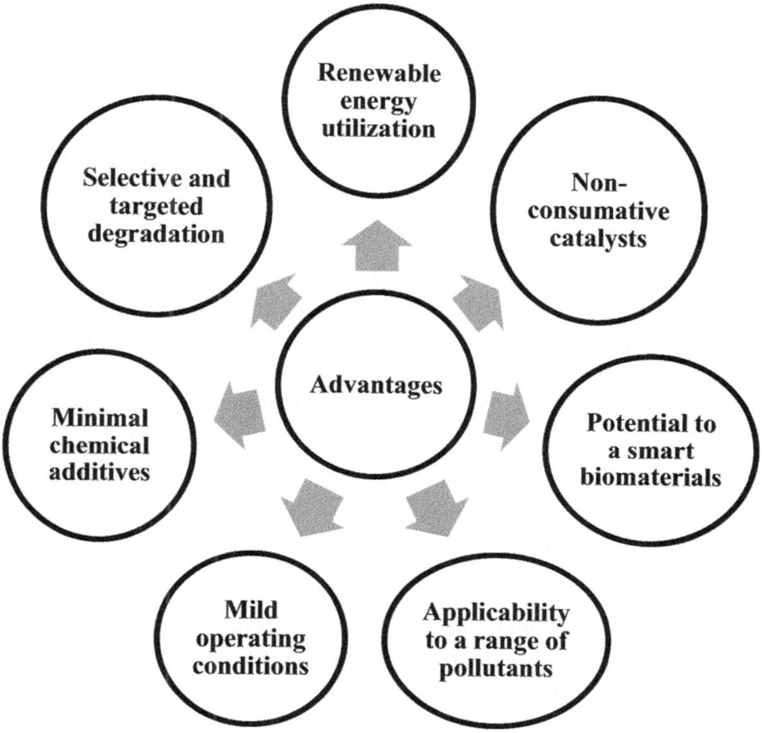

Figure 11.1. *Advantages of photocatalytic degradation*

11.1.2.2. *Role of biomaterials in photocatalysis*

Biomaterials play a crucial role in photocatalysis, contributing to the enhancement, efficiency and specificity of the degradation processes. Their unique properties and versatile nature make them valuable components in developing advanced photocatalytic systems. Here are several key roles that biomaterials play in photocatalysis:

1) *Catalyst support*: biomaterials can serve as excellent supports or carriers for photocatalysts. Their high surface area and porous structures provide ample sites for catalyst immobilization, ensuring efficient utilization of the catalysts during photocatalytic reactions. This immobilization can enhance the stability and reusability of the catalysts.

2) *Enhanced light absorption*: certain biomaterials, such as nanoparticles or nanocomposites derived from natural sources, can exhibit unique optical properties. These materials may enhance light absorption and facilitate the generation of

photoexcited electrons and holes on the catalyst's surface, thereby promoting photocatalytic activity.

3) *Surface functionalization*: biomaterials offer opportunities for surface functionalization, allowing the introduction of specific groups or moieties that can enhance the affinity between the catalyst and the target pollutants. This surface modification helps improve the selectivity and efficiency of photocatalytic degradation processes.

4) *Biocompatibility and environmental friendliness*: utilizing biomaterials in photocatalysis ensures biocompatibility and environmental friendliness. Natural biomaterials, such as chitosan, cellulose, or proteins, are often biodegradable and pose minimal ecological risks. This aligns with the principles of green chemistry, emphasizing sustainability and reduced environmental impact.

5) *Smart biomaterials for targeted degradation*: advanced biomaterials can be engineered as smart materials with specific functionalities. In the context of plastic pollution, for instance, smart biomaterials can be designed to selectively target micro- and nano-plastics for degradation, offering a tailored and effective solution to environmental challenges.

6) *Synergistic effects*: biomaterials can exhibit synergistic effects when combined with photocatalysts. The integration of biomaterials with certain catalysts may enhance the overall performance of the photocatalytic system, promoting more efficient charge separation, reducing electron–hole recombination and improving the overall degradation kinetics.

7) *Biodegradability of photocatalytic products*: biomaterials, especially those with inherent biodegradable properties, can contribute to the eco-friendly nature of photocatalysis. As pollutants undergo photocatalytic degradation, the biomaterials can aid in the breakdown of intermediary products into environmentally benign substances.

8) *Tailored properties*: biomaterials can be tailored to possess specific properties, such as magnetic responsiveness or conductivity. These tailored properties can be advantageous in designing multifunctional photocatalytic systems with enhanced separation and recovery processes.

In summary, the incorporation of biomaterials into photocatalytic systems enhances the overall efficiency, selectivity and sustainability of pollutant degradation processes. Their unique attributes make them invaluable components in the development of advanced and environmentally friendly technologies for tackling various environmental challenges (Sudhakar et al. 2020; Patil et al. 2021; Tariq et al. 2023).

11.2. Smart biomaterials: overview and selection criteria

Smart biomaterials, a class of advanced materials with responsive properties, offer a versatile platform for diverse applications, including photocatalysis. Derived from natural sources or designed to mimic biological structures, these materials exhibit biocompatibility, responsiveness to environmental stimuli and versatility in function. In the realm of photocatalysis, smart biomaterials demonstrate photo-responsive characteristics that can be harnessed for controlled catalytic activity under specific conditions. Key selection criteria for these biomaterials in photocatalysis include their inherent or inducible photocatalytic activity, durability and stability, compatibility with target pollutants, ease of functionalization and sustainability. By meeting these criteria, smart biomaterials can be tailored to initiate and enhance photocatalytic processes, contributing to effective and sustainable solutions for environmental challenges, such as the degradation of pollutants or the remediation of plastic pollution (Holzapfel et al. 2013a; Khan and Tanaka 2017; Montoya et al. 2021).

11.2.1. *Definition and characteristics of smart biomaterials*

Smart biomaterials refer to advanced materials that exhibit responsive and adaptive properties, allowing them to interact intelligently with their environment or external stimuli. These materials are designed with the capability to sense changes in their surroundings and respond accordingly, making them versatile for various applications, including medical devices, drug delivery systems, tissue engineering and environmental technologies. Several key characteristics define smart biomaterials:

1) *Responsiveness*: smart biomaterials exhibit responsiveness to specific stimuli such as temperature, pH, light or the presence of certain chemicals. This responsiveness can trigger changes in the material's properties or behavior.

2) *Adaptability*: these biomaterials can adapt to different environmental conditions or external cues. This adaptability is often programmable or tunable, allowing for controlled responses based on specific requirements.

3) *Biocompatibility*: smart biomaterials are typically biocompatible, meaning they are well tolerated by living tissues and do not elicit harmful reactions. This property is crucial for applications in medicine and biotechnology.

4) *Versatility*: smart biomaterials are versatile in their applications, ranging from controlled drug release and targeted therapy in the biomedical field to pollutant remediation and sensing in environmental applications.

5) *Tailorability*: these materials can be tailored or engineered to have specific properties, allowing researchers to customize their physical, chemical and mechanical characteristics based on the intended use.

6) *Biodegradability*: many smart biomaterials are designed to be biodegradable, breaking down into non-toxic components over time. This characteristic is particularly important for applications where the material may interact with living organisms or the environment.

7) *Functionalization*: smart biomaterials often allow for functionalization or modification of their surface or structure to enhance their performance. This can involve the incorporation of additional functionalities, such as specific ligands or catalytic groups.

8) *Sustainability*: consideration for sustainable sourcing and production methods is integral to smart biomaterials. Using renewable resources and environmentally friendly processes aligns with the broader goals of green and sustainable technology development.

Overall, smart biomaterials integrate the principles of materials science, biology and engineering to create intelligent materials with tailored responses, opening up new possibilities for advancements in healthcare, environmental sustainability and beyond (de Menezes et al. 2019; Amukarimi et al. 2021).

11.2.1.1. *Responsiveness to environmental cues*

The responsiveness of smart biomaterials to environmental cues is a key feature that distinguishes them from conventional materials. These biomaterials are designed to interact intelligently with their surroundings, enabling controlled and adaptive behaviors based on specific environmental triggers. The responsiveness to various cues enhances their versatility in applications such as drug delivery, tissue engineering and environmental technologies. Here are common environmental cues to which smart biomaterials can exhibit responsiveness:

1) Temperature: smart biomaterials can respond to changes in temperature, undergoing alterations in structure, solubility or permeability. This property is particularly valuable for controlled drug release, where temperature variations at specific sites in the body can trigger the release of therapeutic agents.

2) pH: pH-responsive biomaterials undergo changes in their properties in response to variations in acidity or alkalinity. This feature is utilized in drug delivery systems, allowing for targeted release in specific pH environments, such as those found in diseased tissues.

3) Light: photo-responsive biomaterials exhibit changes in their behavior upon exposure to light. This property is exploited in various applications, including

photodynamic therapy in medicine and photocatalysis for environmental remediation.

4) Chemical stimuli: smart biomaterials can respond to specific chemical cues or the presence of particular molecules. This responsiveness is advantageous for applications such as biosensing, where the material can detect and react to specific analytes in the environment.

5) Mechanical forces: some biomaterials respond to mechanical forces or stress. This property is beneficial in tissue engineering, where materials can adapt to the mechanical demands of native tissues, promoting better integration and functionality.

6) Electrical or magnetic fields: biomaterials can be designed to respond to electrical or magnetic fields, allowing for controlled movement or changes in their properties. This responsiveness is utilized in fields such as controlled drug delivery and tissue engineering.

7) Environmental conditions: smart biomaterials can adapt to general environmental conditions, such as changes in humidity or oxygen levels. This adaptability is relevant in various applications, including responsive coatings and sensors.

8) Biological signals: biomaterials can be engineered to respond to specific biological signals within the body, such as enzymatic activity or the presence of specific cells. This responsiveness is exploited for applications like targeted drug delivery in response to biological cues.

By tailoring the design and composition of smart biomaterials, researchers can precisely control their responsiveness to these environmental cues. This programmable adaptability enables the development of advanced materials that can be finely tuned for specific applications, offering innovative solutions in healthcare, biotechnology and environmental science (Khan and Tanaka 2017; Sudhakar et al. 2020; Amukarimi et al. 2021).

11.2.1.2. *Biocompatibility and sustainability*

Biocompatibility and sustainability are two critical aspects that define the utility and ethical considerations of smart biomaterials, particularly in medical and environmental applications.

1) *Biocompatibility*: this refers to the ability of a material to interact with biological systems without causing adverse reactions, inflammation or toxicity. For smart biomaterials, biocompatibility is essential, especially when they come into contact with living tissues or are used for medical purposes.

– Importance: smart biomaterials must be biocompatible to ensure they can be safely integrated into biological environments. This is crucial for applications like implantable medical devices, drug delivery systems and tissue engineering, where interactions with living tissues are common.

– Design considerations: the design of smart biomaterials includes selecting components that are inherently biocompatible or can be modified to ensure compatibility. This involves minimizing potential toxicity, inflammation and immune responses. Surface modifications and coatings may also be employed to enhance biocompatibility.

– Applications: in medicine, biocompatible smart biomaterials are used for controlled drug delivery, bio-responsive implants, tissue scaffolds and other therapeutic interventions. Ensuring that these materials do not elicit harmful responses in the body is critical for their success in clinical applications.

2) *Sustainability*: sustainability in the context of smart biomaterials involves considering the environmental impact of the materials throughout their lifecycle. Sustainable biomaterials are sourced responsibly, manufactured using eco-friendly processes, and are designed to minimize adverse effects on the environment.

– Importance: the increasing emphasis on sustainability reflects a global awareness of environmental issues. Smart biomaterials that are produced sustainably contribute to reducing the ecological footprint of various applications, aligning with broader goals of environmental conservation.

– Design considerations: sustainable smart biomaterials are often derived from renewable resources, utilize green manufacturing processes and may be designed for biodegradability. The selection of materials and production methods should prioritize minimizing energy consumption, waste generation and the use of hazardous chemicals.

– Applications: sustainable smart biomaterials find applications in eco-friendly drug delivery systems, biodegradable implants and environmental technologies. The use of renewable resources and environmentally conscious manufacturing processes is particularly important for minimizing the environmental impact of these materials.

Balancing biocompatibility and sustainability in the design and implementation of smart biomaterials is crucial for responsible and ethical advancements in healthcare and environmental science. The integration of these principles ensures that biomaterials not only perform effectively in biological systems but also contribute to a more sustainable and eco-friendly future (Holzapfel et al. 2013; Montoya et al. 2021).

11.2.2. *Selection criteria for smart biomaterials*

Selecting smart biomaterials involves considering several critical criteria to ensure their effectiveness, safety and applicability for specific applications. The choice of biomaterials is influenced by the intended use, the characteristics required for the application and the desired responses to environmental stimuli. Here are key selection criteria for smart biomaterials:

1) *Responsiveness to stimuli*: we need to choose biomaterials that respond to the relevant environmental stimuli, such as temperature, pH, light or specific chemicals. The responsiveness should align with the requirements of the intended application.

2) *Biocompatibility*: tissue interaction: we need to ensure that the biomaterials are biocompatible, meaning they do not cause adverse effects or immune reactions when in contact with living tissues. This is crucial for medical applications like drug delivery, tissue engineering and implants.

3) *Specificity and selectivity*: we need to opt for biomaterials that exhibit specificity and selectivity in their responses. This is particularly important when designing materials for applications such as targeted drug delivery, where precise interactions are desired.

4) *Tailorability*: we need to select biomaterials that can be easily tailored to meet specific requirements. This may involve adjusting physical, chemical or mechanical properties to optimize their performance for a particular application.

5) *Durability and stability*: we need to choose biomaterials that maintain their structural integrity and functionality over the required duration. This is especially critical for applications like implants, where long-term stability is essential.

6) *Ease of functionalization*: we need to consider biomaterials that can be easily functionalized or modified to enhance their properties. This allows for the incorporation of additional functionalities, such as specific ligands or catalytic groups.

7) *Biodegradability*: for applications where environmental impact is a concern, we selected biomaterials that are biodegradable. Biodegradable materials break down into non-toxic components over time, minimizing their environmental footprint.

8) *Sustainability*: we need to prioritize biomaterials derived from renewable resources or those produced using sustainable and environmentally friendly processes. Sustainable biomaterials contribute to reducing the overall ecological impact.

9) *Cost-effectiveness*: we need to consider the cost-effectiveness of the biomaterials for large-scale production. The materials chosen should be

economically viable for the intended application, especially in fields like healthcare, where affordability is crucial.

10) *Regulatory compliance*: we need to ensure that the biomaterials meet regulatory standards and comply with guidelines for safety and efficacy. This is particularly important for biomaterials intended for medical use.

11) *Scalability*: we need to evaluate the scalability of biomaterial production processes. The selected materials should be producible on a large scale without compromising quality or consistency.

The careful consideration of these criteria ensures that smart biomaterials are tailored to meet the specific needs of their intended applications, whether in medicine, environmental science or other fields, while also adhering to ethical, safety and sustainability standards (Dumbili and Henderson 2020; Zulfikar et al. 2021).

11.3. Design principles for enhanced photocatalysis

Designing smart biomaterials for enhanced photocatalysis involves a strategic integration of key principles to optimize their performance in degrading pollutants and facilitating targeted chemical reactions under light exposure. The incorporation of photoactive components, such as semiconductor nanoparticles like titanium dioxide or zinc oxide, is fundamental to initiate and facilitate the photocatalytic reactions within the biomaterial structure. Maximizing light absorption capacity through tailored composition ensures efficient utilization of wavelengths relevant to catalyst activation. Increasing surface area and porosity, achieved through nanostructuring or porous structures, provides more active sites for photocatalytic reactions, enhancing overall efficiency.

Efficient charge carrier separation, preventing the recombination of photo-generated electrons and holes, is crucial for optimizing the photocatalytic process. The inclusion of redox mediators within the biomaterial facilitates electron transfer, promoting effective reactions with target pollutants. Functional group modification on the biomaterial's surface introduces specific groups that enhance interaction with pollutants, enabling selective and efficient photocatalytic degradation. Light-triggered responsiveness, an engineered feature, allows for controlled activation and deactivation of the photocatalytic process in response to specific stimuli.

For applications with medical or biological relevance, ensuring biosafety and biocompatibility is paramount, making biocompatible biomaterials a priority. Designing for durability and stability ensures biomaterials withstand the conditions of photocatalysis, ensuring longevity and consistent performance over multiple cycles of light exposure. Leveraging nanotechnological approaches, such as incorporating nanoparticles, nanocomposites or nano-coatings, enhances the biomaterial's properties, contributing to improved photocatalytic performance.

Sustainability is a key consideration, encouraging the use of components derived from sustainable sources or those produced using environmentally friendly processes. Incorporating real-time monitoring systems into the biomaterial design allows for precise tracking of photocatalytic reactions, enabling fine-tuning and optimization. Overall, these design principles collectively contribute to the development of smart biomaterials that harness light energy effectively, offering innovative solutions for environmental remediation, wastewater treatment and other applications where controlled and efficient pollutant degradation is essential (Bard 1979; Wang et al. 2021; Fang and Hu 2022).

11.3.1. *Tailoring optical properties*

11.3.1.1. *UV-visible absorption range*

Tailoring the optical properties of smart biomaterials in the UV-visible range involves strategic design considerations to optimize their behavior in response to light within this spectrum. The UV-visible range typically spans wavelengths from approximately 100–750 nm. Here are key strategies for tailoring the optical properties of smart biomaterials within this range:

1) *Material selection*: we need to choose or engineer biomaterials with inherent optical properties in the UV-visible range. Select materials with absorption or emission spectra that match the desired wavelengths for the intended application.

2) *Chromophore integration*: we need to integrate specific chromophores or organic dyes into the biomaterial's structure. These molecules can absorb and emit light within the UV-visible range, allowing for tunable optical properties.

3) *Quantum dots and nanoparticles*: we need to incorporate semiconductor nanoparticles, such as quantum dots, that exhibit size-dependent optical properties. Quantum dots can be engineered to absorb or emit light at specific UV-visible wavelengths.

4) *Surface plasmon resonance*: we need to leverage plasmonic effects by introducing metallic nanoparticles, such as gold or silver, to induce surface plasmon resonance. This can enhance light absorption and scattering in the UV-visible range.

5) *Nanostructuring*: we need to engineer nanostructured surfaces or architectures within the biomaterial to manipulate light interactions. Nanostructuring can influence optical properties, such as enhancing absorption or creating photonic effects in the UV-visible range.

6) *Doping with UV-visible absorbers*: we need to dope the biomaterial with specific compounds that absorb light in the UV-visible range. This can modify the material's absorption spectrum, making it responsive to certain wavelengths.

7) *Optical coatings*: we need to apply coatings with tailored optical properties to the biomaterial's surface. Coatings can modify reflectance, transparency or other optical characteristics within the UV-visible spectrum.

8) *pH-sensitive materials*: we need to integrate pH-sensitive materials that undergo changes in optical properties in response to variations in pH. pH-sensitive dyes or indicators can be employed for applications such as biosensing.

9) *Real-time monitoring with UV-visible probes*: we incorporated optical probes or sensors designed to monitor changes in UV-visible absorption or emission in real-time. This enables dynamic control and adjustment of the biomaterial's optical properties.

10) *Photochromic molecules*: we introduced photochromic molecules that undergo reversible changes in color or absorbance upon exposure to UV or visible light. This property is valuable for applications like smart windows or light-responsive materials.

11) *Optical filters and bandpass structures*: we designed optical filters or bandpass structures within the biomaterial to selectively transmit or block specific UV-visible wavelengths. This can be useful in creating materials with controlled optical responses.

12) *Multiphoton absorption*: we explored biomaterials with multiphoton absorption properties, allowing for absorption of multiple photons simultaneously. This can enable unique optical effects and applications in imaging or sensing.

By strategically implementing these approaches, researchers can tailor the optical properties of smart biomaterials specifically within the UV-visible range, opening up opportunities for diverse applications in fields such as sensing, imaging and light-responsive technologies (Wang et al. 2021; Fang and Hu 2022).

11.3.1.2. *Bandgap engineering*

Bandgap engineering is a critical aspect in tailoring the electronic and optical properties of smart biomaterials. The bandgap refers to the energy difference between the valence band and the conduction band in a material. Here are key strategies for bandgap engineering in smart biomaterials:

1) *Material selection*: we need to choose or design biomaterials with inherent bandgaps suitable for the intended application. Different materials have distinct electronic structures, influencing their bandgap.

2) *Quantum dot integration*: we incorporated semiconductor quantum dots into the biomaterial. Quantum dots exhibit size-dependent bandgaps, and by controlling their size, the bandgap can be tuned for specific electronic and optical properties.

3) *Nanostructuring*: we engineered nanostructures within the biomaterial to influence its bandgap. Nanostructuring can lead to quantum confinement effects, impacting the material's electronic structure and bandgap.

4) *Doping*: we introduced specific dopants or additives to the biomaterial to modify its bandgap. Doping can alter the electronic properties, enabling bandgap engineering for enhanced functionality.

5) *Alloying*: we created alloys by combining different materials to achieve a composite with a tailored bandgap. Alloying offers the ability to fine-tune the electronic properties of the biomaterial.

6) *Strain engineering*: we applied mechanical strain to the biomaterial to modify its electronic structure and bandgap. Strain engineering is particularly relevant for 2D materials and nanomaterials.

7) *Surface functionalization*: we modified the surface of the biomaterial with functional groups or coatings to alter its electronic properties, influencing the bandgap. Surface functionalization can provide additional control over the material's behavior.

8) *Optical excitation*: we utilized optical excitation to induce changes in the bandgap. Illuminating the biomaterial with light can create photo-induced effects, altering its electronic structure temporarily.

9) *Temperature control*: we adjusted the temperature to influence the bandgap. Some materials exhibit temperature-dependent bandgap changes, allowing for control over their electronic properties.

10) *Electric field modulation*: we applied an external electric field to modulate the bandgap of the biomaterial. This approach is particularly relevant for certain 2D materials and semiconductors.

11) *Layered structures*: we designed biomaterials with layered structures to control electronic interactions and bandgap modulation. Layering can provide a platform for engineering electronic properties in a controlled manner.

12) *Integration with nanomaterials*: we incorporated other nanomaterials, such as nanotubes or nanowires, to influence the bandgap of the biomaterial. The interaction between different nanomaterials can lead to synergistic effects in bandgap engineering.

The successful implementation of these bandgap engineering strategies in smart biomaterials can significantly impact their electronic and optical functionalities.

This customization is vital for applications such as sensors, electronic devices and light-responsive systems, where precise control over bandgap properties is essential (Wang et al. 2021).

11.3.2. *Surface functionalization for targeted activity*

11.3.2.1. *Attachment of photocatalytic moieties*

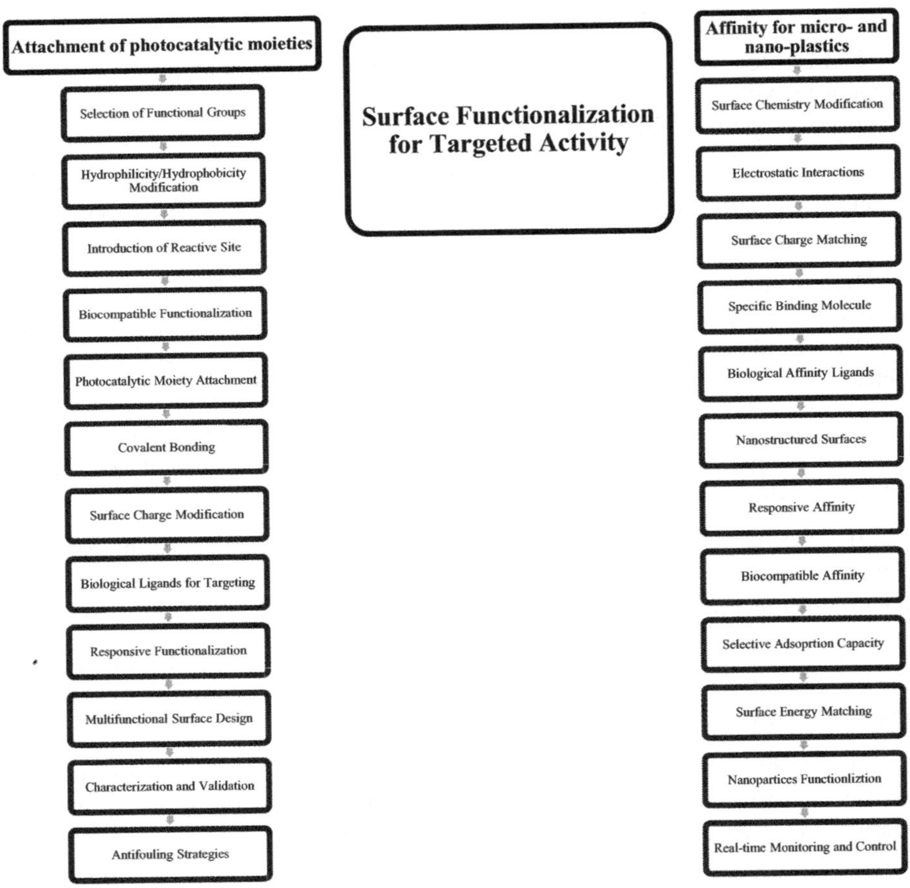

Figure 11.2. *Surface functionalization for targeted activity*

Surface functionalization is a powerful strategy employed in smart biomaterial design to tailor the material's surface properties for targeted activity and attachment of photocatalytic moieties. This process involves modifying the biomaterial's surface with specific functional groups, molecules or coatings to achieve desired characteristics. In the context of photocatalysis, surface functionalization plays a crucial role in enhancing catalytic activity, promoting selective attachment of photocatalytic moieties and facilitating targeted interactions. Here are key considerations and approaches for surface functionalization in smart biomaterials:

1) *Selection of functional groups*: we need to choose functional groups based on the specific requirements of the photocatalytic process. Common functional groups include hydroxyl (-OH), amino (-NH_2), carboxyl (-COOH) and thiol (-SH) groups, which can participate in various chemical interactions.

2) *Hydrophilicity/hydrophobicity modification*: we adjusted the hydrophilicity or hydrophobicity of the biomaterial's surface through functionalization. This influences the material's wettability and interactions with water, impacting the efficiency of photocatalytic reactions.

3) *Introduction of reactive sites*: we introduced reactive sites on the biomaterial's surface for subsequent attachment of photocatalytic moieties. These reactive sites can include double bonds, epoxy groups or other chemically active entities.

4) *Biocompatible functionalization*: we ensured that the selected functionalization is biocompatible, especially if the smart biomaterial is intended for applications in medicine or biology. Biocompatible functionalization is critical for minimizing adverse effects on living systems.

5) *Photocatalytic moiety attachment*: we attached specific photocatalytic moieties to the functionalized surface. This can include semiconductor nanoparticles like titanium dioxide (TiO_2) or other materials with photocatalytic properties.

6) *Covalent bonding*: we employed covalent bonding strategies for robust attachment of photocatalytic moieties. Covalent bonds provide stability and durability, ensuring the moieties remain attached during photocatalytic reactions.

7) *Surface charge modification*: we modified the surface charge of the biomaterial to influence the electrostatic interactions with photocatalytic moieties. This can affect the distribution and orientation of moieties on the surface.

8) *Biological ligands for targeting*: we used biological ligands or targeting molecules during functionalization to achieve site-specific interactions. This is particularly relevant for targeted photocatalytic activity in biological systems.

9) *Responsive functionalization*: we designed responsive functionalization that can change its properties in response to specific stimuli, such as light or changes in pH. Responsive functionalization adds an extra layer of control over the photocatalytic process.

10) *Multifunctional surface design*: we considered multifunctional surface designs that incorporate multiple types of functional groups. This approach allows

for versatile interactions and can enhance the overall performance of the smart biomaterial.

11) *Characterization and validation*: we characterized the functionalized surface to confirm the successful attachment of desired moieties. Validation through techniques like spectroscopy, microscopy or surface analysis ensures the effectiveness of the surface modification.

12) *Antifouling strategies*: we implemented antifouling strategies during functionalization to prevent unwanted interactions or adsorption of biomolecules on the surface. Antifouling coatings can maintain the material's efficiency over extended periods.

By carefully tailoring the surface properties of smart biomaterials through functionalization, researchers can optimize their photocatalytic performance, achieve targeted activity and enhance their applicability in various fields, including environmental remediation, healthcare and energy conversion (Perez et al. 2013; Zamuner et al. 2017; Safavi et al. 2023).

11.3.2.2. *Affinity for micro- and nano-plastics*

Designing smart biomaterials with an affinity for micro- and nano-plastics involves tailoring the material's surface properties to selectively attract, capture or interact with plastic particles at the microscopic and nanoscopic scales. This affinity is crucial for applications such as plastic pollution remediation and environmental cleanup. Here are key considerations for enhancing the affinity of smart biomaterials for micro- and nano-plastics:

1) *Surface chemistry modification*: we tailored the surface chemistry of the biomaterial to introduce functional groups that exhibit specific interactions with plastic surfaces. Functional groups such as hydroxyl (-OH), amino (-NH_2) or carboxyl (-COOH) can enhance affinity.

2) *Electrostatic interactions*: we utilized electrostatic interactions by modifying the surface charge of the biomaterial. Attractive electrostatic forces can enhance the affinity for charged micro- and nano-plastics.

3) *Surface charge matching*: we designed the biomaterial's surface charge to match the charge distribution on micro- and nano-plastics. This can improve the overall affinity through electrostatic complementarity.

4) *Specific binding molecules*: we introduced specific binding molecules or ligands on the biomaterial's surface that have an inherent affinity for plastic particles. Molecular recognition can enhance selectivity.

5) *Biological affinity ligands*: we incorporated biological affinity ligands, such as antibodies or aptamers, that are designed to selectively recognize and bind to plastic surfaces. This approach adds a high degree of specificity.

6) *Nanostructured surfaces*: we engineered nanostructures on the biomaterial's surface to create high surface area and intricate topographies that enhance the physical interactions with micro- and nano-plastics.

7) *Responsive affinity*: we designed responsive biomaterials that can adapt their affinity based on external stimuli. For example, a material that adjusts its surface properties in response to changes in environmental conditions.

8) *Biocompatible affinity*: we ensured that the affinity modifications are biocompatible, especially if the smart biomaterial is intended for environmental applications. Compatibility with the surrounding ecosystem is essential.

9) *Selective adsorption capacity*: we optimized the adsorption capacity of the biomaterial specifically for micro- and nano-plastics. This involves understanding the size, shape and chemical characteristics of the targeted plastic particles.

10) *Surface energy matching*: we matched the surface energy of the biomaterial with that of micro- and nano-plastics. Surface energy considerations can influence the degree of contact and adhesion.

11) *Nanoparticle functionalization*: we functionalized nanoparticles within the biomaterial structure to enhance the overall affinity. Nanoparticles with specific properties can contribute to improved interactions with plastics.

12) *Real-time monitoring and control*: we integrated real-time monitoring and control mechanisms to assess the affinity in situ. This allows for adjustments and optimization based on changing environmental conditions.

13) *Regenerability*: we designed the smart biomaterial to be regenerable, allowing for the recovery and reuse of the material after capturing micro- and nano-plastics. This promotes sustainability in environmental cleanup efforts.

By addressing these considerations, researchers can create smart biomaterials with a heightened and selective affinity for micro- and nano-plastics, contributing to effective and targeted solutions for plastic pollution mitigation (Perez et al. 2013; Safavi et al. 2023).

11.4. Structural modifications for improved efficiency

Structural modifications are instrumental in enhancing the efficiency of smart biomaterials for diverse applications. By introducing nanostructuring, biomaterials can exhibit increased surface area, promoting heightened reactivity and improved performance in catalytic processes. Incorporating a porous architecture enhances mass transfer, allowing for efficient adsorption or controlled release in applications like drug delivery. Hierarchical structures, with multiple organizational levels, provide versatility and efficiency by combining functionalities at different scales. Surface functionalization introduces specific chemical groups, tailoring the biomaterial's reactivity and expanding its capabilities. Core–shell structures integrate distinct features within a single material, optimizing its multifunctionality. Cross-linking enhances stability and mechanical strength, ensuring durability in various environmental conditions. Doping with nanoparticles can impart enhanced properties, such as improved conductivity or catalytic activity. Molecular imprinting creates selective binding sites for recognition, beneficial in applications like sensors. Layer-by-layer assembly offers precise control over structure and function, while responsive polymers adapt to external stimuli for controlled applications. Hybrid materials and nanocomposites leverage the strengths of different components, promoting synergistic effects. Optimizing morphology, employing photonic crystal design and drawing inspiration from biological systems further contribute to the tailored efficiency of smart biomaterials. These structural modifications collectively enhance the biomaterials' performance across a spectrum of applications, from catalysis and sensing to drug delivery and environmental remediation (Zhang et al. 2018; Safavi et al. 2023).

11.4.1. *Nanocomposite formation*

Nanocomposite formation represents a cutting-edge strategy in the design of smart biomaterials, leveraging the synergistic properties of nanoscale components to enhance overall performance. In this process, nanoparticles, such as quantum dots, carbon nanotubes, or metal nanoparticles, are seamlessly integrated into a biomaterial matrix. This integration imparts unique features to the smart biomaterial, combining the inherent properties of the biomaterial with the specific attributes of the nanoparticles. Nanocomposites in smart biomaterials can exhibit improved mechanical strength, enhanced conductivity and superior catalytic or sensing capabilities. The intimate interaction between the biomaterial and nanoparticles at the nanoscale facilitates precise control over material properties, contributing to tailored solutions for applications ranging from biomedical devices to environmental sensors. The versatility of nanocomposite formation enables the development of advanced materials with multifunctionality and responsiveness, paving the way for innovations in diverse fields such as medicine, environmental science and technology (de Menezes et al. 2019).

11.4.1.1. *Integration of nanoparticles*

The integration of nanoparticles into smart biomaterials represents a pivotal advancement in material science, offering a versatile approach to tailoring properties for a wide array of applications. By seamlessly incorporating nanoparticles, such as quantum dots, metallic nanoparticles or carbon nanotubes, into the biomaterial matrix, a synergistic composite is formed. This integration enhances the overall functionality of the biomaterial by leveraging the unique characteristics of the nanoparticles. For instance, metallic nanoparticles may confer improved conductivity, while quantum dots can impart enhanced optical and electronic properties. The controlled dispersion of nanoparticles within the biomaterial structure allows for fine-tuning of mechanical, electrical and catalytic properties. This integration is particularly significant in biomedical applications, where nanocomposites can facilitate targeted drug delivery, imaging or diagnostics. Additionally, the integration of nanoparticles in smart biomaterials plays a crucial role in environmental remediation and sensing technologies. The resulting nanocomposites showcase the potential to revolutionize various industries by providing materials with tailored properties and functionalities (Shodeinde et al. 2020; Triantafyllopoulos and Papaioannou 2022).

11.4.1.2. *Synergistic effects*

Synergistic effects in smart biomaterials refer to the amplified and cooperative outcomes that result from the strategic integration of multiple components within the material. By combining diverse elements such as nanoparticles, polymers or biological entities, smart biomaterials can exhibit properties and functionalities that surpass those of individual components in isolation. The synergistic effects arise from the enhanced interactions and interplay between the integrated components, leading to improved mechanical strength, responsiveness to stimuli or specific biological activities. For example, the combination of responsive polymers and nanoparticles can result in a smart biomaterial that not only adapts to environmental changes but also exhibits enhanced catalytic or sensing capabilities. In biomedical applications, synergistic effects can be harnessed for advanced drug delivery systems that provide targeted release and therapeutic efficacy. Overall, understanding and harnessing synergistic effects in smart biomaterials paves the way for the development of highly efficient, multifunctional materials with applications spanning medicine, environmental science and technology (Shodeinde et al. 2020; Sadeghi et al. 2023).

11.4.2. *Porosity enhancement*

Porosity enhancement in smart biomaterials is a crucial strategy aimed at improving their overall performance, particularly in applications where high surface area and increased reactivity are desirable. By introducing controlled pores or void spaces within the biomaterial structure, various properties such as adsorption capacity, drug delivery efficiency and tissue regeneration capabilities can be significantly enhanced. This porous architecture facilitates improved mass transfer, allowing for better accessibility of reactants to active sites within the material. Strategies for porosity enhancement include nanostructuring, where nanoscale features are incorporated to increase surface area, and the use of sacrificial templates that are later removed to create pores. The resulting biomaterial with enhanced porosity exhibits improved functionality, making it highly adaptable for applications ranging from drug delivery systems in medicine to catalysis in environmental remediation. Porosity enhancement is a pivotal aspect in tailoring smart biomaterials for specific functions and ensuring their effectiveness in diverse fields.

11.4.2.1. *Mesoporous and microporous structures*

Mesoporous and microporous structures are integral features of smart biomaterials, playing a pivotal role in shaping their functional characteristics for diverse applications. Mesopores, typically ranging from 2 to 50 nm, contribute to enhanced surface area, facilitating increased interaction with surrounding molecules. This property is particularly advantageous in drug delivery systems, catalytic applications and adsorption processes. Mesoporous structures excel in controlled release and targeted delivery, making them valuable for encapsulating and delivering therapeutic agents. On the other hand, micropores, with dimensions <2 nm, enable selective adsorption based on size and shape, offering applications in gas separation and molecule capture. The high surface energy of microporous structures enhances adsorption capabilities, contributing to storage, stability and controlled release functions in biomaterials. The synergistic integration of mesoporous and microporous features empowers smart biomaterials to be versatile platforms, adaptable to specific requirements in fields such as medicine, environmental science and biotechnology.

11.4.2.2. *Increased surface area for catalytic reactions*

The incorporation of increased surface area is a fundamental design strategy in the development of smart biomaterials, especially for catalytic applications. By introducing nanostructuring or mesoporous structures within the material, a larger surface area becomes available, providing an extensive platform for catalytic reactions. This heightened surface area allows for more active sites, enabling efficient interactions between the biomaterial and reactants. In catalysis, where the effectiveness of a material relies on its ability to facilitate chemical transformations, the amplified surface area of smart biomaterials enhances reactivity and catalytic efficiency. This design approach proves crucial in various environmental and industrial applications, such as water purification and pollutant degradation, where catalytic biomaterials can effectively harness their increased surface area to accelerate reaction rates and improve overall performance. The optimization of surface area in smart biomaterials exemplifies a strategic approach to boost catalytic prowess, advancing their utility across diverse fields of chemical engineering and environmental science (Maksoud et al. 2022; Intravaia et al. 2023).

11.5. Case studies and applications

Smart biomaterials have found diverse applications across various fields, showcasing their versatility and potential to address complex challenges. The following are a few case studies highlighting the practical use of smart biomaterials.

11.5.1. *Titanium dioxide nanomaterials*

Titanium dioxide (TiO_2) nanomaterials have gained significant attention for their versatile properties, including photocatalytic activity, biocompatibility and unique optical features. In this case study, TiO_2 nanomaterials are employed as smart biomaterials with a focus on their application in drug delivery systems.

The objective is to utilize TiO_2 nanomaterials to develop a drug delivery platform that responds to external stimuli, such as light, enabling controlled and targeted release of therapeutic agents.

1) *TiO_2 nanoparticles synthesis*: TiO_2 nanoparticles are synthesized with controlled size and surface properties to optimize their biocompatibility and drug-loading capacity.

2) *Drug encapsulation*: a model drug is encapsulated within the TiO_2 nanoparticles, taking advantage of their porous structure and high surface area for efficient drug loading.

3) *Photocatalytic responsiveness*: TiO_2's photocatalytic activity is harnessed for responsive drug release. When exposed to specific wavelengths of light, the TiO_2 nanomaterial generates reactive oxygen species, triggering the release of the encapsulated drug.

4) *In vitro studies*: the smart drug delivery system is tested in vitro using cell cultures. The responsiveness of drug release is evaluated under controlled light conditions, simulating external stimuli.

5) *In vivo studies*: efficacy and safety are assessed in animal models, where the TiO_2-based smart biomaterial is implanted or administered, and drug release is triggered externally using light exposure.

RESULT.–

The TiO_2-based smart biomaterial demonstrates:

– controlled drug release in response to light exposure;

– enhanced therapeutic efficacy with targeted drug delivery;

– biocompatibility and minimal cytotoxicity in vitro and in vivo;

– versatility in responding to different light wavelengths for tailored drug release profiles.

APPLICATIONS.–

1) *Cancer therapy*: it includes targeted drug delivery to cancer cells with spatial and temporal control through light exposure, minimizing side effects on healthy tissues.

2) *Infectious disease treatment*: it includes responsive drug release for antimicrobial agents, providing on-demand treatment for infections.

3) *Chronic disease management*: it includes tailored drug release profiles for chronic conditions, offering personalized and controlled therapeutic interventions.

CONCLUSION.–

This case study showcases the potential of TiO_2 nanomaterials as smart biomaterials in drug delivery systems. Their photocatalytic responsiveness, coupled with biocompatibility, opens avenues for innovative and effective therapeutic strategies with applications across various medical fields (Al Othman et al. 2013; Wang et al. 2016; Mydin et al. 2018; Palafox González et al. 2023).

11.5.1.1. *Modification strategies*

TiO_2 modification strategies play a pivotal role in the development of smart biomaterials, enhancing their functionality and responsiveness for a range of applications. Doping TiO_2 with foreign elements refines its electronic structure, expanding its photocatalytic activity into the visible light spectrum. Surface functionalization introduces tailored coatings to enhance biocompatibility and foster specific interactions, crucial in drug delivery systems and biomedical devices. Nanostructuring, through techniques like mesoporous formation, increases surface area, improving reactivity and making TiO_2 suitable for catalysis and sensors. Coupling TiO_2 with other photocatalysts, depositing metal nanoparticles, or creating composites with graphene enhances its performance in applications such as environmental remediation and energy storage. The formation of heterojunctions and the use of hydrothermal synthesis or sol-gel methods allow precise control over TiO_2 properties, ensuring adaptability for diverse smart biomaterial applications. These modification strategies collectively empower TiO_2 to contribute significantly to the advancement of smart biomaterials with tailored functionalities for targeted and responsive applications in medicine, environmental science and beyond.

11.5.1.2. *Performance in micro- and nano-plastic degradation*

TiO_2 has demonstrated promising performance in the degradation of micro- and nano-plastics when incorporated into smart biomaterials. The photocatalytic properties of TiO_2, particularly its ability to generate reactive oxygen species (ROS) upon exposure to ultraviolet (UV) or visible light, make it an effective catalyst for breaking down plastic pollutants. The mechanism involves the photocatalytic oxidation of the plastic surface, leading to the fragmentation of larger plastic particles into smaller, more easily degradable fragments. The high surface area of TiO_2, especially when nanostructured or in mesoporous form, enhances its contact with plastic pollutants, increasing the efficiency of the degradation process. Moreover, TiO_2's ability to initiate photochemical reactions in the presence of light enables targeted and controlled degradation, minimizing environmental impact. This photocatalytic degradation capability positions TiO_2 as a promising component in smart biomaterials designed for environmental remediation, contributing to the sustainable management of micro- and nano-plastic pollution (Wang et al. 2016; Mydin et al. 2018; Palafox González et al. 2023).

11.5.2. *Graphene-based smart biomaterials*

Graphene, a single layer of carbon atoms arranged in a hexagonal lattice, has garnered considerable attention for its exceptional mechanical, electrical and biocompatible properties. In this case study, graphene is employed as a key component in the development of smart biomaterials with a focus on biomedical applications.

The objective is to harness the unique properties of graphene to design smart biomaterials capable of targeted drug delivery, imaging and sensing within the human body.

1) *Graphene synthesis and modification*:

– high-quality graphene is synthesized using methods like chemical vapor deposition (CVD);

– functional groups are introduced to the graphene surface through chemical modifications to enhance its biocompatibility and responsiveness.

2) *Drug delivery system*:

– graphene is incorporated into a drug delivery system where therapeutic agents are loaded onto the graphene surface or within its layers;

– smart functionality is introduced through stimuli-responsive coatings that enable controlled drug release in response to specific triggers, such as pH, temperature or external stimuli.

3) *Imaging and diagnostics*:

– graphene's exceptional conductivity and optical properties are exploited for the development of imaging agents;

– functionalized graphene is used as a contrast agent in imaging techniques like photoacoustic imaging or magnetic resonance imaging (MRI).

4) *Biosensing platform*:

– graphene's high surface area and electrical conductivity are leveraged for the creation of biosensors;

– specific biomolecules are immobilized on graphene surfaces, and changes in electrical conductivity or optical properties are monitored for real-time detection.

RESULTS.–

The graphene-based smart biomaterial exhibits:

– responsive drug delivery: controlled release of therapeutic agents in response to specific stimuli, improving precision and reducing side effects;

– imaging capabilities: enhanced contrast in imaging modalities, providing clearer and more detailed diagnostic information;

– biosensing precision: highly sensitive and selective detection of target biomolecules, offering potential applications in disease diagnosis and monitoring.

APPLICATIONS.–

1) *Cancer therapy*: it includes targeted drug delivery to cancer cells with controlled release, minimizing damage to healthy tissues.

2) *Diagnostic imaging*: it includes improved contrast agents for more accurate and detailed medical imaging.

3) *Point-of-care diagnostics*: it includes portable biosensing devices for rapid and sensitive detection of biomarkers in clinical settings.

CONCLUSION.–

This case study illustrates the multifaceted applications of graphene-based smart biomaterials in the biomedical field. By capitalizing on graphene's unique properties, these materials hold great promise for advancing drug delivery systems, diagnostic imaging and biosensing technologies with potential transformative impacts on healthcare practices (Daneshmandi et al. 2021; Kumar et al. 2021; Sharma et al. 2022).

11.5.2.1. *Unique properties for catalysis*

Graphene-based smart biomaterials exhibit unique properties that make them exceptional candidates for catalysis. These distinctive features leverage graphene's two-dimensional structure and chemical properties, offering advantages in various catalytic applications:

1) *High surface area*: graphene's single layer of carbon atoms arranged in a hexagonal lattice provides an extensive surface area, facilitating a larger number of active sites for catalytic reactions. This property enhances the catalyst's efficiency and reactivity.

2) *Exceptional conductivity*: graphene's excellent electrical conductivity enables rapid electron transfer during catalysis, making it particularly advantageous in electrocatalytic processes. This property enhances the kinetics of redox reactions and supports efficient charge transport.

3) *Chemical inertness*: graphene is chemically inert, contributing to the stability of the catalyst under diverse reaction conditions. This inertness ensures the durability of the catalyst and its suitability for a wide range of catalytic applications.

4) *Tailorable surface chemistry*: the surface of graphene can be modified and functionalized with various groups, allowing for precise tuning of its catalytic properties. Functionalization introduces specific catalytic functionalities, enhancing selectivity and catalytic activity.

5) *Graphene oxide's oxygen-containing groups*: graphene oxide, a derivative of graphene, contains oxygen-containing functional groups (hydroxyl, epoxy, carboxyl), providing additional catalytic sites. These groups participate in redox reactions, expanding the range of catalytic capabilities.

6) *Photocatalytic activity*: graphene exhibits excellent photocatalytic activity, particularly in the visible and near-infrared regions. This property allows graphene-based smart biomaterials to catalyze reactions under light exposure, offering opportunities for light-triggered catalysis.

7) *Two-dimensional structure*: the planar, two-dimensional structure of graphene ensures efficient mass transfer, enabling easy access of reactants to active sites. This

structure promotes faster reaction kinetics and contributes to the overall catalytic efficiency.

8) *Biocompatibility*: in the context of smart biomaterials, the biocompatibility of graphene is crucial for catalytic applications within biological systems. This property allows for applications in biocatalysis and bioelectrocatalysis, expanding the possibilities in the biomedical field.

9) *Synergistic effects in nanocomposites*: graphene's integration into nanocomposites with other catalytically active materials, such as metal nanoparticles, can result in synergistic effects. This synergy enhances catalytic performance by combining the unique properties of each component, leading to improved efficiency.

10) *Stimuli-responsive behavior*: graphene-based smart biomaterials can be engineered to exhibit stimuli-responsive behavior, allowing for on-demand activation or deactivation of catalytic functions in response to specific environmental cues. This responsiveness enhances control and precision in catalytic applications. These unique properties position graphene-based smart biomaterials as versatile and efficient catalysts, with applications ranging from environmental remediation and energy conversion to targeted drug delivery and biomedical catalysis.

11.5.2.2. *Challenges and potential solutions*

1) *Biocompatibility concerns*: ensuring the biocompatibility of graphene-based materials for in vivo applications remains a challenge, as some forms of graphene can induce cytotoxicity. Surface modifications and functionalizations can enhance biocompatibility. Biologically inspired coatings and thorough biocompatibility assessments are crucial before clinical translation.

2) *Large-scale production*: scaling up the production of high-quality graphene can be economically challenging, hindering widespread application.

Advances in scalable synthesis methods, such as CVD and liquid-phase exfoliation, are addressing this challenge. Continued research into cost-effective production methods is essential.

3) *Limited water solubility*: graphene's limited water solubility hampers its dispersion in biological fluids and can affect its interaction with biomolecules. Surface functionalization with hydrophilic groups can enhance water solubility. Amphiphilic polymer coatings or the use of graphene oxide derivatives improve dispersibility and bioavailability.

4) *Long-term stability*: graphene can agglomerate over time, leading to reduced stability and potential loss of functionality. Implementing stabilizing agents, such as

polymers or surfactants, during synthesis can prevent agglomeration. Further research on long-term stability and degradation mechanisms is essential.

5) *Immunological response*: the potential for immune system activation and inflammatory responses to graphene-based materials poses a challenge for biomedical applications. Careful consideration of material design, size and surface modifications can mitigate immunological responses. In-depth immunotoxicity studies are necessary for safe clinical translation.

6) *Regulatory approval*: achieving regulatory approval for graphene-based biomaterials faces hurdles due to concerns about safety and lack of standardized testing protocols. Collaboration between researchers, regulatory agencies and industry stakeholders is essential to establish clear guidelines and standardized testing methods for graphene-based biomaterials.

7) *Environmental impact*: the environmental impact of graphene-based materials, especially during disposal, is a growing concern. Developing strategies for responsible disposal and recycling of graphene-based materials. Research on the environmental fate and impact can guide sustainable practices.

8) *Cost barriers*: the production cost of high-quality graphene can be a barrier to its widespread use. Continued research into cost-effective production methods, such as innovative synthesis techniques and recycling approaches, can help overcome cost barriers.

9) *Lack of standardization*: the lack of standardized protocols for synthesizing and characterizing graphene-based materials hinders reproducibility and comparability of research findings. Establishing standardized protocols for synthesis, characterization and toxicity testing will enhance the reliability and applicability of graphene-based biomaterials.

Addressing these challenges requires a multidisciplinary approach involving materials scientists, biologists, clinicians and regulatory bodies to ensure the safe and effective translation of graphene-based smart biomaterials into practical applications (Bellet et al. 2021; Kumar et al. 2021).

11.6. Challenges and future perspectives

Smart biomaterials face several challenges that include ensuring long-term biocompatibility, stability, regulatory approval, cost-effectiveness and seamless integration with biological systems. Achieving regulatory approval requires standardized protocols, and the lack of such standards hampers progress. Future perspectives in the field involve the development of multifunctional biomaterials with advanced responsiveness, precision medicine applications tailored for individualized treatments and the integration of nanotechnology for enhanced precision. Biohybrid materials combining synthetic and biological components, a focus on sustainability, and the integration of artificial intelligence (AI) for real-time monitoring and predictive modeling are anticipated. Overcoming these challenges and embracing future perspectives will drive innovations, expanding the applications of smart biomaterials in areas such as medicine, biotechnology and environmental science, ultimately contributing to advancements in healthcare, sustainability and technology.

11.6.1. *Overcoming biocompatibility concerns*

Overcoming biocompatibility concerns is essential for the successful implementation of smart biomaterials in medical applications. To address this challenge, researchers are actively engaged in refining surface modifications and functionalizations that enhance the biocompatibility of these materials. Strategies involve coating biomaterial surfaces with bioinspired molecules or biocompatible polymers to mitigate potential cytotoxicity. Additionally, rigorous biocompatibility assessments, including in vitro and in vivo studies, are being conducted to evaluate the safety and performance of smart biomaterials within biological systems. Advances in nanotechnology enable the design of biomaterials with tailored physicochemical properties, promoting compatibility with surrounding tissues. Collaboration between materials scientists, biologists and clinicians is crucial for ensuring comprehensive biocompatibility, with the goal of developing smart biomaterials that seamlessly integrate with biological environments, while minimizing adverse reactions or long-term toxicity concerns. Through these efforts, researchers aim to establish a robust foundation for the clinical translation of smart biomaterials, unlocking their full potential in medical applications.

11.6.1.1. *Impact on aquatic and terrestrial ecosystems*

The impact of smart biomaterials on aquatic and terrestrial ecosystems is a critical consideration in their development and application. While these materials offer innovative solutions for various fields, including medicine and environmental remediation, their introduction into ecosystems raises concerns. In aquatic environments, smart biomaterials could potentially interact with aquatic organisms, altering their behavior, growth or reproduction. The release of nanoparticles or other components during degradation might lead to unintended consequences on aquatic life, affecting the balance of ecosystems. Similarly, in terrestrial ecosystems, the introduction of smart biomaterials may influence soil composition, microbial communities and plant health. Understanding the long-term environmental fate, persistence and potential toxicity of these materials is crucial for mitigating adverse effects. Researchers and policymakers need to collaboratively establish guidelines for the responsible use, disposal and monitoring of smart biomaterials to minimize ecological impacts and ensure sustainable practices (Silva 2019).

11.6.1.2. *Strategies for eco-friendly biomaterials*

Developing eco-friendly biomaterials involves implementing a range of strategies aimed at minimizing environmental impact and promoting sustainable practices. One key approach is sourcing raw materials from renewable resources, such as plant-based polymers, to reduce dependency on finite fossil fuels. Emphasizing biodegradability in biomaterial design ensures that these materials can naturally break down without accumulating in the environment. Recycling-friendly biomaterials contribute to a circular economy by allowing for the recovery and reuse of valuable components. Green synthesis methods, incorporating principles of green chemistry, minimize the use of hazardous substances and reduce energy consumption during manufacturing. Waste valorization further enhances sustainability by utilizing by-products as feedstocks for biomaterial production. Attention to non-toxic components, reduced water usage and optimizing energy-efficient production processes contribute to overall environmental stewardship. Lifecycle assessments guide comprehensive evaluations of biomaterials, aiding in the identification of areas for improvement. By embracing these strategies, the development of eco-friendly biomaterials becomes a concerted effort to balance technological advancements with environmental responsibility, fostering a more sustainable future.

11.6.2. *Scalability and cost-effectiveness*

Achieving scalability and cost-effectiveness is paramount for the practical implementation of smart biomaterials across diverse applications. The scalability of production processes ensures that these advanced materials can be manufactured in quantities necessary for widespread utilization without compromising quality. Innovative synthesis techniques, such as scalable fabrication methods and continuous manufacturing, play a crucial role in efficiently meeting demand. Concurrently, addressing cost-effectiveness involves optimizing production processes to minimize resource consumption, energy requirements and overall production expenses. Leveraging sustainable raw materials, exploring economically viable synthesis routes and integrating recycling strategies contribute to cost-efficient biomaterial development. Collaborative efforts across academia, industry and regulatory bodies are essential to streamline production, reduce costs and enhance the overall accessibility of smart biomaterials for applications ranging from medicine to environmental remediation. As these materials move from the laboratory to real-world scenarios, striking a balance between scalability and cost-effectiveness becomes a key determinant for their successful integration into diverse industries.

11.6.2.1. *Industrial production challenges*

The industrial production of smart biomaterials faces several challenges that require strategic solutions to ensure efficiency and scalability. One significant challenge lies in the complexity of synthesizing biomaterials with precisely controlled properties on a large scale. Achieving reproducibility and maintaining the desired characteristics, especially when transitioning from laboratory-scale to industrial-scale production, poses a formidable hurdle. Additionally, the sourcing of high-quality raw materials and the potential variation in their properties can impact the consistency of the final biomaterial product. The need for specialized equipment and the development of cost-effective, large-scale manufacturing processes are further challenges. Balancing these technical challenges with regulatory compliance, ensuring biocompatibility and implementing sustainable practices in the industrial production of smart biomaterials, requires interdisciplinary collaboration and innovative approaches. Overcoming these challenges is crucial for realizing the full potential of smart biomaterials across various industrial applications, from healthcare to environmental technologies.

11.6.2.2. *Economic feasibility*

The economic feasibility of smart biomaterials is a critical aspect that influences their widespread adoption in various industries. Challenges associated with the high cost of raw materials, intricate manufacturing processes and specialized equipment present obstacles to achieving cost-effectiveness. To enhance economic viability, researchers and industry professionals must explore scalable production methods, optimize manufacturing efficiency and seek sustainable sourcing alternatives. Moreover, establishing robust recycling and waste management systems can contribute to overall cost reduction. Collaborative efforts across sectors, including academia, industry and regulatory bodies, are essential to develop innovative strategies for economic feasibility, making smart biomaterials more accessible and commercially viable. As advancements in technology and manufacturing techniques progress, addressing economic considerations will be pivotal in unlocking the full potential of smart biomaterials for applications in healthcare, environmental solutions and beyond (Holzapfel et al. 2013; Cecen et al. 2023).

11.6.3. *Integration with other remediation techniques*

The integration of smart biomaterials with other remediation techniques offers a synergistic and comprehensive approach to address environmental challenges. By combining the unique properties of smart biomaterials with traditional remediation methods such as phytoremediation, electrokinetic remediation or chemical treatments, a more efficient and tailored remediation strategy can be devised. For instance, smart biomaterials can be designed to selectively target specific contaminants, enhancing the overall effectiveness of remediation processes. Additionally, the adaptability and responsiveness of these biomaterials to environmental cues allow for real-time adjustments, optimizing remediation efforts based on changing conditions. The integration of smart biomaterials with existing techniques not only improves remediation efficiency but also provides opportunities for minimizing environmental impact and reducing the use of conventional, sometimes harsh, remediation agents. As research in this field progresses, the integration of smart biomaterials with other remediation approaches holds great promise for developing sustainable and highly effective solutions to mitigate environmental pollution.

11.6.3.1. *Synergy with biological degradation processes*

Smart biomaterials can exhibit a remarkable synergy with biological degradation processes, offering a novel and sustainable approach to environmental remediation. By integrating into ecosystems and leveraging the inherent capabilities of microorganisms, these biomaterials can enhance and accelerate the natural degradation of contaminants. The materials can be engineered to provide a conducive environment for microbial activity, promoting the growth of pollutant-degrading microorganisms. Additionally, the smart features of these biomaterials, such as responsiveness to environmental cues or controlled release of active agents, enable precise modulation of degradation processes. This synergy not only augments the efficiency of biological degradation but also allows for targeted and programmable interventions. As a result, smart biomaterials offer a promising avenue for the development of eco-friendly and adaptive solutions that work in harmony with nature to combat environmental pollution.

11.6.3.2. *Multi-faceted approaches for plastic removal*

Smart biomaterials contribute to multi-faceted approaches for plastic removal by offering innovative solutions to address the diverse challenges associated with plastic pollution. These biomaterials can be designed with tailored properties for efficient plastic degradation, complementing traditional methods like mechanical recycling and waste collection. Through photocatalytic degradation or enzymatic pathways, smart biomaterials exhibit the capability to break down plastics into environmentally benign components. Additionally, their responsive nature allows for targeted action, optimizing plastic removal processes based on environmental conditions. Integrating smart biomaterials into broader strategies, such as river and ocean cleanup initiatives, presents a holistic and adaptive approach to combat plastic pollution. By combining the strengths of these biomaterials with existing techniques, a multi-faceted and synergistic framework emerges, providing a more comprehensive solution to the global challenge of plastic waste in diverse ecosystems (Silva 2019; Cecen et al. 2023).

11.7. Conclusion

In conclusion, the design and structural modifications of smart biomaterials for photocatalytic degradation of micro- and nano-plastics represent a promising avenue in the quest for sustainable solutions to plastic pollution. Throughout this chapter, we have delved into the intricacies of developing biomaterials with tailored properties, leveraging their photocatalytic capabilities to address the challenges posed by plastic waste at the micro- and nanoscales. The integration of photoactive components, optimization of light absorption capacity and strategic modifications to enhance surface area and porosity have emerged as key factors in maximizing the efficiency of photocatalytic degradation processes. The consideration of charge carrier separation, incorporation of redox mediators and introduction of functional groups showcase the multifaceted nature of these biomaterials, ensuring their adaptability to the complex nature of plastic pollutants. Furthermore, this chapter has underscored the importance of factors such as biocompatibility, sustainability and scalability in the practical implementation of these biomaterials for large-scale plastic removal. As research advances, and interdisciplinary collaboration thrives, the potential for smart biomaterials to play a pivotal role in mitigating the impact of micro- and nano-plastics on our environment becomes increasingly evident. The journey toward a cleaner and healthier planet necessitates the continuous refinement of these biomaterials, their integration into real-world applications and a concerted global effort toward a circular and sustainable approach to plastic management.

11.8. References

Ahmad, A., Sharif, H., Luque, R., Alsaiari, M., Harraz, F.A. (2023). Biomaterials for photocatalysis: An overview. In *Biomaterials for Photocatalysis: Promising New Materials*. De Gruyter, Berlin.

Al Othman, Z.A., Alam, M.M., Naushad, M., Inamuddin, I., Khan, M.F. (2013). Inorganic nanoparticles and nanomaterials based on titanium (Ti): Applications in medicine. *Materials Science Forum*, 754. Trans Tech Publications Ltd, Seestrasse.

Aleksandrov, V. (2024). Nanoremediation: A new and emerging technology. In *Phytoremediation and Biofortification*, Singh, N.K., Afzal, S., Aftab, T. (eds). Apple Academic Press, Palm Bay, FL.

Amukarimi, S., Ramakrishna, S., Mozafari, M. (2021). Smart biomaterials – A proposed definition and overview of the field. *Current Opinion in Biomedical Engineering*, 19, 100311.

Bard, A.J. (1979). Photoelectrochemistry and heterogeneous photo-catalysis at semiconductors. *Journal of Photochemistry*, 10(1), 59–75.

Bellet, P., Gasparotto, M., Pressi, S., Fortunato, A., Scapin, G., Mba, M., Menna, E., Filippini, F. (2021). Graphene-based scaffolds for regenerative medicine. *Nanomaterials*, 11(2), 404.

Cecen, B., Hassan, S., Li, X., Zhang, Y.S. (2023). Smart biomaterials in biomedical applications: Current advances and possible future directions. *Macromolecular Bioscience*, 24, 2200550.

Daneshmandi, L., Barajaa, M., Tahmasbi Rad, A., Sydlik, S.A., Laurencin, C.T. (2021). Graphene-based biomaterials for bone regenerative engineering: A comprehensive review of the field and considerations regarding biocompatibility and biodegradation. *Advanced Healthcare Materials*, 10(1), 2001414.

Deng, J. (2022). Degradable pollen as potential candidates to replace plastic packages. Master's Thesis, Nanyang Technological University.

Dumbili, E. and Henderson, L. (2020). The challenge of plastic pollution in Nigeria. In *Plastic Waste and Recycling*, Letcher, T. (ed.). Academic Press, Cambridge, MA.

Fang, S. and Hu, Y.H. (2022). Thermo-photo catalysis: A whole greater than the sum of its parts. *Chemical Society Reviews*, 51(9), 3609–3647.

Helaïli, N., Boudjamaa, A., Kebir, M., Bachari, K. (2017). Efficient photo–catalytic degradation of malachite green using nickel tungstate material as photo–catalyst. *Environmental Science and Pollution Research*, 24(7), 6481–6491.

Holzapfel, B.M., Reichert, J.C., Schantz, J.T., Gbureck, U., Rackwitz, L., Nöth, U., Jakob, F., Rudert, M., Groll, J., Hutmacher, D.W. (2013). How smart do biomaterials need to be? A translational science and clinical point of view. *Advanced Drug Delivery Reviews*, 65(4), 581–603.

Intravaia, J.T., Graham, T., Kim, H.S., Nanda, H.S., Kumbar, S.G., Nukavarapu, S.P. (2023). Smart orthopedic biomaterials and implants. *Current Opinion in Biomedical Engineering*, 25, 100439.

Iroegbu, A.O.C., Ray, S.S., Mbarane, V., Bordado, J.C., Sardinha, J.P. (2021). Plastic pollution: A perspective on matters arising: Challenges and opportunities. *ACS Omega*, 6(30), 19343–19355.

Kathalingam, A., Santhoshkumar, P., Ramesh, S., Sivanesan, I., Kim, H.S. (2023). Biogenic polymer nanoparticles to remove hydrophobic organic contaminants from water. *Water Environment Research*, 95(10), e10935.

Khan, F. and Tanaka, M. (2017). Designing smart biomaterials for tissue engineering. *International Journal of Molecular Sciences*, 19(1), 17.

Kumar, R., Rauti, R., Scaini, D., Antman-Passig, M., Meshulam, O., Naveh, D., Ballerini, L., Shefi, O. (2021). Graphene-based nanomaterials for neuroengineering: recent advances and future prospective. *Advanced Functional Materials*, 31(46), 2104887.

Maksoud, F.J., de la Paz, M.F.V., Hann, A.J., Thanarak, J., Reilly, G.C., Claeyssens, F., Green, N.H., Zhang, Y.S. (2022). Porous biomaterials for tissue engineering: A review. *Journal of Materials Chemistry B*, *10*(40), 8111–8165.

Mazhandu, Z.S. and Muzenda, E. (2019). Global plastic waste pollution challenges and management. In *2019 7th International Renewable and Sustainable Energy Conference (IRSEC)*. IEEE, Agadir.

de Menezes, B.R.C., Rodrigues, K.F., da Silva Fonseca, B.C., Ribas, R.G., do Amaral Montanheiro, T.L., Thim, G.P. (2019). Recent advances in the use of carbon nanotubes as smart biomaterials. *Journal of Materials Chemistry B*, 7(9), 1343–1360.

Montoya, C., Du, Y., Gianforcaro, A.L., Orrego, S., Yang, M., Lelkes, P.I. (2021). On the road to smart biomaterials for bone research: Definitions, concepts, advances, and outlook. *Bone Research*, 9(1), 12.

Mydin, R.B.S.M.N., Hazan, R., FaridWajidi, M.F., Sreekantan, S. (2018). *Titanium Dioxide Nanotube Arrays for Biomedical Implant Materials and Nanomedicine Applications*, volume 23. IntechOpen, London.

Nair Sreekala, G., Abdullakutty, F., Beena, B. (2019). Green synthesis, characterization, and photo catalytic degradation efficiency of trimanganese tetroxide nanoparticle. *International Journal of Nano Dimension*, 10(4), 400–409.

Okeke, E.S., Ejeromedoghene, O., Addey, C.I., Atakpa, E.O., Bello, S.F., Ezeorba, T.P.C., Chukwudozie, K.I., Okoye, C.O. (2022). Panacea for the nanoplastic surge in Africa: A state-of-the-art review. *Heliyon*, 8, e11562.

Palafox González, P., Rosas Orta, L.R., Contreras, D., Ruíz Camacho, B., Zarraga, R., Martínez Alfaro, M., Alegría Torres, J., Vallejo Montesinos, J. (2023). Study of the side group effect on silane coupling agents, used in titanium dioxide. Case study: Polylactic acid composites. *Polymer-Plastics Technology and Materials*, 62, 1–24.

Patil, P.D., Nadar, S.S., Marghade, D.T. (2021). Photo-enzymatic green synthesis: The potential of combining photo-catalysis and enzymes. In *Advances in Green Synthesis: Avenues and Sustainability*, Inamuddin, Boddula, R., Ahamed, M.I., Khan, A. (eds). Springer, Cham.

Perez, R.A., Won, J.E., Knowles, J.C., Kim, H.W. (2013). Naturally and synthetic smart composite biomaterials for tissue regeneration. *Advanced Drug Delivery Reviews*, 65(4), 471–496.

Querebillo, C.J. (2023). A review on nano Ti-based oxides for dark and photocatalysis: From photoinduced processes to bioimplant applications. *Nanomaterials*, 13(6), 982.

Ratre, P., Nazeer, N., Soni, N., Kaur, P., Tiwari, R., Mishra, P.K. (2024). Smart carbon-based sensors for the detection of non-coding RNAs associated with exposure to micro (nano) plastics: An artificial intelligence perspective. *Environmental Science and Pollution Research*, 31, 1–24.

Sadeghi, A., Afshari, E., Hashemi, M., Kaplan, D., Mozafari, M. (2023). Brainy biomaterials: Latest advances in smart biomaterials to develop the next generation of neural interfaces. *Current Opinion in Biomedical Engineering*, 25, 100420.

Safavi, M.S., Bordbar-Khiabani, A., Walsh, F.C., Mozafari, M., Khalil-Allafi, J. (2023). Surface modified NiTi smart biomaterials: Surface engineering and biological compatibility. *Current Opinion in Biomedical Engineering*, 25, 100429.

Sharma, S.S.A., Bashir, S., Kasi, R., Subramaniam, R.T. (2022). The significance of graphene based composite hydrogels as smart materials: A review on the fabrication, properties, and its applications. *FlatChem*, 33, 100352.

Shodeinde, A.B., Murphy, A.C., Oldenkamp, H.F., Potdar, A.S., Ludolph, C.M., Peppas, N.A. (2020). Recent advances in smart biomaterials for the detection and treatment of autoimmune diseases. *Advanced Functional Materials*, 30(37), 1909556.

Shukla, S., Khan, R., Hussain, C.M. (2020). Nanoremediation. In *The Handbook of Environmental Remediation: Classic and Modern Techniques*, C.M. Hussain (ed.). The Royal Society of Chemistry.

Silva, L.P. (2019). Current trends and challenges in biofabrication using biomaterials and nanomaterials: Future perspectives for 3D/4D bioprinting. In *3D and 4D Printing in Biomedical Applications: Process Engineering and Additive Manufacturing*, Maniruzzaman, M. (ed.). Wiley-VCH Verlag, 373–421.

Sudhakar, M.S., Aggarwal, A., Sah, M.K. (2020). Engineering biomaterials for the bioremediation: Advances in nanotechnological approaches for heavy metals removal from natural resources. In *Emerging Technologies in Environmental Bioremediation*, Shah, M.P., Rodriguez-Couto, S., Sevinc Sengor, S. (eds). Elsevier, Amsterdam.

Tariq, S., Ahmad, A., Luque, R., Alsaiari, M., Harraz, F.A. (2023). 3 recent trends in biomaterials for photodegradation of dyes. In *Biomaterials for Photocatalysis: Promising New Materials*, Ahmad, A., Sharif, H., Luque, R., Alsaiari, M., Harraz, F.A. (eds). De Gruyter, Berlin, 45.

Triantafyllopoulos, I.K. and Papaioannou, N.A. (2022). Application of nanotechnology in medicine. Smart biomaterials and biosensors. *Acta Orthopaedica Et Traumatologica Hellenica*, 73(3).

Wang, Q., Huang, J.Y., Li, H.Q., Chen, Z., Zhao, A.Z.J., Wang, Y., Zhang, K.Q., Sun, H.T., Al-Deyab, S.S., Lai, Y.K. (2016). TiO_2 nanotube platforms for smart drug delivery: A review. *International Journal of Nanomedicine*, 4819–4834.

Wang, S., Wang, C., Pan, W., Sun, W., Yang, D. (2021). Two-dimensional silicon for (photo) catalysis. *Solar RRL*, 5(2), 2000392.

Zamuner, A., Brun, P., Scorzeto, M., Sica, G., Castagliuolo, I., Dettin, M. (2017). Smart biomaterials: Surfaces functionalized with proteolytically stable osteoblast-adhesive peptides. *Bioactive Materials*, 2(3), 121–130.

Zhang, K., Wang, S., Zhou, C., Cheng, L., Gao, X., Xie, X., Sun, J., Wang, H., Weir, M.D., Reynolds, M.A. et al. (2018). Advanced smart biomaterials and constructs for hard tissue engineering and regeneration. *Bone Research*, 6(1), 31.

Zulfikar, M., Nadeem, R., Javed, T., Jilani, M.I., Javed, I. (2021). Green synthesis of Fe nanoparticles by using *Mangifera indica* extract and its application in photo-catalytic degradation of dyes. *Water Science and Technology*, 83(7), 1739–1752.

12

Nanocomposites: Sustainable Resources for Photodegradation of Micro- and Nano-Plastics

The issue of plastic pollution, especially the spread of micro-plastics (MPs) and nano-plastics (NPs), has generated an urgent demand for innovative and sustainable solutions. Conventional plastic disposal methods like burning and burying produce harmful MPs and NPs that can accumulate and transport toxins, posing environmental and health risks. Scientific research on this issue started in the early 21st century, with recent efforts gaining momentum. Within the context of tackling MP and NP waste, there are noteworthy strategies to consider. These strategies encompass the use of light-sensitive techniques to convert plastics into low-density hydrocarbons and the promotion of recycling. This chapter delves into the cutting-edge of science and technology, exploring the realm of photocatalytic nanocomposites. Nanocomposites bring together the distinct characteristics of nanomaterials with customized matrices, encompassing a range of materials. These include classic inorganic semiconductor-oxide nanoparticles such as TiO_2 and ZnO, nanocomposites with AgNPs, Fe_2O_3, MoS_2, polymeric nanocomposites, quantum dots and carbon-based nanotubes. These materials offer promising methods to enhance the decomposition of MP and NP pollutants through photodegradation.

This chapter discusses the science behind processing nanocomposites and modifying their surfaces, explores their origins, development, and characterization in order to strengthen their ability to drive photocatalysis effectively, while also addressing eco-friendly and sustainable techniques for producing these materials. This research significantly contributes to the advancement of advanced nanocomposites and technologies aimed at combatting the global environmental issue of plastic pollution.

Chapter written by Nisha SHANKHWAR, Pinki SINGH, Jewel THOMAS and Satyendra SINGH.

12.1. Introduction

12.1.1. *Addressing environmental challenges with nanocomposites*

In the realm of environmental stewardship, the emergence of micro- and nano-pollutants, characterized by their minuscule measurements in micrometers or nanometers, presents a pressing concern. These pollutants originate from various sources including industrial activities, agricultural practices and the gradual degradation of larger materials. Within this intricate landscape, micro-pollutants span from 0.1 to 100 μm, whereas nano-pollutants fall within the nanoscale range of 1 to 100 nm. Notably, within the scope of plastic pollution, MPs and NPs represent a subset of these pollutants, with MPs ranging from 0.1 μm to 5 mm and NPs existing at the nanoscale, often resulting from the breakdown of larger plastic materials or the degradation of MPs. The production of synthetic plastics has seen exponential growth in various sectors globally, with diverse applications ranging from textile fiber production, to packaging in the food industry, aimed at reducing wastage and transportation costs. With a market boasting over 30,000 different natural and synthetic polymers, the identification of potentially harmful polymers and their impact on human health is pivotal for responsible usage. Notably, advancements have led to the development of biodegradable plastics (BPs) to mitigate environmental concerns. However, despite these efforts, over 250,000 tons of plastic waste continue to accumulate in marine environments worldwide.

The escalating issue of plastic pollution in marine ecosystems, underscored by the distressing sight of dead marine animals laden with plastic waste, has become increasingly apparent in recent years. Micro-plastics (5–1,000 μm) have been detected in commercially important fish from the Mediterranean, and even in drinking water sourced from plastic bottles. This article delves into the global challenge posed by MPs and NPs, emphasizing their pervasive presence in oceans, seas, rivers and lakes. Notably, with global plastic production reaching nearly 350 million tonnes in 2017, China emerged as the leading producer, accounting for 29.4%, followed by Europe with 18.5%.

Experimental findings indicate that three weeks of UV irradiation can yield nanoscale plastic particles. Nanoparticles (1–100 nm) exhibit distinct chemical and physical properties compared to their larger counterparts, highlighting the need for comprehensive understanding and proactive measures. Recent breakthroughs in the field include the successful degradation of low-density polyethylene film (LDPE) in water, utilizing visible light-excited heterogeneous zinc oxide photocatalysts, along with polyethylene (PE) plastic, employing a polypyrrole/TiO_2 (PPy/TiO_2) nanocomposite as a photocatalyst. Despite the limited knowledge about nanoplastics in the environment, they are recognized as an emerging concern of significant magnitude.

Furthermore, the behavior of MPs and NPs warrants careful examination. Microplastics can infiltrate organisms through ingestion, potentially posing adverse effects on various tissues and cells. Studies have revealed the contamination of shellfish and commercially important fish with MPs, serving as a concerning route of exposure for human consumers. Notably, consumers in European countries with high shellfish consumption ingest significant amounts of MPs annually. Seafood species consumed whole, such as mollusks and crustaceans, pose a greater contamination risk compared to gutted fish or peeled shrimp. Additionally, MPs have been identified in drinking water stored in plastic bottles and other packaging materials, underscoring the pervasiveness of this issue and the imperative for proactive mitigation strategies.

12.2. Photocatalytic degradation of micro- and nano-plastics

This chapter offers an extensive examination of various approaches to plastic degradation, with a particular focus on photocatalytic degradation. It categorizes plastic recycling methods into primary, secondary, tertiary and quaternary categories. Of particular note is the discussion on tertiary methods, or chemical recycling, which break down plastics into smaller molecular weight polymers for their use in the chemical industry. The discourse also encompasses thermal and mechanical degradation methods, delving into the thermal degradation behaviors of PE and polypropylene (PP). Furthermore, the article discusses the biodegradation of plastics in seawater, illustrating the release of dissolved organic carbon and the stimulation of microbial activity. Notable studies on the utilization of carbon-based molecules for breaking down marine MPs are also highlighted.

The spotlight shines on photocatalytic methods in the review, exemplified by successful cases such as the degradation of LDPE using visible light-excited zinc oxide photocatalysts and PE plastic with a polypyrrole/TiO_2 nanocomposite as a photocatalyst. The potential of photocatalytic processes, leveraging renewable energy sources like solar energy, emerges as a promising solution to plastic pollution.

This review presents an overview of the latest catalytic and photocatalytic methods for degrading MPs. Additionally, it addresses the global plastic production quantities and their impact on human health, underscoring the emerging concern of MPs and NPs.

12.3. Nanocomposites in environmental remediation

Nanocomposites herald a paradigm shift in materials science, melding nanotechnology with composite materials to unlock unprecedented possibilities. The fundamental principles governing nanocomposites, including synthesis, composition and structure, underpin their exceptional mechanical, thermal, electrical and optical properties. Since they are precision-engineered, these nanocomposites find versatile applications, notably reshaping environmental remediation efforts against air, water and soil pollution challenges. Engineered nanoscale components, especially carbon-based nanomaterials, contribute to advanced water disinfection, pollutant detection and removal. Exemplified by materials like graphene and metal–organic frameworks, nanocomposites showcase multifaceted capabilities, providing innovative solutions to combat pressing environmental issues.

This section provides a comprehensive overview of the pervasive environmental challenges posed by MP and NP pollution, accentuating their integration into daily life. It underscores the imperative to address these challenges and introduces the role of nanocomposites as a potential solution within the broader context of environmental remediation.

12.3.1. *Understanding nanocomposites*

Nanocomposites epitomize the fusion of nanotechnology principles with composite materials to fabricate structures with enhanced properties. At their core, nanocomposites comprise a matrix material reinforced with nanoscale particles. Typically composed of two or more materials with disparate characteristics, with at least one in a microscopic scale, a composite can be classified as a nanocomposite when at least one of the reinforcement dimensions falls within the nanometer range (from 10 to 200 nm). This amalgamation yields materials exhibiting superior mechanical, thermal, electrical and optical properties compared to traditional composites (Hamd et al. 2022).

Delving into the foundational aspects of nanocomposites, this section explores their composition, structure and the unique properties that render them conducive to catalyze the photodegradation process. It elucidates how nanocomposites, with nanoscale filler materials dispersed in a polymer matrix, exhibit enhanced mechanical, thermal and optical characteristics crucial for effective photodegradation.

12.3.2. *Enhanced mechanical, thermal, electrical and optical properties*

Nanocomposites stand out in the domain of MP and NP photodegradation due to their exceptional mechanical, thermal, electrical and optical attributes, distinguishing them from conventional materials. One of their key advantages lies in their enhanced mechanical properties. By incorporating nanoscale reinforcements, nanocomposites gain superior strength, stiffness and toughness. This enhancement is attributed to the high surface area and aspect ratio of the nanofillers, facilitating efficient stress transfer within the composite structure. Additionally, the small size of the nanoparticles reduces the likelihood of defects, bolstering the overall mechanical robustness of nanocomposites (Fa et al. 2008). Besides mechanical improvements, nanocomposites exhibit remarkable thermal conductivity owing to the unique properties of nanoscale fillers. The integration of these fillers not only enhances mechanical strength, stiffness and toughness but also reinforces resistance to deformation and fracture, essential factors in addressing the challenges posed by MP and NP pollution. Within the context of photodegradation, nanocomposites featuring specialized nanofillers demonstrate unparalleled thermal properties, promoting efficient heat transfer – an advantageous trait for developing advanced materials tailored for MP and NP degradation across diverse applications. Moreover, the utilization of carbon-based nanomaterials in nanocomposites enhances electrical conductivity, broadening their utility in creating sensors and electronic components aimed at detecting and mitigating MP and NP pollutants. Additionally, the optical properties of these nanocomposites, influenced by carefully selected nanofillers, offer tunability in color, transparency and light interactions, thereby contributing to the development of innovative solutions for identifying and addressing MP and NP particles. The synergistic amalgamation of these diverse properties positions nanocomposites as versatile materials with significant applications in advancing technologies related to MP and NP photodegradation. Through ongoing research, nanocomposites hold the promise of further breakthroughs and expanded applications in the continuous efforts to combat plastic pollution in the realms of technology and engineering.

12.3.3. *Nanocomposite composition and structure*

Nanocomposites represent engineered materials that fuse a matrix substance with nanoscale fillers to achieve enhanced properties. The composition and structure of nanocomposites play a pivotal role in determining their overall performance and functionality. These composites consist of fundamental building blocks, with the matrix material serving as the continuous phase that envelops and supports the nanofillers. Polymers, metals, ceramics and various combinations thereof commonly serve as matrix materials, offering a versatile platform for tailoring nanocomposite properties to specific application requirements. The second major component, nanoscale fillers, imparts distinct characteristics to nanocomposites and can manifest in forms such as nanoparticles, nanotubes, nanofibers or other nanosized structures. Carbon-based materials such as graphene and carbon nanotubes, metal oxides like titanium dioxide (TiO_2) and zinc oxide (ZnO), as well as clay nanoparticles such as montmorillonite, are common nanofillers. The selection of nanofillers depends on the desired properties and applications of the nanocomposite. The structure of nanocomposites is intricately tied to the arrangement of nanofillers within the matrix, with uniform dispersion being crucial for optimizing their properties. Achieving this uniformity is essential for realizing the full potential benefits, as agglomeration or poor dispersion can compromise material performance. Factors such as surface energy, matrix compatibility and processing conditions influence nanofiller dispersion. Functionalization with surfactants or chemical treatments enhances nanofiller compatibility with the matrix, facilitating better dispersion and maximizing their overall contribution to material properties.

12.4. Synthesis of nanocomposites

Focusing on the practical aspects, this section examines various methods for synthesizing nanocomposites, such as in situ polymerization, melt blending and solution mixing. It underscores the importance of optimizing synthesis parameters to achieve desired properties, ensuring the effectiveness of nanocomposites in the photodegradation of MPs and NPs.

An essential aspect of harnessing nanocomposites for photodegradation involves understanding their synthesis. The chapter delves into various synthesis methods, including in situ polymerization, melt blending and solution mixing, each influencing the final properties of the nanocomposite. Special attention is given to optimizing synthesis parameters to achieve desired characteristics, such as improved light absorption and efficient degradation capabilities.

The synthesis of nanocomposites entails dispersing nanoscale fillers, such as nanoparticles or nanotubes, within a matrix material. Achieving a uniform distribution of these fillers is crucial for optimizing the material's performance (Llorente-García et al. 2020). Foundational methods like in situ polymerization, melt blending and solution mixing are complemented by newer approaches such as layer-by-layer assembly, chemical vapor deposition and electrospinning. The choice of synthesis method depends on the specific requirements of the desired nanocomposite, emphasizing the importance of a nuanced approach in materials science.

12.4.1. *Synthesis techniques*

– In situ polymerization: simultaneously forming the polymer matrix and incorporating nanofillers, in situ polymerization involves mixing monomers and nanofillers, initiating polymerization to enhance control over nanofiller loading and distribution. Catalyst systems and techniques like ultrasonication may be employed for improved efficiency.

– Melt blending: widely used melt blending involves mixing nanofillers with a molten polymer matrix, promoting dispersion through mechanical mixing and shear forces. Despite challenges in achieving uniform distribution, its simplicity and scalability make it prevalent for various polymer nanocomposites.

– Solution mixing: involving dispersal in a solvent, solution mixing allows for precise control over nanofiller dispersion. Challenges include environmental impact and limited solubility of certain nanofillers, but it offers a uniform distribution, minimizing agglomeration.

– Layer-by-layer assembly: alternating layers of nanofillers and matrix create a stratified structure, offering fine control over properties at the nanoscale. Electrostatic interactions or hydrogen bonding drive the layering process, commonly used for thin films and coatings.

– Chemical vapor deposition (CVD): CVD synthesizes nanocomposites in thin film form by depositing nanomaterials through chemical reactions of precursor gases. This technique provides precise control over film thickness and composition, commonly applied in electronic device and sensor fabrication.

– Electrospinning: utilized for nanocomposite fiber fabrication, electrospinning involves subjecting a polymer solution with nanofillers to an electric field. This results in the formation of nanofibers with enhanced properties, widely used in fields such as tissue engineering and filtration.

12.4.2. *Optimization of synthesis parameters*

The optimization of synthesis parameters stands as a pivotal step in crafting nanocomposites with tailored properties, ensuring efficiency, reproducibility and precise control. These parameters encompass a spectrum of variables including temperature, reaction time, precursor concentrations and mixing techniques. Temperature exerts a significant influence on reaction kinetics, phase formation and nanofiller dispersion, necessitating careful adjustment to achieve optimal results. Balancing the reaction time is paramount to steer clear of under-reaction or over-reaction, striking a delicate equilibrium for attaining desired nanocomposite properties. Precursor concentrations wield a direct impact on composition, nanofiller loading and distribution, demanding meticulous optimization to uphold stability and reproducibility. The choice of mixing technique, whether it be mechanical stirring or ultrasonication, plays a critical role in fostering uniform dispersion and enhancing overall material homogeneity. Moreover, optimization of pH and chemical environment proves vital for effective nanocomposite formation, where pH control becomes instrumental in sensitive reactions, and the selection of solvents and additives influences nanofiller stability and matrix interactions. Fine-tuning these conditions heightens compatibility and facilitates desired reactions. In methods such as chemical vapor deposition, factors like pressure and atmospheric conditions influence nanomaterial growth, necessitating meticulous optimization for reproducibility and property adjustments. Subsequent post-treatment processes, such as annealing or heat treatment, further augment nanocomposite properties by optimizing crystallinity and interfacial interactions. Striking a harmonious balance between conflicting factors, such as maximizing dispersion while minimizing agglomeration, emerges as the central goal of optimization. This iterative process guarantees reproducibility, efficiency and cost-effectiveness in both research and industrial applications.

12.5. Photodegradation mechanisms

Offering profound insights into the underlying mechanisms, this section illuminates the intricate process of photodegradation, wherein nanocomposites harness light energy to engender reactive species that catalyze the breakdown of plastic polymers. The synergistic interplay between nanocomposites and light exposure is dissected to underscore how nanocomposites augment the efficacy of the photodegradation process.

The mechanisms underpinning photodegradation in nanocomposites entail intricate processes propelled by the interaction between nanomaterials and light energy. The primary objective revolves around leveraging the photocatalytic properties of nanocomposites to initiate and expedite the degradation of targeted pollutants, such as MPs and NPs. Herein lie the key mechanisms involved in the photodegradation process. The photodegradation of plastics constitutes a multifaceted mechanism reliant on the interaction among plastics, light and suitable photocatalysts. Plastics, when exposed solely to ultraviolet (UV) or sunlight irradiation, undergo processes such as chain scission, branching, cross-linking and oxidation reactions directly within their polymer macromolecules. However, due to the inherently slow natural degradation of plastics, which spans hundreds of years, achieving efficient photodegradation necessitates the involvement of catalysts.

Catalysts suited for the task play a pivotal role in facilitating and expediting the photodegradation process.

12.5.1. *Mechanism of photocatalytic reaction*

Upon exposure to light, these catalysts absorb photons, thereby promoting electrons from the valence band to the conduction band, ultimately generating electron–hole pairs. This phenomenon occurs in close proximity to the plastics, thereby initiating the photodegradation process.

12.5.2. *Energy absorption and electron–hole pair generation*

Plastics, renowned for their poor light absorption, benefit from the presence of photocatalysts. Upon exposure to light, the photocatalyst seizes photons, consequently propelling electrons from the valence band to the conduction band. This transition instigates the formation of electron–hole pairs within the interior of the catalyst, which is pivotal for setting off subsequent reactions.

12.5.3. *Charge aggregation and surface migration*

The electrons (eCB^-) and holes (hVB^+) generated within the photocatalyst coalesce and migrate toward the surface. This crucial process aims to forestall the swift recombination of electron–hole pairs, ensuring their intact arrival at the surface, primed to engage in the degradation reactions.

12.5.4. *Redox reactions at the interface*

As the aggregated charges reach the surface of the photocatalyst, they incite redox reactions at the interface with the plastics. These reactions entail the transfer of electrons and holes between the photocatalyst and the plastic, precipitating various chemical transformations. This step constitutes a pivotal juncture for the actual degradation of the plastic.

12.5.5. *Oxygen evolution reaction (OER) in an oxygen-rich atmosphere*

In an oxygen-rich atmosphere, molecular oxygen (O_2) reacts with free electrons in the valence band, yielding the formation of superoxide anion radicals. Concurrently, holes in the valence band migrate to the surface and engage in reactions with water (H_2O), generating hydroxyl radicals or directly interacting with the plastics. This OER process engenders highly oxidizing radical species, thereby facilitating nonselective oxidation reactions on the plastic.

12.5.6. *Hydrogen evolution reaction (HER) in an inert atmosphere*

Conversely, in an inert atmosphere typically comprising nitrogen (N_2) or argon (Ar), another form of photocatalytic reaction, termed the hydrogen evolution reaction (HER), transpires. In this scenario, the generated holes catalyze the transformation of plastics, yielding selective value-added chemicals. Simultaneously, photoinduced electrons partake in reactions with protons to yield hydrogen gas. This HER process enables the production of valuable compounds from plastic waste.

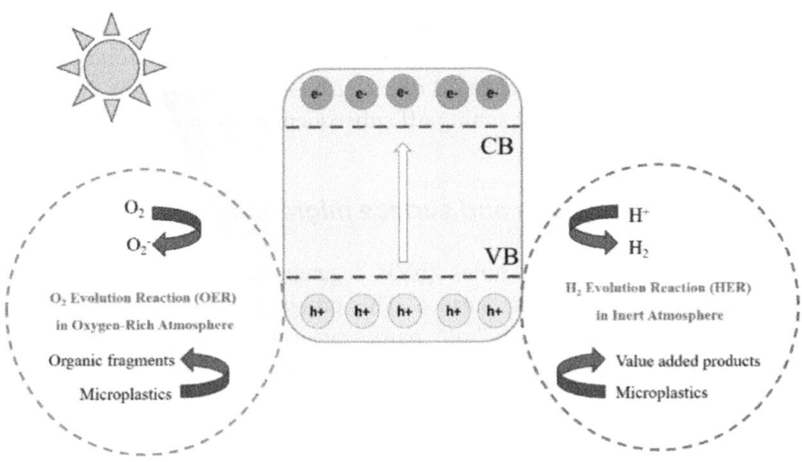

Figure 12.1. *Schematic representation of Mechanism of photocatalytic reaction in different environments. For a color version of this figure, see www.iste.co.uk/singh/photodegradation.zip*

12.6. Nanocomposites for micro- and nano-plastic degradation

The growing concern surrounding MP pollution has spurred extensive research into effective degradation methods, with catalytic photodegradation emerging as a promising avenue. Various catalysts have been investigated, each showcasing unique capabilities in breaking down these persistent pollutants. Let us delve into the diverse range of catalysts and comprehend their applications and mechanisms in addressing the environmental impact of MPs.

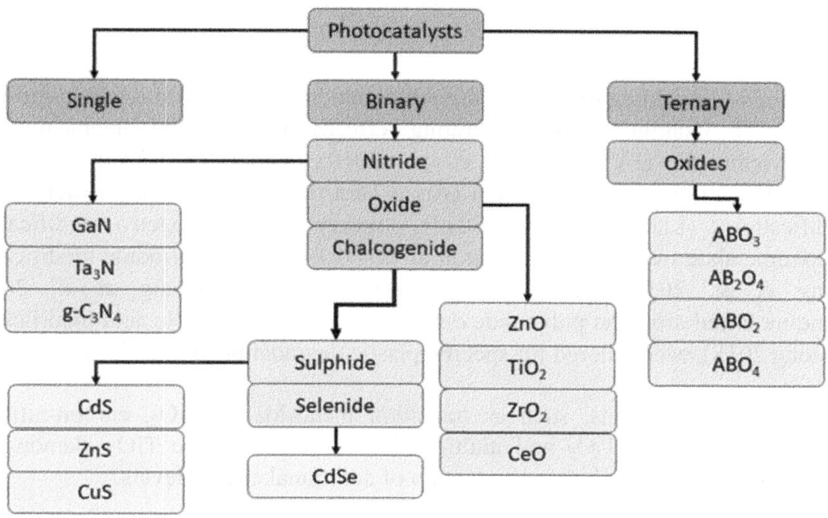

Figure 12.2. *Flow chart representation of some binary and ternary semiconductor photocatalysts (Hamd et al. 2022). For a color version of this figure, see www.iste.co.uk/singh/photodegradation.zip*

12.6.1. *Titanium dioxide and modified composites*

Titanium dioxide (TiO_2), a transition metal oxide, has emerged as a vital photocatalyst with wide-ranging applications, from environmental remediation to photovoltaic cells. Its role in the photodegradation of MPs is particularly noteworthy, with successful applications in decomposing various pollutants, including high-density polyethylene (HDPE) MPs (Llorente-García et al. 2020). Operating primarily under the UV region, TiO_2 catalyzes light-induced redox reactions and hydrophilic conversion, presenting new possibilities for applications, especially in building materials (Ma et al. 2019). The timeline of studies on modified TiO_2 variants showcases ongoing efforts to enhance efficiency in MPs degradation.

Modified TiO_2-based nanocomposites have been extensively explored for their efficiency as photocatalysts in various environmental applications. One notable modification involves the use of phthalocyanine, an artificial pigment comprising four isoindole components in a large ring. The $FePc/TiO_2$ composite demonstrated efficacy in degrading polystyrene (PS) (Fa et al. 2008). Similarly, $CuPc/TiO_2$ was synthesized through two methods, one involving commercial TiO_2 in ethanol-CuPc solution and the other utilizing sol-gel synthesized TiO_2 with added ethanol-CuPc solution. These modifications enhance the catalytic activity for the decomposition of PE and PS.

Vitamin C modification entails mixing commercial TiO_2 with a tetrahydrofuran solution of vitamin C, demonstrating effectiveness in the degradation of polyvinylchloride (PVC) (Yang et al. 2010). Other modifications include protein-based nitrogen modification (Ariza-Tarazona et al. 2019), polyacrylamide modification (Liang et al. 2013), mesoporous nitrogen modification, polyoxometalate modification (Yang et al. 2011b), bismuth oxyiodide modification (Yang et al. 2011a), graphene oxide-reduced TiO_2 (Zhang et al. 2020), amine-modified aromatic polyamide dendrimer TiO_2 and ascorbic acid modification (Soitong 2018), each tailored for specific plastic degradation applications.

Doped TiO_2 variants, such as transition metal-loaded TiO_2, carbon-nitrogen TiO_2, graphite-doped TiO_2 and multi-walled carbon nanotube TiO_2, demonstrate enhanced activity through the introduction of additional energy levels.

12.6.2. *Zinc oxide and modified composites*

While titanium dioxide (TiO_2) is widely recognized as a photocatalyst, substantial interest lies in the extensive exploration of zinc oxide (ZnO) nanoparticles due to their remarkable photocatalytic efficacy against diverse contaminants. ZnO stands out for its excellent stability, nontoxic nature and cost-effectiveness, making it a compelling alternative to TiO_2. ZnO exhibits versatility by being applicable for both UV and visible irradiation, displaying excellent absorptivity across a significant portion of the solar spectrum. Furthermore, ZnO boasts favorable mechanical, optical and electrical characteristics, along with inherent antifouling and antibacterial properties.

In investigations focusing on solid-phase degradation, ZnO modified with PVP exhibited enhanced performance, demonstrating improved photocatalytic efficiency and tensile strength in LDPE films (Lam et al. 2022). Another approach involved the use of iron-doped ZnO, which resulted in heightened mass loss and displayed antibacterial activity in LDPE films (Lam et al. 2021). These outcomes underscore the potential of engineered ZnO composites in solid-phase MP degradation.

Shifting to aqueous-phase degradation, UV-assisted thermo-photocatalysis with ZnO was employed to break down PP, emphasizing the significance of reaction temperature and MP size (Razali et al. 2020). Additionally, the utilization of ZnO nanorods for visible-light degradation of PP led to a significant reduction in particle volume (Uheida et al. 2021). Studies comparing ZnO nanorods and platinum-doped

ZnO in degrading LDPE film highlighted the crucial role of catalyst surface area (Tofa et al. 2019).

Collectively, these investigations emphasize the versatility and effectiveness of ZnO-based nanocomposites as catalysts for both solid and aqueous-phase MP degradation, offering promising solutions for environmental remediation strategies.

12.6.3. *Zirconium dioxide and modified composites*

While not as commonly associated with photocatalysis as TiO_2 or ZnO, zirconium dioxide (ZrO_2) and its composites have displayed potential in certain applications and composite forms, showcasing catalytic and photocatalytic properties (Bandara et al. 2017).

12.6.4. *Tungsten trioxide and modified composites*

As a semiconductor oxide, tungsten trioxide (WO_3) and its composites have shown promise in degrading environmental pollutants through its photocatalytic activity (Wang et al. 2023). Ongoing research explores its potential for environmental remediation.

12.6.5. *Carbon nitride-based composites*

Carbon nitride (C_3N_4) and its composites, such as C_3N_4-metal hybrids or C_3N_4-graphene composites, have been explored for their potential in degrading MPs, aiming to enhance photocatalytic activity (Cao et al. 2022; Chai et al. 2023).

12.6.6. *Perovskite-like materials*

Metal halide perovskites, known for their high absorption coefficients and tunable bandgaps, have been explored for their potential in photocatalysis, including MP degradation (Nkwachukwu and Arotiba 2021; Venkataramana et al. 2021; Cao et al. 2022).

12.7. Photodegradation efficiency

The efficiency of photocatalysts in degrading MPs is significantly influenced by various operating parameters and material characteristics. The bandgap of photocatalysts plays a crucial role, where a narrower bandgap enhances degradation efficiency, broadening the range of suitable materials (Fonseca-Cervantes et al. 2020). Activation energies during the degradation process vary based on plastic properties, with different plastics exhibiting distinct activation energies for the formation of degradation products. Thermodynamics, as indicated by Gibbs energy changes (ΔG0), offer insights into the feasibility of degradation reactions. Doped semiconductors like Fe/Ag-doped TiO_2 act as electron–hole pair traps, preventing rapid recombination and enhancing photocatalytic efficiency. Surface modifications in hybrid catalysts, such as CuPc–TiO_2 and FePc–TiO_2, extend light absorption into the visible spectrum, promoting more efficient reactions. Composite photocatalysts, combining different materials, achieve uniform dispersion, higher stability and enhanced degradation rates compared to single-component catalysts. The microstructure of composites prevents semiconductor aggregation on polymers, contributing to their effectiveness in plastic degradation.

Various factors impact photocatalytic degradation efficiency, including solution pH, termination rate and the generation of free radicals essential for extracting hydrogen atoms from polymer chains. pH influences reactive species generation and MP adsorption onto photocatalyst surfaces, with acidic conditions typically favoring degradation. Temperature affects the reaction rate and reactive species generation, influencing overall efficiency (Ariza-Tarazona et al. 2020). MP concentration influences available active sites on photocatalyst surfaces, affecting morphology and performance. MP size and shape also influence adsorption on photocatalyst surfaces. Smaller MPs tend to adsorb better, while larger MPs may face challenges due to their substantial particle size (Joo et al. 2021). The shape of MPs impacts dispersion behavior and contact with the semiconductor, affecting photocatalytic systems aiming for complete mineralization to CO_2 and H_2O. Visible light photocatalysis studies using mesoporous N-TiO_2 coatings on HDPE and LDPE MPs indicate that smaller MPs experience greater degradation, while film-shaped MPs exhibit reduced degradation (Llorente-García et al. 2020). Understanding these factors is vital for optimizing photocatalytic systems and enhancing their performance in plastic waste remediation.

12.7.1. *Light absorption*

Nanocomposites are designed with specific nanomaterials, often semiconductors, that have the ability to absorb light in the UV or visible range. The absorbed photons impart energy to electrons within the nanomaterial, elevating them to higher energy states.

12.7.2. *Electron–hole pair generation*

The absorbed energy creates electron–hole pairs, where electrons are excited to higher energy levels (conduction band) and leave behind positively charged holes in the valence band. This separation of charges is a crucial step in generating reactive species for subsequent reactions.

12.7.3. *Reactive oxygen species formation*

Electrons in the conduction band and holes in the valence band participate in redox reactions, leading to the formation of reactive oxygen species (ROS) such as superoxide radicals ($\bullet O2^-$), hydroxyl radicals (•OH) and singlet oxygen ($^1O_2$). ROS are highly reactive and play a key role in breaking down the chemical structure of MPs and NPs.

12.7.4. *Interaction with micro- and nano-plastics*

The generated ROS interact with the surface of MPs and NPs, initiating oxidation and fragmentation processes. ROS attack the polymer chains, breaking them into smaller fragments and ultimately leading to the degradation of plastic particles.

12.7.5. *Mineralization*

In the final stages of photodegradation, the smaller plastic fragments undergo further oxidation until they are mineralized into harmless by-products, such as carbon dioxide, water and other simple compounds.

The overall goal is to convert complex plastic structures into environmentally benign substances.

12.8. Applications and case studies

Real-world applications take center stage in this section, showcasing how nanocomposites are deployed in diverse environmental settings. The discussion includes water treatment and soil remediation applications, along with case studies from various industries, demonstrating the versatility and efficacy of nanocomposites in addressing MP and NP pollution.

12.8.1. *Nanocomposites for micro- and nano-plastic pollution control*

Nanocomposites, featuring nanoscale constituents embedded within a matrix, are pivotal in combatting environmental pollution, excelling in remediation efforts across air, water and soil. In the realm of MP and NP photodegradation, they offer promising avenues for mitigating plastic pollution in diverse environmental contexts. The integration of nanoscale components, derived from carbon-based nanomaterials like carbon nanotubes, graphene and their derivatives, imbues nanocomposites with exceptional physical, electronic and chemical properties. This unique amalgamation of attributes renders nanocomposites highly versatile for addressing a myriad of environmental challenges. Notably, carbon-based nanomaterials, particularly graphene-based nanocomposites, play a vital role in pollutant quantification and detection, owing to their outstanding surface area, mechanical robustness and electron mobility. In the domain of water purification, nanocomposites serve as indispensable tools in disinfection processes, where carbon and metal-based nanomaterials function as advanced electrodes and surfaces, facilitating effective disinfection. The distinctive electrical properties of graphene, combined with the antimicrobial potential of metal nanoparticles, contribute to the advancement of water disinfection technologies. Furthermore, in tackling the challenge of pollutant removal from water resources, nanocomposites offer innovative solutions. Graphitic carbon nitride (g-C_3N_4), a two-dimensional organic polymeric material, demonstrates efficacy as both an adsorbent and photocatalyst for eliminating and degrading various contaminants in water. Exploring the versatile capabilities of nanocomposites, including Bi-based oxide nanostructures, metal–organic frameworks (MOFs) and 2D materials like graphene and MOFs, unlocks their full potential in the ongoing pursuit of sustainable environmental solutions.

12.8.2. *Application in photodegradation*

The application of nanocomposites in photodegradation extends to water and air purification processes. In water treatment, nanocomposite-based photocatalysts prove effective in degrading organic pollutants such as dyes, pesticides and pharmaceutical residues, offering a sustainable and efficient solution for water remediation (Liang et al. 2013).

This chapter highlights real-world applications of nanocomposites in the photodegradation of MPs and NPs, featuring case studies from diverse industries and environmental settings. These applications underscore the versatility and efficacy of nanocomposites in addressing challenges posed by plastic pollution. Encompassing scenarios ranging from water treatment to soil remediation, these case studies illustrate the potential of nanocomposites in diverse environmental contexts.

12.9. Challenges and considerations/future directions

While acknowledging the promise of nanocomposites, this section delves into the challenges and considerations associated with their utilization. It examines potential environmental impacts, scalability issues and ethical considerations to provide a balanced perspective on the deployment of nanocomposites for photodegradation.

Despite the promise of nanocomposites, challenges and considerations are discussed, including potential environmental impacts and scalability issues. The chapter concludes by outlining future directions in research and development, emphasizing the need for continued innovation in nanocomposite technology to enhance their efficiency, sustainability and broader implementation in combating MP and NP pollution through photodegradation.

Looking ahead, this section outlines future directions for research and development in nanocomposite technology. It emphasizes the need for continuous innovation to enhance efficiency and sustainability, encouraging broader implementation strategies to tackle the ongoing issue of MP and NP pollution.

12.9.1. *Future vistas and emerging trends*

Exploring future vistas and emerging trends in scientific research offers insight into potential pathways for the field in the coming years. This section delves into the evolving knowledge landscape, envisioning innovative directions and breakthroughs that hold promise for addressing existing challenges and uncovering new frontiers. Researchers scrutinize emerging trends to anticipate the trajectory of their field, fostering a proactive approach to advancements that can shape the future of scientific inquiry and application.

12.9.2. *The power of cross-disciplinary collaboration*

Recognizing the interconnected nature of modern scientific challenges. The power of cross-disciplinary collaboration explores the synergy that emerges when experts from diverse fields join forces. By fostering interdisciplinary collaboration, researchers aim to create a holistic approach to problem-solving, transcending traditional boundaries and catalyzing advancements that draw on the collective expertise of varied scientific domains.

12.10. Conclusion

This chapter addresses the pressing issue of plastic pollution, with a specific focus on the proliferation of MPs and NPs, exploring cutting-edge strategies to combat MP and NP waste. Highlighting two noteworthy approaches, the chapter delves into the utilization of light-sensitive techniques for converting plastics into low-density hydrocarbons and the promotion of recycling. A pivotal aspect of this exploration is the introduction of photocatalytic nanocomposites, which amalgamate nanomaterials with customized matrices. Various nanomaterials, including inorganic semiconductor oxide nanoparticles and polymeric nanocomposites, offer promising avenues for enhancing the decomposition of MP and NP pollutants through photodegradation. The chapter delves into the science behind processing and modifying nanocomposites, addressing their origins, development and characterization. Additionally, it explores eco-friendly and sustainable techniques for producing these materials, contributing significantly to the advancement of advanced nanocomposites and technologies aimed at combating the global environmental issue of plastic pollution.

12.11. Acknowledgments

The authors wish to express their sincere gratitude to the Department of Biotechnology (DBT), Government of India, for the generous financial assistance provided under project No. DBT/2022-23/July/N/2240 as part of the DBT-RA Program.

12.12. References

Ariza-Tarazona, M.C., Villarreal-Chiu, J.F., Barbieri, V., Siligardi, C., Cedillo-González, E.I. (2019). New strategy for microplastic degradation: Green photocatalysis using a protein-based porous N-TiO_2 semiconductor. *Ceramics International*, 45(7), 9618–9624. doi: 10.1016/j.ceramint.2018.10.208.

Ariza-Tarazona, M.C., Villarreal-Chiu, J.F., Hernández-López, J.M., Rivera De la Rosa, J., Barbieri, V., Siligardi, C., Cedillo-González, E.I. (2020). Microplastic pollution reduction by a carbon and nitrogen-doped TiO_2: Effect of pH and temperature in the photocatalytic degradation process. *Journal of Hazardous Materials*, 395, 122632. doi: 10.1016/j.jhazmat.2020.122632.

Bandara, W.R.N., de Silva, R.M., de Silva, K.M.N., Dahanayake, D., Gunasekara, S., Thanabalasingam, K. (2017). Is nano ZrO_2 a better photocatalyst than nano TiO_2 for degradation of plastics? *RSC Adv.*, 7(73), 46155–46163. doi: 10.1039/C7RA08324F.

Cao, R., Zhang, M.Q., Hu, C., Xiao, D., Wang, M., Ma, D. (2022). Catalytic oxidation of polystyrene to aromatic oxygenates over a graphitic carbon nitride catalyst. *Nature Communications*, 13(1), 4809. doi: 10.1038/s41467-022-32510-x.

Chai, C., Liang, H., Yao, R., Wang, F., Song, N., Wu, J., Li, Y. (2023). Photocatalytic degradation of polyethylene and polystyrene microplastics by α-Fe_2O_3/g-C_3N_4. *Environmental Science and Pollution Research International*, 30(58), 121702–121712. doi: 10.1007/s11356-023-31000-x.

Fa, W., Zan, L., Gong, C., Zhong, J., Deng, K. (2008). Solid-phase photocatalytic degradation of polystyrene with TiO_2 modified by iron (II) phthalocyanine. *Applied Catalysis B*, 79(3), 216–223. doi: 10.1016/j.apcatb.2007.10.018.

Fonseca-Cervantes, O.R., Pérez-Larios, A., Romero Arellano, V.H., Sulbaran-Rangel, B., Guzmán González, C.A. (2020). Effects in band gap for photocatalysis in TiO_2 support by adding gold and ruthenium. *Processes*, 8(9), 1032. doi: 10.3390/pr8091032.

Hamd, W., Daher, E.A., Tofa, T.S., Dutta, J. (2022). Recent advances in photocatalytic removal of microplastics: Mechanisms, kinetic degradation, and reactor design. *Frontiers in Marine Science*, 9, 885614. doi: 10.3389/fmars.2022.885614.

Joo, S.H., Liang, Y., Kim, M., Byun, J., Choi, H. (2021). Microplastics with adsorbed contaminants: Mechanisms and treatment. *Environmental Challenges*, 3, 100042. doi: 10.1016/j.envc.2021.100042.

Lam, S.M., Sin, J.C., Zeng, H., Lin, H., Li, H., Chai, Y.Y., Choong, M.K., Mohamed, A.R. (2021). Green synthesis of Fe-ZnO nanoparticles with improved sunlight photocatalytic performance for polyethylene film deterioration and bacterial inactivation. *Materials Science in Semiconductor Processing*, 123, 105574. doi: 10.1016/j.mssp.2020.105574.

Lam, S.M., Chew, K.C., Sin, J.C., Zeng, H., Lin, H., Li, H., Lim, J.W., Mohamed, A.R. (2022). Ameliorated photodegradation performance of polyethylene and polystyrene films incorporated with ZnO-PVP catalyst. *Journal of Environmental Chemical Engineering*, 10(3), 107594. doi: 10.1016/j.jece.2022.107594.

Liang, W., Luo, Y., Song, S., Dong, X., Yu, X. (2013). High photocatalytic degradation activity of polyethylene containing polyacrylamide grafted TiO_2. *Polymer Degradation and Stability*, 98(9), 1754–1761. doi: 10.1016/j.polymdegradstab.2013.05.027.

Llorente-García, B.E., Hernández-López, J.M., Zaldívar-Cadena, A.A., Siligardi, C., Cedillo-González, E.I. (2020). First insights into photocatalytic degradation of HDPE and LDPE microplastics by a mesoporous N–TiO_2 coating: Effect of size and shape of microplastics. *Coatings*, 10(7), 658. doi: 10.3390/coatings10070658.

Ma, D., Zhai, S., Wang, Y., Liu, A., Chen, C. (2019). TiO_2 photocatalysis for transfer hydrogenation. *Molecules*, 24(2), 330. doi: 10.3390/molecules24020330.

Nkwachukwu, O.V. and Arotiba, O.A. (2021). Perovskite oxide-based materials for photocatalytic and photoelectrocatalytic treatment of water. *Frontiers in Chemistry*, 9, 634630. doi: 10.3389/fchem.2021.634630.

Razali, N., Abdullah, W.R.W., Mohd Zikir, N. (2020). Effect of thermo-photocatalytic process using zinc oxide on degradation of macro/micro-plastic in aqueous environment. *Journal of Sustainable Science and Management*, 15(6), 1–14. doi: 10.46754/jssm.2020.08.001.

Soitong, T. (2018). Photo-degradation of polypropylene-ascorbic acid TiO_2 composite films. *International Polymer Processing*, 33(1), 29–34. doi: 10.3139/217.3315.

Tofa, T.S., Ye, F., Kunjali, K.L., Dutta, J. (2019). Enhanced visible light photodegradation of microplastic fragments with plasmonic platinum/zinc oxide nanorod photocatalysts. *Catalysts*, 9(10), 819. doi: 10.3390/catal9100819.

Uheida, A., Mejía, H.G., Abdel-Rehim, M., Hamd, W., Dutta, J. (2021). Visible light photocatalytic degradation of polypropylene microplastics in a continuous water flow system. *Journal of Hazardous Materials*, 406, 124299. doi: 10.1016/j.jhazmat.2020.124299.

Venkataramana, C., Botsa, S.M., Shyamala, P., Muralikrishna, R. (2021). Photocatalytic degradation of polyethylene plastics by $NiAl_2O_4$ spinels-synthesis and characterization. *Chemosphere*, 265, 129021. doi: 10.1016/j.chemosphere.2020.129021.

Wang, X., Zhu, Z., Jiang, J., Li, R., Xiong, J. (2023). Preparation of heterojunction C_3N_4/WO_3 photocatalyst for degradation of microplastics in water. *Chemosphere*, 337, 139206. doi: 10.1016/j.chemosphere.2023.139206.

Yang, C., Gong, C., Peng, T., Deng, K., Zan, L. (2010). High photocatalytic degradation activity of the polyvinyl chloride (PVC)-vitamin C (VC)–TiO_2 nano-composite film. *Journal of Hazardous Materials*, 178(1–3), 152–156. doi: 10.1016/j.jhazmat.2010.01.056.

Yang, C., Deng, K., Peng, T., Zan, L. (2011a). Enhanced solid-phase photocatalytic degradation activity of a poly(vinyl chloride)-TiO_2 nanocomposite film with bismuth oxyiodide. *Chemical Engineering and Technology*, 34(6), 886–892. doi: 10.1002/ceat.201000450.

Yang, C., Tian, L., Ye, L., Peng, T., Deng, K., Zan, L. (2011b). Enhancement of photocatalytic degradation activity of poly(vinyl chloride)-TiO_2 nanocomposite film with polyoxometalate. *Journal of Applied Polymer Science*, 120(4), 2048–2053. doi: 10.1002/app.33316.

Zhang, Y., Sun, T., Zhang, D., Shi, Z., Zhang, X., Li, C., Wang, L., Song, J., Lin, Q. (2020). Enhanced photodegradability of PVC plastics film by codoping nano-graphite and TiO_2. *Polymer Degradation and Stability*, 181, 109332. doi: 10.1016/j.polymdegradstab.2020.109332.

13

Fabrication of Plant/Biogenic-based Metallic Nanomaterials for Degradation of Micro- and Nano-Plastics

The extensive use of plastics is an alarming issue for our environment nowadays. Among the different types of plastics, micro-plastics (MPs) have a particle size of <5 μm and are formed by the fragmentation of large plastics particles. However, nano-plastics (NPs) have a particle size of <1 μm and are accidentally produced by the fragmentation of larger particles. Degradation of these plastics is another revolutionary problem that has to be handled by science. Nanoscience has provided a unique path to resolve such issues. Plant-mediated metallic nanomaterial (NM) has gained greater attention over the chemical method. This is due to its ecofriendly synthesis process. This chapter focuses on the various synthesis methods for metallic nanoparticles using plant extracts and biogenic materials. The number of properties of these NMs is also discussed. Biogenic NMs used for the degradation of MPs are discussed in detail. This chapter also summarizes the resources of NPs and their adverse effect on human health. It discusses the degradation of these NPs using various plant-mediated nanoparticles.

13.1. Introduction

Since its invention in 1907, plastic has become a crucial part of human life. Its durability, inertness, lightweight nature and corrosion resistance make it a superior substitute for many other materials. Despite these advantages, plastics are often regarded as one of humanity's worst creations. They do not degrade naturally and require energy-intensive chemical or thermal methods for disposal. These methods not only demand significant resources, but also produce hazardous by-products that pose risks to both human health and the environment, making them impractical solutions (Andrady and Neal 2009; Cole et al. 2011; Jakubowicz et al. 2021).

Chapter written by Preeti GUPTA and Geeta SINGH.

Nowadays, the extensively used synthetic polymers are polypropylene (PP), polystyrene (PS), polyethylene (PE) in its various forms (low-density polyethylene (LDPE), linear low-density polyethylene (LLDPE) and high-density polyethylene (HDPE)), polyethylene terephthalate (PET), polyamides (PAs), polylactic acid (PLA), polyvinyl chloride (PVC), polycarbonate (PC), acrylonitrile butadiene styrene (ABS) and polyurethanes (PU) in various sectors (Sun et al. 2019; Rima 2020).

Contamination of plastic has emerged as a critical issue in environmental research. As per the recent report, global plastic production reached 400.3 million tons in 2022 (Rafey and Siddiqui 2021; Nayanathara Thathsarani Pilapitiya and Ratnayake 2024). In recent years, the focus has not only been on MPs, but also on NPs, which are increasingly recognized as significant contaminants of concern (Wang et al. 2017; Koelmans 2019; Zhang et al. 2020; Kokalj et al. 2021; Kiran et al. 2022).

Synthetic polymer particles with a size $\leq$ 5 mm are known as MPs, and first became a concern in 2014. These MPs are the result of the breakdown of larger plastic waste and subsequently enter the environment (Andrady and Neal 2009). Importantly, MPs are not the final stage of plastic degradation; they can decompose further into even smaller particles known as NPs (Coppock et al. 2017; Novotna et al. 2019; Kokalj et al. 2021). NPs are particles with sizes ranging from 1 nm to 1 μm, which distinguishes them from engineered nanomaterials (ENMs) (Mason et al. 2016; Coppock et al. 2017; Hanpeng et al. 2022; Keerthana Devi et al. 2022).

MPs can be broadly categorized into two main types: primary MPs and secondary MPs. Primary MPs are plastics that are already small, typically measuring around 2–5 mm in diameter. In contrast, secondary MPs are tiny plastic fragments that originate from the breakdown of larger plastic items, occurring both in marine and terrestrial environments. Over time, a mix of physical, biological and chemical processes can weaken the structure of these larger plastics, leading to their fragmentation into smaller pieces (Bessa et al. 2018).

Secondary MPs result from the gradual breakdown of macro-plastics in water and involve three key mechanisms: (a) biodeterioration: refers to changes in the mechanical, chemical and physical properties of the polymer caused by microorganisms growing on or within its surface; (b) biofragmentation: process involves microorganisms breaking down polymers into smaller units, such as oligomers and monomers; (c) assimilation: microorganisms obtain carbon, energy and nutrients from the fragmented polymers, converting the carbon in plastic into CO_2, water and biomass. Another degradation mechanism of MPs is photo-degradation (exposed to prolonged periods of sunlight). In this type of mechanism, UV radiations causes oxidation of the polymer matrix, which leads to bond cleavage (Emadian et al. 2017; Magalhães et al. 2020; Thakur et al. 2023). Some biodegradable and non-biodegradable MPs are listed in Table 13.1 (Thakur et al. 2023).

Non-biodegradable micro-plastic	Biodegradable micro-plastic
Polyethylene Polystyrene Polypropylene Polyethylene teraphthalate (PET) Polyurethane	Polylactic acid (PLA) Polyhydroxyalkanoates (PHA) Polycapro lactone (PCL)

Table 13.1. *Different examples of micro-plastics: biodegradable and nonbiodegradable (Thakur et al. 2023)*

Similar to MPs, NPs can be divided into primary and secondary sources based on how they are released from products and applications (Table 13.2) (Keerthana Devi et al. 2022). Primary NPs are generated during manufacturing, but their use is relatively limited due to the associated challenges and costs. Secondary sources, on the other hand, are more hypothetical and refer to the potential creation of NPs from the ongoing breakdown of MPs in the environment, though this process has yet to be confirmed.

Scientists have begun using the term NPs to differentiate them from MPs due to growing concerns about the NPs' potential environmental impact. Despite these concerns, there is still limited information in the scientific literature regarding the abundance, distribution and biological effects of NPs, even though this category is believed to have significant implications for toxicity and ecotoxicity.

Types of nano-plastics	Name of nano-plastics	Particle size	Source
Primary nano-plastics	Polyethylene (PE)	300 nm	Cosmetic industries, plastic
	Polyethylene terephthalate (PET)	<100 nm	Plastic bag industries
	Polymethyl methacrylate (PMMA)	140 nm	Chemical industries
	Polyvinyl chloride (PVC)	120 nm	Chemical industries and metabolic
	Polystyrene (PS)	60 nm	Teabag chemical industries
Secondary nano-plastics	PET	800 nm, 60–80 nm, 20 nm	Bottles and plastic bags
	Polystyrene (PS)	100, 40 and 20 nm	From used plastic bottles
	Polysulfone (PSo)	<150 nm	Plastic bags

Table 13.2. *Nano-plastics, their source and particle sizes*

To aid risk assessment agencies and policymakers in managing the issue of MPs and NPs, several recommendations and research gaps need to be addressed. The suggestions included improving the evaluation of the sources (product-specific) of plastic waste and conducting studies on the potential development of MPs and NPs resulting from the degradation and fragmentation of discarded plastics (Alexy et al. 2020). The current state of research on plant biogenic metallic NMs for the degradation of MPs is an emerging area of study. While there is limited information available on this specific topic, we can gather insights from relevant research areas.

This chapter focuses on the impact of MPs and NPs in our environment. The authors emphasized the synthesis of NMs extracted from different parts of plants and its effect on the degradation of the plastic waste. Beyond that, some directions of future research to better comprehend or solve MPs and NPs issues have simultaneously been suggested.

13.2. Environment and micro- and nano-plastics

The widespread use of plastics, MPs and NPs has recently emerged as a major topic of global discussion, primarily due to their potential harmful effects on various ecosystems rather than their societal benefits. Growing evidence indicates that MPs negatively affect different water bodies and other components of the ecosystem.

About eight million metric tons of plastic waste enter aquatic ecosystems each year, and this figure is expected to increase fourfold by 2050 (Indhur et al. 2023). This alarming trend has raised significant concerns within the scientific community, prompting efforts to mitigate the devastating effects of plastic pollution. Most plastic products are nondegradable, with a lifespan of around 100 years, which is vastly disproportionate to current disposal rates. One of the critical issues stemming from this pollution is the emergence of MPs and NPs (MPs and NPs) as contaminants in water bodies, posing a substantial challenge to the availability of clean and safe water. MP and NP ubiquity has placed them in almost all ecosystems, including the oceans, rivers, lakes, freshwater, groundwater, seashores and soil. Growing concerns associated with these MPs and NPs stems from their bioaccumulation in the food chain, adverse side effects, and potential toxicity to different ecological systems (Abuwatfa et al. 2021; Rosa et al. 2021; Indhur et al. 2023).

The presence of plastic in terrestrial environments, such as soils, is a significant concern. Estimates suggest that the amount of mismanaged plastic waste can be 4–23 times greater than that reported for marine environments (Horton et al. 2017; Pérez-Reverón et al. 2023). In agricultural soils specifically, plastics can originate from soil management practices and, to a lesser extent, from atmospheric deposition. Conventional farming utilizes a substantial amount of plastic products, including mulch, greenhouse coverings, packaging, irrigation pipes and seedbeds, among others. When it comes to MPs and NPs, irrigation with treated wastewater and the application of sewage sludge or compost are particularly noteworthy (Yong et al. 2020; You et al. 2022).

A report from various sources (Peng et al. 2020; Soares et al. 2020) indicates that the prevalence of various types of MPs found in marine environments or sediments is as follows: PE at 79%, PP at 64% and PS at 40%. Additionally, Eunomia Research & Consulting Ltd (2016) reports that over 80% of the annual plastic litter entering the sea originates from land-based sources.

NPs have raised concerns regarding the potential risks they pose to the environment, but further research is needed to clarify these risks. First and foremost, we lack comprehensive data on the quantity of NPs present in the environment. This information is crucial for estimating their likelihood of occurrence, which is essential for effective risk assessment. Additionally, it is important to consider the level of certainty in these assessments.

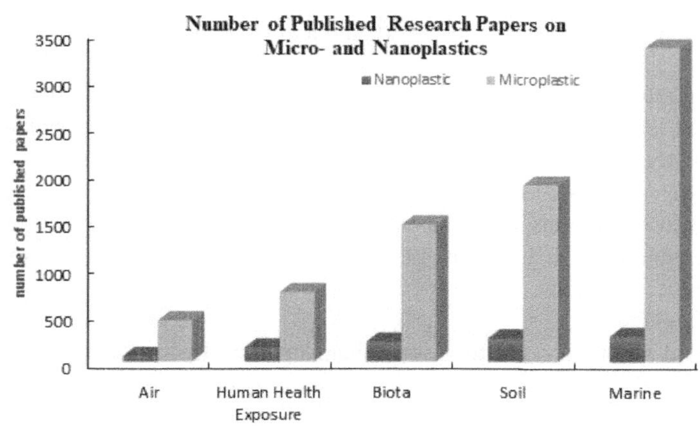

Figure 13.1. *Publications on micro- and nano-plastics affecting different environmental components. For a color version of this figure, see www.iste.co.uk/singh/photodegradation.zip*

Figure 13.1 represents the literature related to research in the field of MPs and NPs in the various compartments of the ecosystem. Allena et al. (2022) note that there has been limited research on the presence of NPs and MPs in water and soil, which significantly disrupts environmental balance. Approximately 159 research articles have been published on MP contamination in soil, while around 74 articles focus on NPs and their environmental impact (Jakubowicz et al. 2021). The number of research papers on MPs and NPs has significantly increased between 2000 and 2020, with a growth rate of 73.0% from 2012 to 2021. China has the largest number of published papers on MPs and NPs, accounting for 25.48% of the total (Keerthana Devi et al. 2022; Kiran et al. 2022).

13.3. Role of nanomaterials in micro- and nano-plastics

Nanotechnology is a diverse field that integrates physics, chemistry, biology and various engineering disciplines. It involves materials with external dimensions measuring 100 nm or less, which can take forms such as tubes, particles, rods and fibers. NMs are utilized across multiple domains, including agriculture, medicine, environmental science and energy production. They play crucial roles in areas such as aerospace and chemical engineering, solar hydrogen production, fuel cells, batteries, power generation, optics, sensors, construction, automotive engineering, consumer electronics, thermoelectric devices, pharmaceuticals and the cosmetics industry. These materials possess unique properties that can be enhanced by adjusting their sizes and morphologies. Typically, NMs have significantly larger surface areas compared to their bulk counterparts, and the magnetic properties of elements often change at the nanoscale (Chandran et al. 2023).

Plastic pollution represents a serious threat to ecosystems and living organisms. The frequent landfilling and incineration of plastic waste reduce the soil's capacity to absorb water, resulting in soil sterility. Aquatic organisms accumulate in streams where plastic waste is discarded, negatively impacting aquatic ecosystems. While there are various solutions available – physical, chemical and biological – traditional biological methods, such as enzymatic degradation of plastics, often fall short due to environmental limitations.

NMs hold significant promise for enhancing the effectiveness of existing MP removal methods and offering innovative solutions to emerging challenges. Their ability to accelerate degradation processes by promoting the growth of beneficial microorganisms and acting as efficient catalysts makes nanoparticles particularly valuable for polymer breakdown. Additionally, the diverse sizes and shapes of nanoparticles can enhance both their physicochemical stability and degradability (Chandran et al. 2023).

Several examples of NMs have been developed to degrade plastics (LDPE and HDPE) at the micro- and nano-scale (Pol and Thiyagarajan 2010; Sah et al. 2010; Rodrigues et al. 2019; Raghavendra et al. 2022; Remya et al. 2022; Chandran et al. 2023). For instance, gold (Au) nanoparticles were synthesized by combining hydrogen tetrachloroaurate (III) ($HAuCl_4 \cdot 3H_2O$) with aqueous leaf extract from Ananas comosus. The photocatalytic activity of these gold nanoparticles was demonstrated when they interacted with LDPE films, leading to the degradation of the LDPE. Under sunlight exposure, the LDPE film lost 51.4% of its weight within 24 hours, paving the way for new advancements in polymer degradation techniques.

Bimetallic core/shell nanoparticles (Pd_{core}/Au_{shell}) were synthesized from Pd (II) and Au (III) solutions using the leaf extract of Ananas comosus as a reducing agent. The photocatalytic properties of these nanoparticles were activated when they were introduced into the solid-phase degradation process of LDPE films under solar irradiation. After a long time (~10 days) of solar exposure, the weight of the nanoparticle-infused polymer decreased by 55.8 ± 5.9%. Additionally, these nanoparticles were effective in degrading LDPE films in dark conditions, achieving a degradation rate of 1.90 ± 0.3%. Notably, the nanoparticles can be reused for up to five photocatalytic degradation cycles without a loss in their efficiency for degrading LDPE films. This opens up a sustainable setup to degrade polymers by newer methods and technologies.

Several nanoparticles such as zinc oxide nanoparticles and silver nanoparticles are used to degrade LDPE and HDPE (Pol and Thiyagarajan 2010; Sah et al. 2010; Rodrigues et al. 2019; Raghavendra et al. 2022; Remya et al. 2022; Chandran et al. 2023).

Chokejaroenrat et al. (2024) reported the use of nanoscale $CuO/BiVO_4$ photocatalysts to tackle the challenge of efficient MP degradation in wastewater treatment. They demonstrated the effectiveness of a copper oxide/bismuth vanadate nanocomposite ($CuO/BiVO_4$) in degrading MPs by activating H_2O_2 under visible light. The enhanced oxidative performance was attributed to the doping of CuO onto $BiVO_4$ nanocomposites, which increased the generation of reactive oxygen species (ROS) upon H_2O_2 activation. The study focused on three commonly used plastics as representatives of MPs: PP, low-density PE and PA. Additionally, titanium oxide has also been utilized in the degradation of MPs (Nabi et al. 2021; Ge et al. 2022).

13.4. Plant/biogenic metallic nanomaterials

The green synthesis approach for nanoparticles involves creating metal nanoparticles (MNPs) using eco-friendly and sustainable materials and processes. This method utilizes natural or renewable resources, such as plant extracts or microorganisms, instead of synthetic chemicals. Bimetallic nanoparticles (BMNPs) produced through this method are generally less toxic and generate less waste, enhancing sustainability. A significant advantage of the green synthesis approach is its use of natural materials or biocatalysts, including enzymes, fungi and microorganisms, often without requiring high temperatures for the reduction of metal ions (Alsaiari et al. 2023; Kumar et al. 2023).

There are several methods for producing NMs, which can be categorized into two main approaches. The first is the bottom-up approach, which includes chemical methods such as hydrothermal, sonochemical, electrochemical, photochemical, microwave, microemulsion, pyrolysis-redox, co-precipitation and sol-gel techniques. In this approach, NMs are created by assembling single atoms or molecules (Mason et al. 2016). The second is the top-down approach, which encompasses physical methods like lithography. Additionally, the biological method, also a bottom-up approach, involves synthesizing NMs using biopolymers, biomolecules, viruses, bacteria, yeasts, plants and fungi (Saratale et al. 2018).

While green synthesis of NMs holds great promise, several challenges hinder its broader adoption in industrial manufacturing. Key issues include product quality control, synthesis conditions, material selection and application suitability. One major limitation is the variability in the quality of green-synthesized NMs. Different extracts and microorganisms can produce particles of varying shapes and sizes, often leading to inconsistent properties. Recent studies have highlighted significant discrepancies in particle size, which complicates large-scale production and control during manufacturing. Additionally, the resulting NMs can differ widely in morphology, including non-spherical shapes, amorphous structures and irregular clusters. Such variability restricts the practical applications of these materials and makes mass production difficult.

The use of plant biogenic metallic NMs for the degradation of MPs has several advantages. These includes increased rate of biodegradation, that is, cost-effective and environmental friendly. Plant biogenic metallic NMs are biodegradable and non-toxic, making them a more environmentally friendly option. Recent advancements in biogenic metallic NMs for environmental applications include the biogenic synthesis of MNPs, environmental remediation, bio-reduction of metal ions, bioremediation of contaminated soil, water treatment and bio-reduction of carbon dioxide.

Another challenge is the low yield and conversion rates of green-synthesized nanoparticles, further complicating their industrial viability. These factors collectively limit the effectiveness and scalability of green synthesis methods in producing NMs.

13.4.1. *Characterization technique involved in nanomaterials*

Before NMs can be utilized in any application, it is essential to characterize their properties. Various characterization techniques are employed to assess the physical, chemical and other attributes of NMs, tailored to the specific application. These methods help identify key properties such as topography, roughness, surface chemistry, stability and distribution, as well as surface area, shape, size and composition.

The characterization of metallic nanoparticles employs a range of techniques, including UV-Vis absorption, transmission electron microscopy (TEM), field emission scanning electron microscopy (FESEM), dynamic light scattering (DLS), X-ray diffraction (XRD), inductively coupled plasma mass spectrometry (ICP-MS), differential scanning calorimetry, thermogravimetric analysis (TGA) and surface-enhanced Raman spectroscopy (SERS). These methods are essential for measuring and analyzing properties such as size, shape, composition and surface characteristics (Figures 13.1 and 13.2).

Understanding these properties is crucial for optimizing MNP performance across various applications. TEM, for example, utilizes a high-energy electron beam to capture high-resolution images of nanoparticles, enabling precise measurement of their size and shape. DLS, on the other hand, assesses nanoparticle size by analyzing the light scattering behavior of MNPs in solution. This technique is particularly effective for determining size distribution, a key factor influencing MNP performance. DLS can also evaluate the stability of nanoparticle suspensions and their behavior in applications.

13.4.2. *Properties of nanomaterials*

The review highlights various properties of magnetic nanoparticles (MNPs), including electronic, optical, magnetic, mechanical, thermal and antimicrobial characteristics, which are relevant for diverse biomedical applications. The electronic properties of MNPs are influenced by factors such as size, surface area, chemical composition and modifications. Additionally, several biogenic NMs have been synthesized, showing potential applications as adsorbents, photocatalysts and disinfectants.

Zero-valent metallic nanoparticles, either reduced and capped to a neutral charge by biological macromolecules during green synthesis or chemically reduced, are derived from metal and metal oxide salts. These nanoparticles exhibit unique morphological characteristics, strong affinities and high adsorptive efficiency. Examples include zero-valent nanoparticles of silver (AgNPs), gold (AuNPs), silver–gold alloys (Ag–AuNPs), zinc oxide (ZnONPs), copper (CuNPs), iron (FeNPs), iron oxide (Fe_2O_3NPs) and magnesium oxide (MgONPs). These nanoparticles have diverse applications in water treatment, agriculture, medicine and the food industry. As adsorbents, they offer a wealth of reactive sites, large surface areas, surface plasmon resonance and catalytic properties, making them effective for comprehensive water remediation.

Biogenic nanoparticles of selenium (SeNPs), palladium (PdNPs) and platinum (PtNPs) have demonstrated significant capabilities in environmental applications. SeNPs, in particular, effectively adsorb heavy metals such as zinc (Zn), nickel (Ni), copper (Cu), mercury (Hg) and cadmium (Cd) due to their amorphous structure, which facilitates a ligand-like adsorption process (Jain et al. 2015). Additionally, the photocatalytic properties of metallic nanoparticles, including ZnO NPs, NiO NPs, TiO_2 NPs and Fe_2O_3 NPs, enable them to interact with light photons to break down various organic pollutants.

13.5. Degradation of micro- and nano-plastics

Plastic recycling can be classified into four categories: primary, secondary, tertiary and quaternary. Primary recycling refers to re-extrusion, where plastics are returned to the production process while retaining the same characteristics. Secondary recycling involves mechanical recycling, which uses physical processes to recycle various plastic products. Tertiary recycling focuses on chemical recycling, aiming to produce raw materials for the chemical industry. Quaternary recycling pertains to energy recovery, where plastic waste is partially or completely oxidized to generate heat, gaseous fuels, oils or materials for disposal. For MPs, ozonation and photocatalytic methods for degradation are discussed (Bratovcic 2019; Xiao et al. 2024). The generation degradation mechanism of pollutants by using NMs is shown in Figure 13.2.

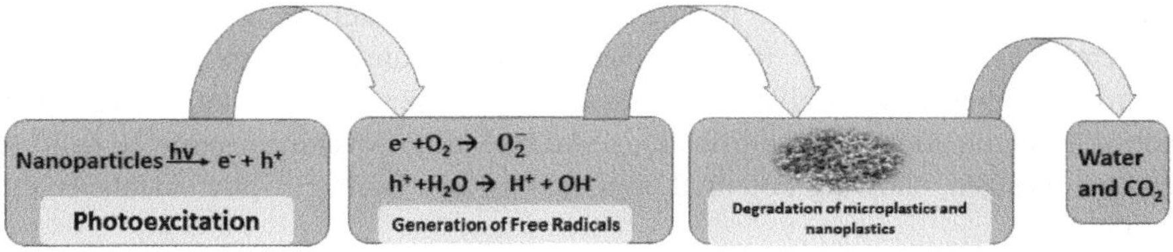

Figure 13.2. *Mechanism of degradation of pollutants by photocatalysis. For a color version of this figure, see www.iste.co.uk/singh/photodegradation.zip*

Photocatalysis can be categorized into two types: photodegradation and photosynthesis. Photodegradation involves the non-selective oxidation of plastics into smaller fragments, which can range from MPs and NPs to small molecules and ultimately carbon dioxide (CO_2). This degradation typically occurs in the presence of air or oxygen, and is facilitated by photogenerated holes and electrons that create radicals, such as hydroxyl radicals, which drive the oxidation and breakdown of plastic materials. In contrast, photosynthesis is an upcycling process that converts carbon- and hydrogen-rich plastics into valuable target products, such as fuels, chemicals and materials. However, this process is less efficient and poses challenges in recovering small, dispersed MPs from the environment.

Current studies on the photocatalytic degradation of plastics indicate that this process generates lower molecular weight intermediates that can be utilized in organic synthesis. Although research on photocatalytic methods for plastic degradation is still limited, the benefits of using solar radiation and photocatalysts to transform waste into more valuable intermediate products are clear. Consequently, developing photocatalytic processes that harness renewable energy sources like solar energy is a promising option from economic, energy and environmental standpoints.

Advanced oxidation processes (AOPs), which generate in situ reactive species, primarily hydroxyl radicals (OH), have emerged as promising technologies for degrading plastic waste. The most studied AOPs include ozonation, combinations of ozone and UV radiation, peroxide (H_2O_2) with UV radiation and the use of UV radiation alongside photocatalysts. However, a significant drawback of all AOPs is their high maintenance costs, which limits their widespread practical application despite their effectiveness. With the development of highly efficient UV radiation sources, visible spectrum-absorbing catalysts and improved reactor designs, technologies utilizing UV and solar radiation show great potential for broader implementation. Nevertheless, further research is needed to fully realize this potential.

Tofa et al. (2019) demonstrated the successful degradation of MP fragments, specifically LDPE film in water using visible light-activated heterogeneous zinc oxide photocatalysts.

Shengying et al. (2010) conducted the degradation of PE plastic under sunlight using a polypyrrole/TiO_2 (PPy/TiO_2) nanocomposite as the photocatalyst. This nanocomposite was prepared using sol-gel and emulsion polymerization methods.

Tian et al. (2019) investigated the degradation and mineralization of PS NPs under ultraviolet (UV) radiation at 254 nm using carbon radioisotope tracer technology.

This pioneering study explores the application of photocatalysis for the degradation of HDPE MPs extracted from a commercially available facial scrub. Two innovative semiconductors based on N-TiO_2 were employed, with one being green synthesized using the extrapallial fluid of fresh blue mussels. This eco-friendly approach demonstrated exceptional capacity for promoting photocatalytic degradation. The key findings of this research are (a) environmental conditions, pollutant/N-TiO_2 interaction and photocatalyst surface area must be carefully optimized to prevent photocatalysis arrest; (b) batch tests indicate that photocatalysis of HDPE microbeads warrants further investigation and adaptation for domestic wastewater treatment plants.

Ongoing research focuses on determining the degradation paths of PE MPs and optimal conditions for scaling this technology. This groundbreaking research paves the way for the development of sustainable solutions for MP degradation, with potential applications in wastewater treatment and environmental remediation.

13.6. Conclusion and future prospectives

Today, MPs and NPs are pervasive in the environment. Their small size leads to unique and significant behaviors. As particles decrease in size, the ratio of surface area to volume increases, making their chemical behavior more influenced by surface properties than by the materials themselves. Consequently, the interactions of MPs and NPs with the environment cannot be inferred from the behavior of macro-plastics.

Understanding the contamination routes of MPs in foods and beverages is crucial for evaluating the extent of food contamination. Additionally, it is important to explore how NPs interact with the environment, particularly with living organisms, as they may pose different health risks compared to larger particles. Efforts to prevent MP pollution should be politically and socioeconomically viable.

NMs possess unique properties that allow them to integrate effectively with existing wastewater treatment technologies, presenting promising opportunities to enhance the removal of MPs and NPs from the ecosystem. Biogenically synthesized metallic nanoparticles have garnered significant attention due to their eco-friendliness, unique characteristics and potential applications in medicine, sensing, food preservation and other consumables. Utilizing biological sources as reducing and capping agents in the synthesis of these metallic nanoparticles offers several advantages over conventional methods, including lower costs, sustainability and the absence of toxic chemicals. Furthermore, biogenic metallic nanoparticles (BMNPs) are widely used for a variety of applications, such as analytical sensing of toxic chemicals, biomedical applications, food preservation, wastewater remediation, coating of surgical devices and in household appliances.

The literature review indicates that several BMNPs have been synthesized and are widely used for treating various pollutants in water, soil and human health hazard systems. However, there remains considerable uncertainty due to the limited number of studies published on the fabrication of plant-based or BMNPs specifically for MP and NP degradation. Additionally, nanoparticles have been explored for their potential to protect plants from MP-induced physical obstruction. To address this knowledge gap and tackle key research questions, there is a pressing need to focus on the synthesis of BMNPs and thoroughly explore their properties. To advance in the field, researchers should investigate the use of plant biogenic metallic NMs for MP degradation, exploring their potential to enhance biodegradation processes. This could involve studying the interactions between plant biogenic NMs, microorganisms and MPs to understand the underlying mechanisms and optimize degradation efficiencies.

13.7. References

Abuwatfa, W.H., Al-Muqbel, D., Al-Othman, A., Halalsheh, N., Tawalbeh, M. (2021). Insights into the removal of microplastics from water using biochar in the era of Covid-19: A mini review. Case studies in chem. *Environmental Engineering*, 4, 100151.

Alexy, P., Anklam, E., Emans, T., Furfari, A., Galgani, F., Hanke, G., Koelmans, A., Pant, R., Saveyn, H., Sokull Kluettgen, B. (2020). Managing the analytical challenges related to micro and nanoplastics in the environment and food: Filling the knowledge gaps. *Food Additives and Contaminants. Part A, Chemistry, Analysis, Control, Exposure and Risk Assessment*, 37, 1–10.

Allena, S., Allen, D., Karbalaei, S., Maselli, V., Walker, T.R. (2022). Micro(nano)plastics sources fate, and effects: What we know after ten years of research. *J. Hazard. Mater. Adv.*, 6, 100057.

Alsaiari, N.S., Alzahrani, F.M., Amari, A., Osman, H., Harharah, H.N., Elboughdiri, N., Tahoon, M.A. (2023). Plant and microbial approaches as green methods for the synthesis of nanomaterials: Synthesis, applications, and future perspectives. *Molecules*, 28, 463.

Andrady, A.L. and Neal, M.A. (2009). Applications and societal benefits of plastics. *Philosophical Transactions of the Royal Society of London. Series B, Biological Sciences*, 364, 1977–1984. doi: 10.1098/rstb.2008.0304.

Ariza Tarazona, M.C., Villarreal, C.J.F., Barbieri, V., Siligardi, C., Cedillo, G.E.I. (2018). New strategy for microplastic degradation: Green photocatalysis using a protein-based porous N-TIO_2, semiconductor. *Ceram. Internat.*, 45, 9618–9624.

Azeez, L., Lateef, A., Olabode, O. (2023). An overview of biogenic metallic nanoparticles for water treatment and purification: The state of the art. *Water Science and Technology*, 88, 851–873.

Bessa, F., Barría, P., Neto, J.M., Frias, J.P.G.L., Otero, V., Sobral, P., Marques, J.C. (2018). Occurrence of microplastics in commercial fish from a natural estuarine environment. *Marine Pollution Bulletin*, 128, 575–584.

Bratovcic, A. (2019). Degradation of micro-and nano-plastics by photocatalytic methods. *Journal of Nanoscience and Nanotechnology Applications*, 3, 1–6.

Chandran, R.R., Thomson, B.I., Natishah, A.J., Mary, J., Nachiyar, V. (2023). Nanotechnology in plastic degradation. *Biosciences Biotechnology Research Asia*, 20, 53–68.

Chokejaroenrat, C., Watcharatharapong, T., Thienprasert, J.T., Angkaew, A., Poompoung, T., Chinwong, C., Chirasatienpon, T., Sakulthaew, C. (2024). Decomposition of microplastics using copper oxide/bismuth vanadate-based photocatalysts: Insight mechanisms and environmental impacts. *Marine Pollution Bulletin*, 201, 116205.

Chu, S., Zhang, B., Zhao, X., Soo, H.S., Wang, F., Xiao, R., Zhang, H. (2022). Photocatalytic conversion of plastic waste: From photodegradation to photosynthesis. *Advanced Energy Materials*, 12, 2200435.

City of Westminster (n.d.). Plastic waste and pollution [everything you need to know in 2020] [Online]. Available at: https://cleanstreets.westminster.gov.uk/plastic-waste-complete-guide/#1.

Cole, M., Lindeque, P., Halsband, C., Galloway, T.S. (2011). Microplastics as contaminants in the marine environment: A review. *Marine Pollution Bulletin*, 62, 2588–2597.

Coppock, R.L., Cole, M., Lindeque, P.K., Queirós, A.M., Galloway, T.S. (2017). A small-scale, portable method for extracting microplastics from marine sediments. *Environmental Pollution*, 230, 829–837.

Emadian, S.M., Onay, T.T., Demirel, B. (2017). Biodegradation of bioplastics in natural environments. *Waste Management*, 59, 526–536.

Eunomia Research & Consulting Ltd (2016). Plastics in the marine environment [Online]. Available at: https://eunomia.eco/reports/plastics-in-the-marine-environment/.

Ge, J., Zhang, Z., Ouyang, Z., Shang, M., Liu, P., Li, H., Guo, X. (2022). Photocatalytic degradation of (micro)plastics using TiO_2-based and other catalysts: Properties, influencing factor, and mechanism. *Environmental Research*, 209, 112729.

Hanpeng, L., Xing, L., Man, Q. (2022). Nanoplastics and human health: Hazard identification and biointerface. *Nanomaterials*, 12, 1298.

Horton, A.A., Walton, A., Spurgeon, D.J., Lahive, E., Svendsen, C. (2017). Microplastics in freshwater and terrestrial environments: Evaluating the current understanding to identify the knowledge gaps and future research priorities. *Science of the Total Environment*, 586, 127–141.

Indhur, R., Amoah, I., Bux, F., Kumari, S. (2023). Nanomaterials for microplastic removal from wastewater: Current state of the art nanomaterials and future prospects. *ACS ES&T Water*, 3, 3741–3754.

Jain, R., Jordan, N., Schild, D., van Hullebusch, E., Weiss, S., Franzen, C., Farges, F., Hübner, R., Lens, P.N.L. (2015). Adsorption of zinc by biogenic elemental selenium nanoparticles. *Chemical Engineering Journal*, 260, 855–863.

Jakubowicz, I., Enebro, J., Yarahmadi, N. (2021). Challenges in the search for nanoplastics in the environment—A critical review from the polymer science perspective. *Polymer Testing*, 93, 106953.

Jayeoye, T.J., Eze, F.N., Singh, S., Olatunde, O.O., Benjakul, S., Rujiralai, T. (2021). Synthesis of gold nanoparticles/polyaniline boronic acid/sodium alginate aqueous nanocomposite based on chemical oxidative polymerization for biological applications. *International Journal of Biological Macromolecules*, 179, 196–205.

Keerthana Devi, M., Karmegam, N., Manikandan, S., Subbaiya, R., Song, H., Kwon, E.E., Sarkar, B., Bolan, N., Kim, W., Rinklebe, J., Govarthanan, M. (2022). Removal of nanoplastics in water treatment processes: A review. *Science of the Total Environment*, 845, 157168.

Kiran, B.R., Kopperi, H., Mohan, V., Micro, S. (2022). Micro/nano-plastics occurrence, identification, risk analysis and mitigation: Challenges and perspectives. *Reviews in Environmental Science and Bio/Technology*, 21, 169–203.

Koelmans, A.A. (2019). Proxies for nanoplastic. *Nature Nanotechnology*, 14, 307–308.

Kokalj, A.J., Hartmann, N.B., Drobne, D., Potthoff, A., Kühnel, D. (2021). Quality of nanoplastics and microplastics ecotoxicity studies: Refining quality criteria for nanomaterial studies. *Journal of Hazardous Materials*, 415, 125751.

Kumar, A., Shah, S.R., Jayeoye, T.J., Kumar, A., Parihar, A., Prajapati, B., Singh, S., Kapoor, D.U. (2023). Biogenic metallic nanoparticles: Biomedical, analytical, food preservation, and applications in other consumable products. *Frontiers in Nanotechnology*, 5, 1175149.

Li, S., Xu, S., He, L., Xu, F., Wang, Y., Zhang, L. (2010). Photocatalytic degradation of polyethylene plastic with polypyrrole/TiO, nanocomposite as photocatalyst. *Polymer-Plastics Technology and Engineering*, 49, 400–406.

Magalhães, S., Alves, L., Medronho, B., Romano, A., da Graça Rasteiro, M. (2020). Microplastics in ecosystems: From current trends to bio-based removal strategies. *Molecules*, 25, 3954. doi: 10.3390/molecules25173954.

Marimuthu, S., Antonisamy, A.J., Malayandi, S., Rajendran, K., Tsai, P.C., Pugazhendhi, A., Ponnusamy, V.K. (2020). Silver nanoparticles in dye effluent treatment: A review on synthesis, treatment methods, mechanisms, photocatalytic degradation, toxic effects and mitigation of toxicity. *Journal of Photochemistry and Photobiology. B, Biology*, 205, 111823.

Mason, S.A., Garneau, D., Sutton, R., Chu, Y., Ehmann, K., Barnes, J., Fink, P., Papazissimos, D., Rogers, D.L. (2016). Microplastic pollution is widely detected in US municipal wastewater treatment plant effluent. *Environmental Pollution*, 218, 1045–1054.

Mustapha, S., Ndamitso, M.M., Abdulkareem, A.S., Tijani, J.O., Shuaib, D.T., Ajala, A.O., Mohammed, A.K. (2020). Application of TiO_2 and ZnO nanoparticles immobilized on clay in wastewater treatment: A review. *Applied Water Science*, 10, 49.

Nabi, I., Bacha, A.U.R., Ahmad, F., Zhang, L. (2021). Application of titanium dioxide for the photocatalytic degradation of macro- and micro-plastics: A review. *Journal of Environmental Chemical Engineering*, 9, 105964.

Nayanathara Thathsarani Pilapitiya, P.G.C. and Ratnayake, A.S. (2024). The world of plastic waste: A review. *Cleaner Materials*, 11, 100220.

Novotna, K., Cermakova, L., Pivokonska, L., Cajthaml, T., Pivokonsky, M. (2019). Drinking water treatment – Current knowledge and research needs. *Science of the Total Environment*, 667, 730–740.

Peng, L., Fu, D., Qi, H., Lan, C.Q., Yu, H., Ge, C. (2020). Micro and nanoplastics in marine environment: Source, distribution and threats – A review. *Science of the Total Environment*, 698, 134254.

Pérez-Reverón,R., Álvarez-Méndez, S.J., González-Sálamo, J., Socas-Hernández, C., Díaz-Peña, F.J., Hernández-Sánchez, C., Hernández-Borges, J. (2023). Nanoplastics in the soil environment: Analytical methods, occurrence, fate and ecological implications. *Environmental Policy and Law*, 317, 120788.

Pol, V.G. and Thiyagarajan, P. (2010). Remediating plastic waste into carbon nanotubes. *Journal of Environmental Monitoring*, 12, 455–459.

Rafey, A. and Siddiqui, F.Z. (2021). A review of plastic waste management in India – Challenges and opportunities. *International Journal of Environment Analytical Chemistry*, 103, 3971–3987.

Raghavendra, V.B., Shankar, S., Govindappa, M., Pugazhendhi, A., Sharma, M., Nayaka, S.C. (2022). Green synthesis of zinc oxide nanoparticles (ZnO NPs) for effective degradation of dye, polyethylene and antibacterial performance in waste water treatment. *Journal of Inorganic and Organometallic Polymers and Materials*, 32, 614–630.

Remya, R.R., Julius, A., Suman, T.Y., Vinayagam, M., Karthick, A., Pazhanimuthu, C., Samrot, A., Muhibbullah, M. (2022). Role of nanoparticles in biodegradation and its importance of environmental and biomedical applications. *Journal of Nanomaterials*. doi: 10.1155/2022/6090846.

Rima, F.J.G.S. (2020). Setting the facts straight on plastics [Online]. Available at: https://www.weforum.org/agenda/2019/10/plastics-what-are-they-explainer/ [Accessed 4 February 2020].

Rodrigues, M.O., Abrantes, N., Gonçalves, F.J.M., Nogueira, H., Marques, J.C., Gonçalves, A.M.M. (2019). Impacts of plastic products used in daily life on the environment and human health: What is known? *Environmental Toxicology and Pharmacology*, 72, 103239.

Rosa, E.V., Fascineli, M.L., da Silva, I.C., Rodrigues, M.O., Chaker, J.A., Grisolia, C.K., Moya, S.E., Campos, A.F., Sousa, M.H. (2021). Carbon nitride nanosheets magnetically decorated with Fe_3O_4 nanoparticles by homogeneous precipitation: Adsorption-photocatalytic performance and acute toxicity assessment. *Environmental Nanotechnology, Monitoring and Management*, 16, 100549.

Sacco, N.A., Zoppas, F.M., Devard, A., González Muñoz, M.P., García, G., Marchesini, F.A. (2023). Recent advances in microplastics removal from water with special attention given to photocatalytic degradation: Review of scientific research. *Microplastics*, 2, 278–303.

Sah, A., Kapri, A., Zaidi, M.G.H., Negi, H., Goel, R. (2010). Implications of fullerene-60 upon in-vitro LDPE biodegradation. *Journal of Microbiology and Biotechnology*, 20, 908–916.

Saratale, R.G., Karuppusamy, I., Saratale, G.D., Pugazhendhi, A., Kumar, G., Park, Y., Ghodake, G.S., Bharagava, R.N., Banu, J.R., Shin, H.S. (2018). A comprehensive review on green nanomaterials using biological systems: Recent perception and their future applications. *Colloids and Surfaces. B, Biointerfaces*, 170, 20–35.

Soares, J., Miguel, I., Venâncio, C., Lopes, I., Oliveira, M. (2020). Perspectives on micro(nano)plastics in the marine environment: Biological and societal considerations. *Water*, 12, 3208.

Sun, J., Dai, X., Wang, Q., van Loosdrecht, M.C.M., Ni, B.J. (2019). Microplastics in wastewater treatment plants: Detection, occurrence and removal. *Water Research*, 152, 21–37.

Thakur, B., Singh, J., Singh, J., Angmo, D., Vig, A.P. (2023). Biodegradation of different types of microplastics: Molecular mechanism and degradation efficiency. *Science of the Total Environment*, 877, 162912.

Tian, L., Chen, Q., Jiang, W., Wang, L., Xie, H., Kalogerakis, N., Ma, Y., Ji, R. (2019). A carbon-14 radiotracer-based study on the phototransformation of polystyrene nanoplastics in water versus in air. *Environmental Science: Nano*, 6, 2907–2917.

Tofa, T.S., Kunjali, K.L., Paul, S., Dutta, J. (2019). Visible light photocatalytic degradation of microplastic residues with zinc oxide nanorods. *Environmental Chemistry Letters*, 17, 1341–1346.

Wang, W., Ndungu, A.W., Li, Z., Wang, J. (2017). Microplastics pollution in inland freshwaters of China: A case study in urban surface waters of Wuhan, China. *Science of the Total Environment*, 575, 1369–1374.

Xiao, Y., Tian, Y., Xu, W., Zhu, J. (2024). Photodegradation of microplastics through nanomaterials: Insights into photocatalysts modification and detailed mechanisms. *Materials*, 17, 2755.

Yong, C.Q.Y., Valiyaveettil, S., Tang, B.L. (2020). Toxicity of microplastics and nanoplastics in mammalian systems. *International Journal of Environmental Research and Public Health*, 17(5), 1–24.

You, X., Wang, S., Li, G., Du, L., Dong, X., (2022). Microplastics in the soil: A review of distribution, anthropogenic impact, and interaction with soil microorganisms based on meta-analysis. *Science of the Total Environment*, 832, 154975.

Zhang, Y., Kang, S., Allen, S., Allen, D., Gao, T., Sillanpää, M. (2020). Atmospheric microplastics: A review on current status and perspectives. *Earth-Science Reviews*, 203, 103118.

14

Efficiency of Hybrid Materials for Photocatalytic Degradation of Micro- and Nano-Plastics

The prevalence of micro-plastics (MPs) and nano-plastics has led to serious concerns about their harmful effects, which include health risks, ecological disruption and the potential for contamination to spread in the environment. The fragmentation of plastics (primary and secondary) produces micro- and nano-plastics (MNPs). MNPs stay for a very long time in the environment, and once they are released, their degradation time is prolonged. Therefore, it is imperative to mitigate and remove these plastics in order to protect the environment and human health. Consequently, the growing issue of MNP contamination calls for creative methods to ensure their effective removal and restoration. Nanotechnology is one of the many upcoming advancements that hybrid materials are now having a significant influence on. Today, multifunctional hybrid materials with precise control over structure and morphology are possible thanks to the enormous amount of basic research completed over the last 20 years. The industrial market has already seen the introduction of some of these hybrid materials. These specific surface alterations have the potential to greatly improve the photogenerated electron–hole pairs separation, expand the reactive surface area and the production of more active sites for the reaction and subsequent breakdown of MNPs. Numerous intriguing characteristics, including miniaturization, hierarchical organization, resistance, recognition, self-repairing qualities and adaptability, will be combined in this new generation of hybrid materials.

This chapter explores the emerging hybrid smart materials as photocatalysts working for the degradation of MNPs, and the production of advanced photocatalytic materials capable of effectively addressing the critical issue of plastic pollution in both macro- and nano-forms.

Chapter written by Vaishali GUPTA and Satyendra SINGH.

14.1. Introduction

There has been an appreciable rise in the number of deceased marine animals over the last few years. Many of them are filled with plastic debris that congregates in the natural waterbodies like seas and oceans. Actually, drinking water in plastic bottles contain MPs (5–1,000 μm), and the commercially significant fish from the Mediterranean Sea have been found containing good fragments of the MNPs. The massive amounts of MNPs that are currently present in rivers, lakes, seas and oceans are a problem for humanity. In 2017, the manufacturing of plastics worldwide nearly topped 350 million tonnes, with Europe producing 18.5% of plastics, and China producing 29.4%. Experimental data indicate that three weeks of UV light cause polymers to form nanometer-sized particles. The chemical and physical characteristics of nanoparticles (sizes ranging from 1 to 100 nm) differ from those of macroscopic objects made of the same material. Many plastics pollutants like polyvinyl chloride (PVC) and polyethylene terephthalate (PET) were discovered at the micro- and nano-scales very later. According to a recent research, carbon-based hybrid compounds have the potential to degrade marine MPs effectively without endangering the nearby microbes. Both polyethylene (PE) plastic and low-density polyethylene (LDPE) film using polypyrrole/TiO_2 (PPy/TiO_2) nanocomposite as photocatalyst successfully degraded polluted water using sunlight induced inorganic zinc oxide photocatalysts. There are currently relatively few scientific publications available and a very limited amount of knowledge regarding nano-plastics present in the environment. Due to this, MNPs are regarded as a developing problem that needs to be closely watched. Lower molecular weight (MW) intermediates are produced when plastics are broken down by photocatalysis. These harmful and toxic intermediates can then be utilized as raw materials by the chemical industry to make secondary petrochemical products, plastics or some non-toxic synthesis. An overview of the most recent catalytic and photocatalytic techniques for breaking down MPs is given in this paper, together with information on the amount of plastic generated worldwide and its effects on human health.

An extended polymer chain possesses a substantial MW, crystallinity, hydrophobicity and other essential properties that contribute to plastics' resistance to biodegradation (Wilkes and Aristilde 2017). It is not possible to transport large compounds with a high MW over the membrane containing bacteria. Therefore, in order for long-chain polymers to get across the cell membrane, they must first depolymerize into smaller monomers (Shah et al. 2008). After that, monomers have the ability to cross the cell membrane and are assimilated by intracellular metabolism (Kolvenbach et al. 2014).

There is an urgent need to focus on the processes by which and how plastics are carried by rivers into the oceans, how they are dumped there and may decompose on the way, how they accumulate in rivers, sediments, soils and organisms, and how these plastics affect ecosystems that support life at all food chain levels. Because plastic garbage builds up in rivers and oceans, it may threaten our ecosystem. The presence of plastic waste in rivers and oceans constitutes a threat to global environmental health that affects ecosystems, people and animals alike. Plastics are thought to contain harmful substances such as bisphenol A, phthalates and polychlorinated biphenyls (Toussaint et al. 2019). As these contaminants can entangle, suffocate or disturb the digestive processes of fish, birds, mammals and turtles, they also have a substantial impact on marine fauna (Derraik 2002). Plastics found in the environment undergo a weathering process by natural attacks over it, and cause physical, chemical and biological alterations, resulting in the formation of MPs (Andrady 2011), which have a diameter of <5 mm, and finally nano-plastics (da Costa et al. 2016), which have a size of <1 µm.

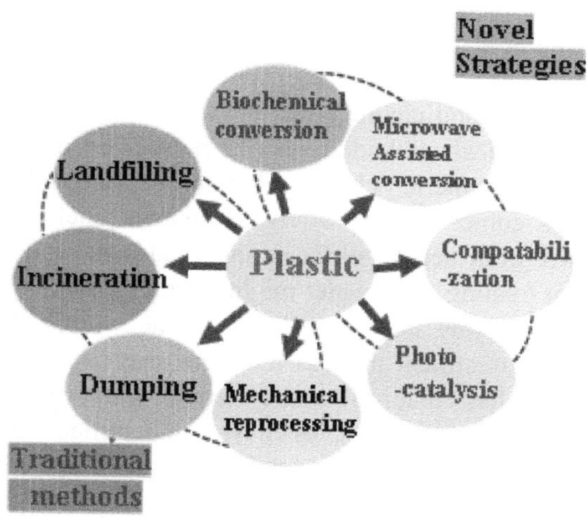

Figure 14.1. *Old and novel strategies for degradation of plastic waste. For a color version of this figure, see www.iste.co.uk/singh/photodegradation.zip*

A critical analysis was conducted on food contamination and edible animal species since 2010. They gathered information that around 200 edible animal species, 23 molluscs, five birds, seven sweet water fish and various other aquatic animals, as well as beer and water get contaminated by MNPs present over there. Dense plastics are dispersed randomly in the river or may be bypassed by light plastics, floating on the surface of the river and influenced by sunlight. These low-density plastics may also cause chemical and physical changes to the edible food. Furthermore, it may take years or even decades for sediment-bound plastics to enter the oceans. During that time, they may interact with various biological materials, such as microbial biofilms, changing their characteristics and behavior. Additionally, pollutants may have absorbed or adsorbed onto their surfaces, exposing vulnerable organisms to long-term exposure (Bratovcic 2019a). There are changes in approach from old time to now, as many novel strategies have been proposed to stop mixing these sediments and for the deduction of these plastic pollutants. Figure 14.1 shows contemporary and novel strategies for degrading MNPs pollutants.

Micro-plastics (MPs) are insoluble particles that are many times smaller than five millimetres and undetectable to the human eye (Kershaw and Rochman 2015; Bratovcic 2019a). MPs have been a growing source of concern, since they endanger both humans and aquatic life. They not only aid in the buildup of plastics in the environment, but they may also aid in the dispersal of micropollutants because of their absorption (Sun et al. 2019). As MPs pass through water, they absorb organic and metal pollutants. When these chemicals are consumed by aquatic animals, they build up along the food chain. When MPs naturally decompose, a variety of dangerous organic chemicals and additives like plasticizers and antitoxicers will be released from the polymeric chain. This process could take decades in manufacturing. MPs come from a variety of sources, including improper waste management, runoff from sealed surfaces and roadways, and wastewater inputs. Despite the fact that treating wastewater might lessen the quantity of plastics, many particles still find their way into rivers all over the world. Moreover, biosolids, which are wastewater treatment process leftovers, can be utilized as fertilizers in agriculture. This implies that MPs may enter soils and possibly precious groundwater resources through other, as of yet unidentified, routes. Nano-plastics are a subset of MPs that have at least one dimension and a size equal to or <100 nm. Despite being made of the same material, nanoparticles typically have chemical and physical properties different from macroscopic items (Ferreira Soares et al. 2020). As a result, when nanoparticles travel throughout the environment, their surface area and other physicochemical characteristics may have a significant impact on how they change and become bioavailable (Bratovcic 2019b). As a result, there are substantial distinctions between MNPs, and there is particular worry about how nano-plastics interact with the environment and living things (Maurer-Jones et al. 2013). To evaluate potential health risks, it is crucial to understand how nano-plastics interact with the environmental bodies, particularly with living things. This is important because behavior of nano-plastics particles is different from their micron-sized counterparts. The most popular and extensively researched plastic polymers include light polyethylene (LDPE), dense polyethylene (high-density polyethylene (HDPE)), polypropylene (PP) and polystyrene (PS). Other soft plastics that are biodegradable include polylactic acid, nylon, acrylic, polycarbonate, polyurethanes, and other plastics. However, at the shorelines of oceans, styrene has been discovered in sand as well as in waters all over the world, and nano-plastics have just been detected in the North Atlantic (Ter Halle et al. 2017). They show that a mixture of plastics was responsible for the anthropogenic fingerprint seen in the colloidal component of seawater. The different size classes have different compositions of polymers. Polyvinyl chloride (PVC) and PS were discovered at the micro- and nano-scales. Additionally, they noticed that when the size of the debris decreased, the pyrolytic signals of PE changed; this refers to plastics' structural changes brought on by the weathering process.

14.2. Behavior of micro- and nano-plastics

MPs can be ingested and, if absorbed, may find their way into various tissues and cells, where they may have a variety of adverse effects (Von Moos et al. 2012; Wright et al. 2013; Pedà et al. 2016; Foley et al. 2018). Studies conducted by Barboza and Gimenez (2015) and Foley et al. (2018) have demonstrated that contamination of shellfish and numerous commercially valuable fish species by MNPs is common, suggesting that MPs may be a possible pathway for human exposure to MPs (Bouwmeester et al. 2015).

Consumers in European nations who consume a lot of shellfish can absorb up to 11,000 MP particles (which range in size from 5 to 1,000 μm) annually, while consumers in nations that consume less shellfish typically swallow 1,800 MPs annually (Van Cauwenberghe and Janssen 2014; Bouwmeester et al. 2015), which is still a significant exposure. Research indicates that approximately 170 MP particles ranging size 200–1,000 μm are consumed by each person when just shrimp are considered (Devriese et al. 2015). In relation to human eating of aquatic lifes such as *Mytilus edulis*, especially persons from Denmark, Norway and the Netherlands were discovered to contain MPs (Vandermeersch et al. 2015).

Furthermore, MPs have been found in the digestive system and liver of various aquatic creatures, which have occasionally eaten it in their entirety (Collard et al. 2017; Compa et al. 2018) as well as in the stomachs of commercially significant fish from the Mediterranean (Vandermeersch et al. 2015). Seafood contamination is typically more of a concern for complete seafood species like mollusks and fish than, for example, peeled shrimp or gutted fish. Additionally, some commercially accessible drinking water comes in glass, plastic and beverage carton bottles.

Furthermore, tap water from other nations or branded drinking water in glass, plastic and beverage cartons that are sold in German supermarkets have been shown to contain MPs (Kosuth et al. 2018; Schymanski et al. 2018). This review's primary focus is on the photocatalytic degradation of polymers, including their benefits and drawbacks in relation to alternative techniques.

14.3. Objective of the chapter

This chapter aims to discuss in detail: (1) systematically present details about the photocatalytic pollutants; (2) investigate the traditional and novel ways to degrade pollutants and mechanisms of the deduction of plastics by various hybrid smart photocatalysts; (3) discuss the role of reactive oxygen species (ROS) during photocatalysis of plastics; and (4) analyze challenges and opportunities that lie ahead in the development of AOPs methods for the deduction of plastics. An overview of the many approaches that could be used to degrade plastics using photocatalysis is given in this chapter.

14.4. Global plastic production

In 2017, the global plastic production was about 350 million tonnes, which did not include PET, polyamide (PA), polyacryl, thermosets, nor PP fibers. Plastics are produced in China at a rate of 29.4%, 18.5% in Europe and 17.7% in NAFTA. based on Figure 14.2 (Bratovcic 2019a).

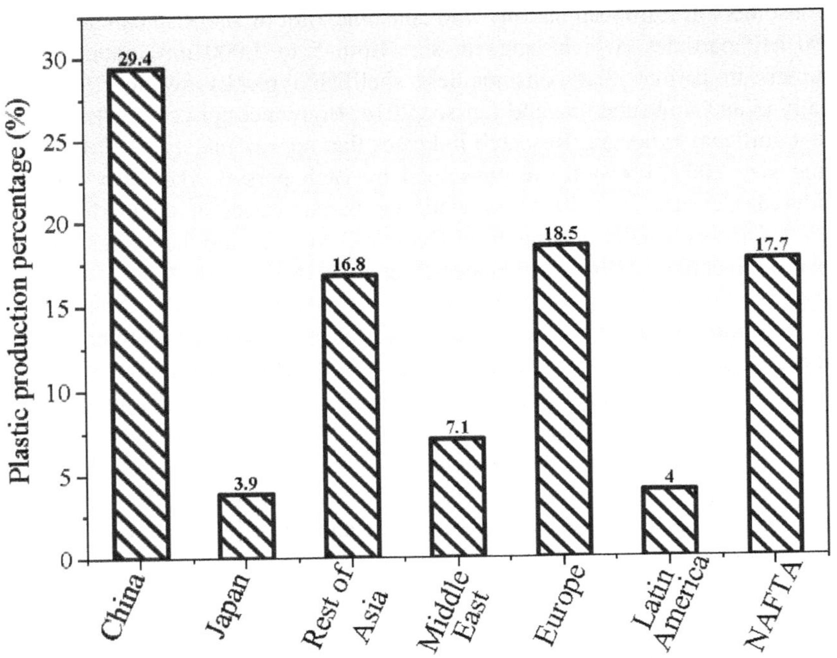

Figure 14.2. *Estimation of global plastics production*

Almost 1.5 million people in Europe directly work in the plastics industry, which employs about 60,000 businesses. In 2017, the plastics sector in Europe generated 355 billion euros in revenue. More than eight million tonnes of plastic garbage was gathered in 2016 for recycling, both inside and outside the European Union.

14.5. Photocatalytic degradation

Novel and alternative chemical techniques for the vanishment of plastics have emerged rapidly over the conventional methods of managing waste plastics. These techniques are favorable, productive, non-toxic products and environmentally friendly, and transform plastics into non-toxic products with harmless secondary products (Luo et al. 2022). These techniques include compatibilization, pyrolysis, gasification, microwave-assisted, biochemical and high-degree water conversion (Zhang et al. 2021). These methods have pros and cons and are similar in addition to being environmentally friendly. However, it can speed up the degradation of plastics to break down using conventional procedures. It takes hundreds of years, compared to a few days or even a few hours using developing strategies (Zhao et al. 2022). Additionally, clean energy and other valuable goods are produced throughout the degradation processes, which helps to obtain higher commercial benefits (Bäckström et al. 2017). However, because of their high cost and low yielding rate, these approaches also have substantial drawbacks. Photocatalytic degradation is thought to be a very efficient and ecologically friendly purification process when compared to other methods (Bhatkhande et al. 2002; Guo et al. 2020; Nasir and Sassani 2021). It produces two electrons and two holes in the redox reaction by using light irradiation to stimulate photocatalysts (Li et al. 2019; Awfa et al. 2020).

Advanced oxidation processes (AOPs) are a recognized novel strategy for the decomposition of plastic waste. These processes are elevated by the formation of in situ reactive species, mainly superoxide (O_2^-) and hydroxyl radicals (•OH). Many redox reactions occur during degradation, which are expressed in the following (Gupta and Singh 2023):

$$\text{Photocatalyst} + \text{h}\nu \rightarrow \text{Photocatalyst}\,(e^- + h^+)$$

$$\text{h}^+ + \text{H}_2\text{O} \rightarrow \text{OH}^\text{o} + \text{H}^+$$

$$\text{h}^+ + \text{OH}^- \rightarrow \text{OH}^\text{o}$$

$$e^- + \text{O}_2 \rightarrow \text{O}_2{}^{\text{o}-}$$

$$\text{OH}^\text{o}/\text{O}_2{}^\text{o} + \text{micro-/nano-plastics} \longrightarrow \text{Intermediates} + \text{CO}_2 + \text{H}_2\text{O}$$

The most researched AOPs include ozonation, which is the coupling of ozone and UV radiation, hydrogen peroxide (H_2O_2) and UV radiation, and UV radiation and photocatalysts. The practical use of these incredible potential technologies is hindered due to their high maintenance charges for all AOPs. However, technologies using ultra-violet radiation or solar radiation have a lot of potential for degradation of water-based pollutants because of high absorptivity of UV radiation, strong penetration power; also, catalysts that absorb in a visible region can greatly improve reactor design. The MPs may break down in this process into smaller inorganic molecules like water and carbon dioxide (Turchi and Ollis 1990). Numerous studies have shown how well photocatalytic technology degrades (micro-)plastics (Turchi and Ollis 1990; Uekert et al. 2019). The two primary forms of photocatalysts are bare and composite. An alternative would be for the initial molecule to partially oxidize into less stable intermediates if intermediates are prone to additional environmental degradation and do not endanger human health or the aquatic environment. The properties of secondary products after photocatalytic degradation of plastics pollutants depends upon many factors such as rate of redox reactions, treatment time and type of catalyst matrix used. Partial degradation of organic pollutants for the formation of such secondary pollutants may be more hazardous than parent pollutants, so analysis of the final product is of more interest, and the mechanism behind the degradation of each type of pollutant is discussed in Figure 14.3. In contrast to most recent photocatalysts (Cui et al. 2019), nothing is known about the most recent developments in the photocatalytic breakdown of MNPs.

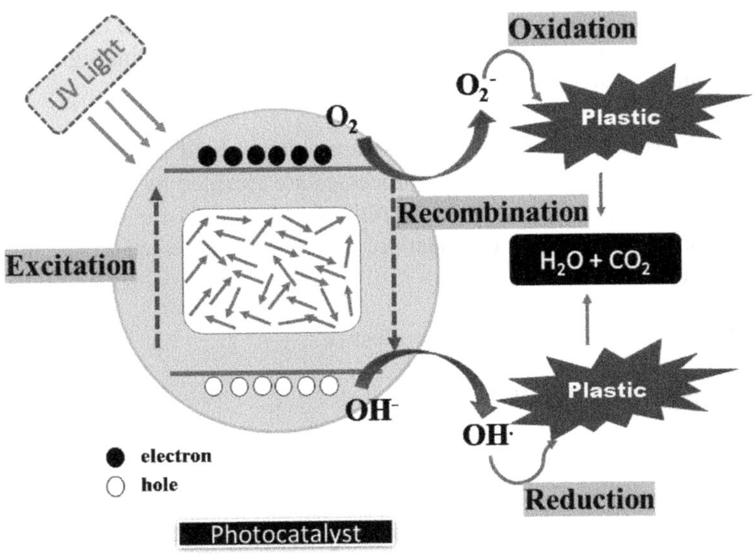

Figure 14.3. *Mechanism of photocatalysis. For a color version of this figure, see www.iste.co.uk/singh/photodegradation.zip*

From 14C radioisotope tracer technology, Tian et al. (2019) investigated the complete mineralization and degradation of PS nano-plastics under exposure to ultraviolet (UV) radiation at 254 nm. 14C-styrene was converted into 14C-PS nano-plastics. Water's role in the photodegradation of PS nano-plastics was investigated by subjecting 14C-PS nano-plastics to airborne UV light or suspending them in water. However, no discernible change was observed from the Fourier-transform infrared spectroscopy (FTIR) analysis. X-ray photoelectron spectroscopy (XPS) analysis demonstrated clearly that C–O groups found on the surface after exposure of 48 hours of UV irradiation, suggesting that partial photo-oxidation only occurs on the upper surface layer of the PS nano-plastics. Following the radiation, the PS nano-plastics' MW had risen in the natural air, indicating good linkage of the PS chains, but it was not altered significantly when water was present. In comparison to air (6.17 ± 0.1%), water had a greater mineralization of PS nano-plastics (17.1 ± 0.55%). When water was exposed to UV light, a notable quantity (11.0 ± 0.1%) of by-products with tiny MW was found; this amount was significantly greater than what washed off the surface of nano-plastics that were exposed to air. The increased photo-activity in water raises the possibility that there are distinct mechanisms behind driving the PS nano-plastics via photo-transformation in the two different matrices. The current investigation offered the first proof of PS nano-plastic photodegradation in aquatic conditions.

14.6. Hybrid smart materials for degradation of micro- and nano-plastics

Human culture and customs have included smart composite materials, in which some are organic, some are inorganic or some are hybrid, made by humans for about 5,000 years. In the second half of the 20th century, the development of the so-called "chimie douce" gave rise to the idea of "hybrid organic–inorganic" nanocomposites, sparking numerous partnerships amongst biologists, physicists and chemists. As a result, a new interdisciplinary idea was generated by the blending of these radically diverse scientific communities. Thanks to the enormous amount of fundamental research done over the past 20 years, custom-made organic–inorganic hybrid materials with exact structure, morphology and topography are now possible. A few of these hybrid materials are now available for purchase in the industrial sector. A growing number of custom multi-scale hybrid materials are having an impact on various fields including energy, optics, nanomedicine, environment and catalysis. We highlight a number of basic and practical features for the hybrid materials in this article, taking various applications and hybrid materials that exist in different forms including nanomaterials, mesostructured thin films, hybrid nanocomposites with chemistry akin to Legos, and some advanced hybrid materials for energy applications. Moreover, many review articles summarize several commercial uses for hybrid materials (Nicole et al. 2014).

The term “Hybrid Organic–Inorganic Materials” (HOIM) refers to materials that combine organic and inorganic component at the nanoscale. These materials are classified as nanocomposite materials or hybrid smart materials. Such close minglings have not historically been carried out by humans, but during billions of years of development, nature has successfully experienced them, giving rise to multifunctional, hierarchical, multi-scale systems (Nicole et al. 2014). Here, we summarize several smart hybrid materials used recently for the application of deduction of MNPs, tabulated in Table 14.1. Photocatalysis degradation of MNPs uses different trigger sources such as UV light lamp, or visible light or sunlight. For wide implementation or practical purposes, UV light degradation is not favorable as it adversely affects environment and lives; therefore, focusing on visible or sunlight degradation is the primary goal. Sunlight contains just 3–5% UV light of total exposure. Use of sunlight as a trigger source would save power and cost.

Catalyst type	Photocatalyst	Plastic	Light used	Duration	Weight loss %	Reference
Hybrid	FePc-TiO_2	PS	Sunlight	12 days	35	Nicole et al. (2014)
	PPy/TiO_2	PE	Sunlight	240 h	35.4	Baddam (2014)
	CuPc/TiO_2	PE	160 h	160 h	–	Li et al. (2010)
	$FePcCl_{16}$-TiO_2	PVC	UV light	240 h	50	Vaschetto et al. (1999)
	PAM-g-TiO_2	LDPE	UV light	520 h	39.85	Fa et al. (2011)
	T@AIH	LDPE	Xenon lamp	72 h		Liang et al. (2013)
	TiO_2-ZnO	PE, PES	UV light	480 h, 624 h	100	Bang et al. (2021)
	NH_2-MIL-88B(Fe)/MoS_2	HDPE	Sunlight	48 h	–	He et al. (2023)
	MXene/$Zn_xCd_{1-x}S$	PET	Solar light	–	–	Feng et al. (2022)

Table 14.1. *List of recent hybrid materials reported for photocatalytic degradation of micro- and nano-plastics*

The goal of initial hybrid photocatalyst investigations was to combine TiO_2 with specific polymers for better use (Zhao et al. 2008; Li et al. 2010). In 2008, Zhao et al. (2008) conducted a research on the heterogeneous photodegradation of PE plastic pollutants when exposed to sun radiation. They synthesized TiO_2/CuPc (copper phthalocyanine/modified TiO_2) as a photocatalyst and observed that the CuPc-modified TiO_2 composite film improved photocatalytic reduction of PE.

The rate of weight loss for the degradation of PE using (TiO_2/CuPc) samples was significantly higher than that of the PE-TiO_2 sample after 160 h, reported by Zhao et al. The degradation rate was at its maximum around 0.23 when the mass ratio of CuPc to TiO_2 was almost equivalent to 0.7 weight percent. The PE-(TiO_2/CuPc) film has a wider response compared to the PE-TiO_2 film. The film exhibited a greater photovoltaic response range, a stronger photovoltaic intensity and a wider absorption range in the solar spectrum.

Li et al. (2010) produced PPy/TiO_2, a highly active photocatalyst, via emulsion polymerization and sol–gel techniques. The PE-PPy/TiO_2 experienced a weight loss rate of 35.4% after 240 h of sunshine irradiation, which was significantly greater than the rates of 11.7% for PPy and 3.2% for TiO_2. TiO_2 crystalline structure was preserved and its capacity to absorb visible light was increased by depositing the PPy conducting polymer as a film, which also explains why PPy/TiO_2's photocatalytic activity was significantly increased (Vaschetto et al. 1999).

In order to facilitate the photodecomposition of a poly(vinyl chloride) (PVC) composite, Fa et al. (2011) employed perchlorinated iron(II) phthalocyanine-modified TiO_2 nanoparticle ($FePcCl_{16}$) as a photocatalyst. Compared to blank TiO_2, which could only absorb UV light, the $FePcCl_{16}$–TiO_2 composite showed exceptional absorption in the visible area beyond 400 nm. It barely absorbed in light longer than 400 nm. After 240 h of UV light irradiation, the PVC–$FePcCl_{16}$–TiO_2 film experienced a 50% weight loss rate, which was significantly greater than the weight loss rate by PVC–TiO_2 film, which is around 30% under the same circumstances. The PVC–$FePcCl_{16}$–TiO_2 film vanished weight at a rate of 81% during the course of a 15-day solar light irradiation trial, which was 80% greater than the PVC–TiO_2 film. When exposed to UV light and sunshine, the PVC–$FePcCl_{16}$-TiO_2 film exhibited a greater rate of photodegradation.

Photocatalytic degradation using polyacrylamide-modified TiO_2 (PAM-g-TiO_2) nanocomposite was produced by Liang et al. (2013) to investigate the weight loss of LDPE/PAM-g-TiO_2. LDPE/TiO_2 underwent a weight loss rate of 30.63% after being exposed to 520 h of UV light. Under the exposure for 520 h of UV light irradiation, the weight loss caused by LDPE/PAM-TiO_2 composite film dropped by 39.85%, and its MW dropped by around 94.60%. It is anticipated that the LDPE matrix will exhibit good dispersion and hydrophilicity upon the introduction of PAM on TiO_2 particles. The breakdown of LDPE was started by a rise in hydroxyl radicals caused by the interaction between TiO_2 and the adsorbed H_2O in the PAM phase.

Zhao et al. (2021) created TiO_2@amylose/polyiodide/hydroxyethyl cellulose (T@AIH), which allowed for the encouragement of self-color changing in addition to the stable stage. It promotes a self-indicating delayed commencement of function and a self-color changing feature. The specific auto-coloring characteristic, which benefits by the inclusion of hydroxyethyl cellulose (HEC), is well matched with stages of degradation, which turns it to white from blue. Following a 72-h exposure to the T@A0.1IN photocatalyst, LDPE-CP underwent oxidation and reduction, resulting in a 0.6556 carbonyl index (CI), and then mineralized into CO_2. Note that 233[thin space (1/6-em)]105 g mol^{-1} to 206[thin space (1/6-em)]370 g mol^{-1} was the change in MW of LEPE-CP. In the same way, the average MW of numbers (Mn) dropped from 26[thin space (1/6-em)]234 to 1462 g mol^{-1}. The decomposition of the polymer is suggested by the drop in MW and molecular numbers, which points to the breakage of a chain reaction in LDPE.

In addition to TiO_2, C_3N_4 is frequently used in hybrid photocatalyst research. The use of benzene sulfonyl chloride hybridized with g-C_3N_4 (BS_5–CN), a new intramolecular donor–acceptor conjugated g-C_3N_4, was suggested by Sun et al. (2021). By extending the π-electron delocalization, this synthesis strengthened the absorption of the visible light range or narrowing the bandgap, which improved the photocatalytic performance considerably. CN and BS_5-CN have indirect band gaps of 2.75 and 2.54 eV, respectively. Additionally, the electron transmission was sped up and the recombination efficiency of electron–hole pairs reduced by grafting the electron acceptor BS onto the CN framework. In order to turn PLA plastic wastes into non-toxic, environment friendly, high-value chemical products, lactate undergoes oxidation simultaneously into a kind of formate and acetate. This solved the energy and environmental pollution problems.

Han et al. (2022) concentrated on the synthesis of H_2 and polymeric carbonized dot-graphitic (CPDs-CN) catalytic under PET upcycling using solar light irradiation. In this article, PET was utilized as a reagent to yield high-value chemicals, primarily ethanol, glycolic acid and glycolaldehyde, along with water-splitting-produced H_2. These results show promise for the sustainable production of solar chemicals or fuels with added value.

As interest in creating sustainable materials and eco-friendly polymers grows, so does the concentration on good strength and practical natural fiber-based smart hybrid materials. By using the different sized fibers ranging from micro- to nano-meter, the multi-scale hybridization phenomenon was obtained from bamboo, and an environmentally friendly long fibrous composite with good mechanical and diffusive properties was created. Consequently, the surface of the micro-bamboo fiber was simultaneously covered and attached to by the cellulose nanofibers. The multi-scale hybrid phenomena between the microfibers and bamboo nanoparticles enhanced the fibrous composite's toughness, Young's modulus, elongation and tensile strength, and results are shown in Figure 14.4.

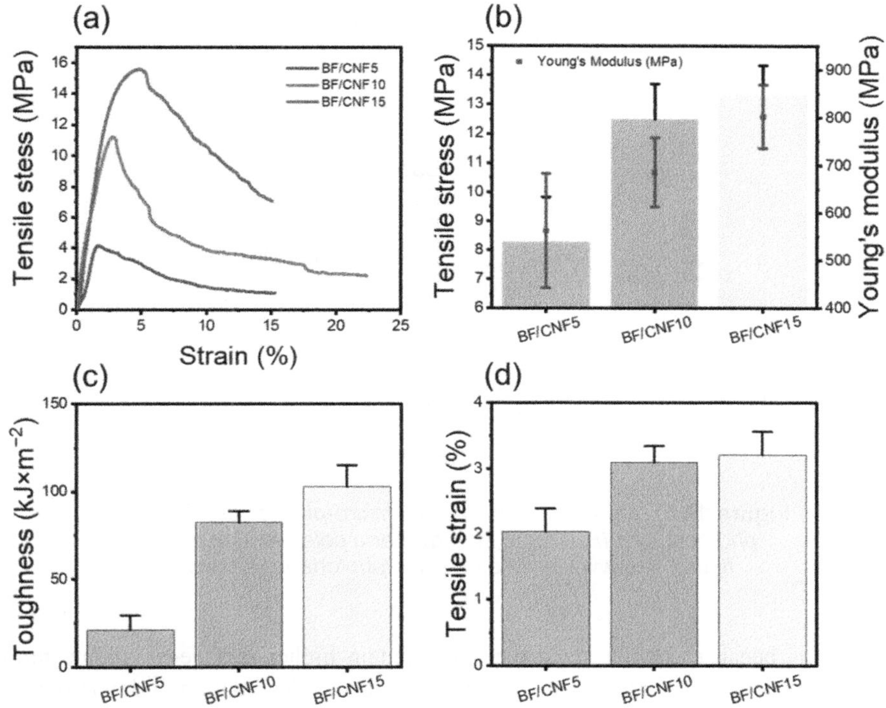

Figure 14.4. *Mechanical properties of BF/CNF fibrous preforms. (a) Representative tensile stress–strain curve; (b) tensile stress and Young's modulus; (c) toughness and (d) tensile strain (Bang et al. 2021). For a color version of this figure, see www.iste.co.uk/singh/photodegradation.zip*

The reinforcing of biodegradable fiber and reinforced plastic (FRP) was also impacted by the improvement of the mechanical properties of the fibrous preform. This environmentally friendly nano-/microfiber preform can be widely used in FRP forced preforms and other applications in the eco-friendly plastics sector (Bang et al. 2021).

Using a TiO_2/ZnO (He et al. 2023) photocatalyst and 365 nm of light, we efficiently demonstrated the photocatalytic destruction of several MPs, including HDPE derived from face cleanser, PES microfibers and environmental MPs. The shape of MPs was discovered to have a significant influence on the deterioration, as these fibers showed saturation of weight loss, resulting in visible destruction to the fibers while preserving their integrity. For MPs with large sizes (PES fibers and plastic pieces), ball milling was necessary to produce smaller sized MPs (~2 μm particles) that were degraded, and results are shown in Figure 14.5.

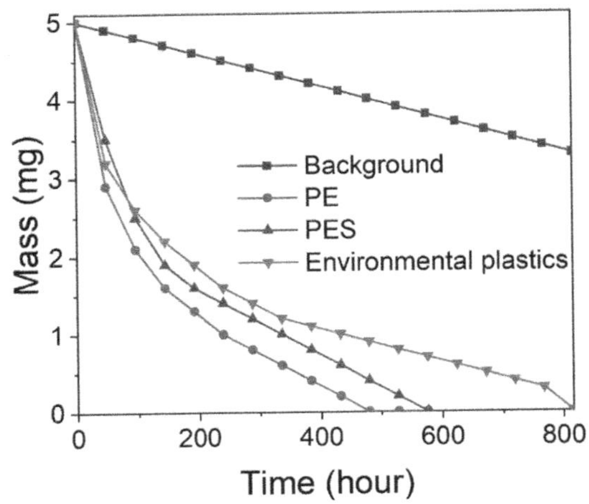

Figure 14.5. *Degradation of different micro-plastics by TiO_2/ZnO photocatalyst (Bratovcic 2019a). For a color version of this figure, see www.iste.co.uk/singh/photodegradation.zip*

This is because smaller sized particles contain higher surface-to-volume ratio, which proposes more active sites to react; so, faster degradation is favorable for nanosized plastics or smaller fragments of plastics. Additionally, the addition of an electron scavenger, $Na_2S_2O_8$, significantly increased ROS generation, which in turn enhanced photocatalytic degradation. These steps were necessary to achieve 100% mass loss of MPs. While varying treatment durations were required for distinct kinds of MPs to achieve 100% mass reduction, the morphologies of photocatalysts also play an important role in determining efficiency of the same material.

A nanoflower novel hybrid photocatalyst, NH_2-MIL-88B(Fe) conjugated with the MoS_2 catalyst (Feng et al. 2022), was synthesized via the hydrothermal method, as shown in Figure 14.6.

It had nano-flower like morphology. Novel photocatalyst NH_2-MIL-88B(Fe)/MoS_2 had an amazing degradation ability for tetracycline hydrochloride using visible-light exposure. Although there are still many obstacles to overcome, NH_2-MIL-88B (Fe) photocatalysts show excellent efficiency in photocatalytic energy conversion and environmental cleanup. It was successfully developed to manufacture a dense novel effective composite (NH_2-MIL-88B(Fe)/MoS_2) by depositing of MoS_2 on NH_2-MIL-88B(Fe) utilizing NH_2-MIL-88B(Fe) as a precursor. Tetracycline hydrochloride (TC) had a photodegradation efficiency of 96.44% and a salinity of 72.36%. HDPE film and MP particles have been broken down in water using the NH_2-MIL-88B(Fe)-MoS_2-2 as a photocatalyst. Several investigations verified that the hybrid system's specific surface area, recombination resistance, carrier lifetime and light absorption were all improved upon attaching NH_2-MIL-88B(Fe) to MoS_2.

Potential degradation pathways were identified and studied by the detection of the emerging intermediates by LC-MS. The intermediate's toxicity assessment revealed that NH_2-MIL-88B(Fe)/MoS_2 might likewise lessen TC's toxicity through a photocatalytic reaction. On the NH_2-MIL-88B (Fe)/MoS_2 heterojunction, more photoinduced active species (radical h^+, radical O_2^- and radical OH^o) are activated, according to calculations using density functional theory and charge transfer mechanism analysis, as shown in Figure 14.7.

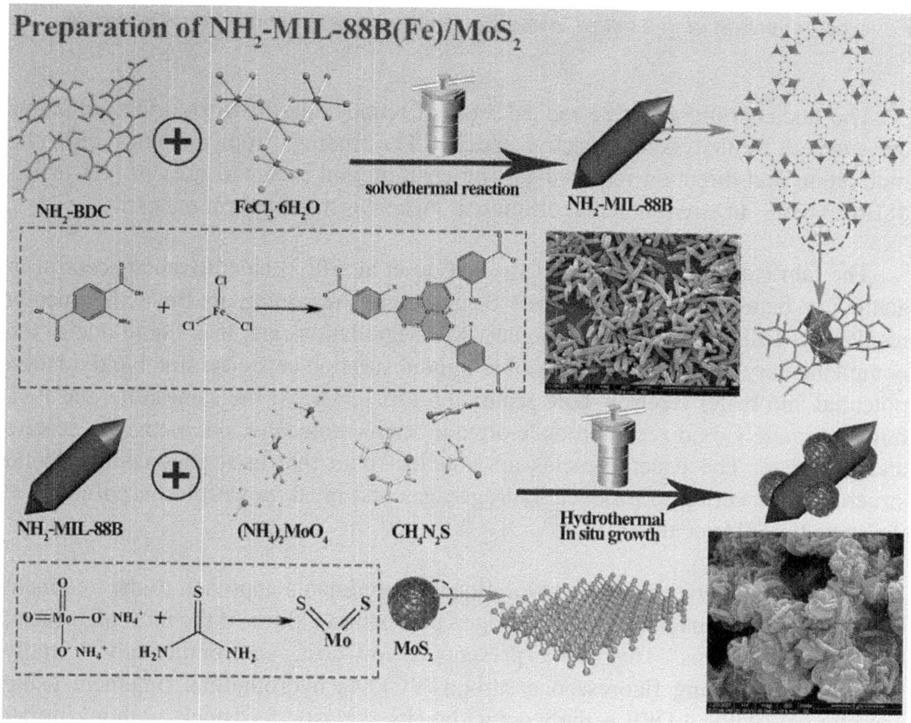

Figure 14.6. *Schematic diagram of the synthesis process of the nanoflower hybrid NH_2-MIL-88B(Fe)/MoS_2. Reprinted from Bratovcic (2019a), copyright (2022) with permission from Elsevier. For a color version of this figure, see www.iste.co.uk/singh/photodegradation.zip*

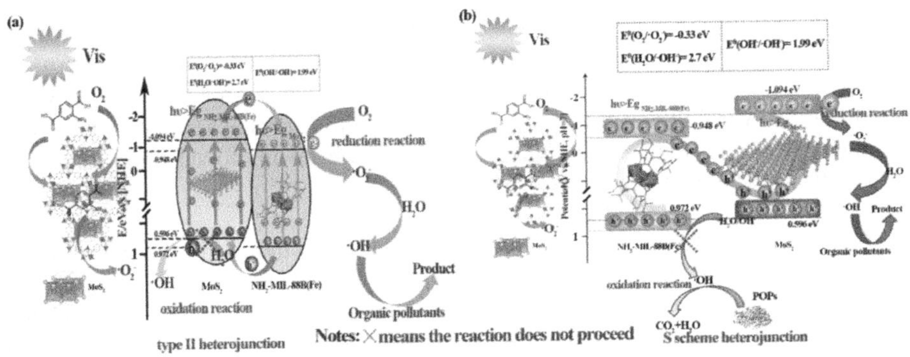

Figure 14.7. *(a) type II photocatalytic mechanism for NH_2-MIL-88B(Fe)/MoS_2 nanocomposite under Xenon lamp irradiation; (b) S-scheme photocatalytic mechanism for NH_2-MIL-88B(Fe)/MoS_2 nanocomposite under Xenon lamp irradiation. Reprinted from Bratovcic (2019a), copyright (2022) with permission from Elsevier. For a color version of this figure, see www.iste.co.uk/singh/photodegradation.zip*

The nucleophilic and free radical sites of tetracycline hydrochloride pollutants are attacked by the excited reactive species. The findings from the study offer the motivation and direction required for the creation and potential uses of NH_2-MIL-88B(Fe)/MoS_2 in environmental cleanup as an active smart hybrid material.

The fabrication of MXene/$Zn_xCd_{1-x}S$ (Cao et al. 2022) has proven successful in achieving better efficiency for MP degradation. According to the experimental results, M~2/$Zn_{0.6}Cd_{0.4}S$ has the highest MP degradation rate. It is found due to the complementary effects of higher visible light absorption capacity, suitable oxidation potential and better electron–hole separation efficiency. At the same time, the PET bottle ultimately oxidizes to produce organic micro-molecules like methanol, acetate and glycolate. The paper presents an excellent plan that uses photocatalyst band structure engineering to produce hydrogen gas and break down plastic pollution at the same time (He et al. 2023).

The current work presents a revolutionary coordinated approach to the treatment of water, waste management and hydrogen production – three of the most significant worldwide concerns. This work presents a low-cost, environmentally friendly method of producing fluorescence carbon (FC) via hydrothermal treatment using commercial LDPE. LDPE is the most widely used plastic feedstock, without further treatment. The produced FCs demonstrated strong fluorescence characteristics with two emission bands, making them suitable for use in the detection of Cu^{2+} ions in aqueous solutions. With a LOD of 86.5 nM, the sensing results were highly successful, indicating their potential use in the analysis of Cu^{2+} ions in industrial effluent. The solution FC@CuO was mixed with TiO_2 to create an

FCs@CuO@TiO_2 composite as a smart photocatalyst (Mehta et al. 2022). The tertiary composite is astonishing and a novel technique. A simple hydrothermal method was utilized to create FCs from LDPE, a material that is widely used.

14.7. Conclusions and suggestions for the future

Amazing work have been applied to investigate new and influential methods of decomposing plastics as environmental consciousness has grown. One approach with tremendous promise is photocatalysis. In this review, catalysts from common pure semiconductors such as TiO_2, ZnO and MnO to composites, that is, smart and hybrid materials are being summarized systematically. Next, we enumerate a few more factors that our analysis revealed to have an impact on catalytic efficiency and help with future photocatalysis optimization. Nonetheless, given that plastics now degrade at a rate of <10%, ubiquitous MPs continue to be a hazard to every human, animal and plant on the planet. We still need to make progress because photocatalysis is only at the theoretical level right now. Here, we outline a few constraints and future directions for development that could encourage the possibility of more useful applications and merit further in-depth investigation.

14.8. References

Andrady, A.L. (2011). Microplastics in the marine environment. *Marine Pollution Bulletin*, 62, 1596–1605.

Awfa, D., Ateia, M., Fujii, M., Yoshimura, C. (2020). Photocatalytic degradation of organic micropollutants: Inhibition mechanisms by different fractions of natural organic matter. *Water Research*, 174, 115643.

Bäckström, E., Odelius, K., Hakkarainen, M. (2017). Trash to treasure: Microwave-assisted conversion of polyethylene to functional chemicals. *Industrial and Engineering Chemistry Research*, 56, 14814–14821.

Baddam, S.R. (2014). *Effect of UV and Amount of Titanium Dioxide on the Degradation of Polystyrene-Titanium Dioxide Composite Films*. Lamar University, Beaumont.

Bang, J., Lee, H., Yang, Y., Oh, J.K., Kwak, H.W. (2021). Nano/micro hybrid bamboo fibrous preforms for robust biodegradable fiber reinforced plastics. *Polymers*, 13, 636.

Barboza, L.G.A. and Gimenez, B.C.G. (2015). Microplastics in the marine environment: Current trends and future perspectives. *Marine Pollution Bulletin*, 97, 5–12.

Bhatkhande, D.S., Pangarkar, V.G., Beenackers, A.A.C.M. (2002). Photocatalytic degradation for environmental applications – A review. *Journal of Chemical Technology & Biotechnology*, 77, 102–116.

Bouwmeester, H., Hollman, P.C.H., Peters, R.J.B. (2015). Potential health impact of environmentally released micro- and nano-plastics in the human food production chain: Experiences from nanotoxicology. *Environmental Science and Technology*, 49, 8932–8947.

Bratovcic, A. (2019a). Degradation of micro- and nano-plastics by photocatalytic methods. *J. Nanosci. Nanotechnol. Appl.*, 3, 206.

Bratovcic, A. (2019b). Different applications of nanomaterials and their impact on the environment. *Int. Journal of Materials Science and Engineering*, 5, 1–7.

Cao, B., Wan, S., Wang, Y., Guo, H., Ou, M., Zhong, Q. (2022). Highly-efficient visible-light-driven photocatalytic H2 evolution integrated with microplastic degradation over MXene/ZnxCd1-xS photocatalyst. *Journal of Colloid and Interface Science*, 605, 311–319.

Collard, F., Gilbert, B., Compère, P., Eppe, G., Das, K., Jauniaux, T., Parmentier, E. (2017). Microplastics in livers of European anchovies (*Engraulis encrasicolus* L.). *Environmental Pollution*, 229, 1000–1005.

Compa, M., Ventero, A., Iglesias, M., Deudero, S. (2018). Ingestion of microplastics and natural fibres in *Sardina pilchardus* (Walbaum, 1792) and *Engraulis encrasicolus* (Linnaeus, 1758) along the Spanish Mediterranean coast. *Marine Pollution Bulletin*, 128, 89–96.

da Costa, J.P., Santos, P.S.M., Duarte, A.C., Rocha-Santos, T. (2016). (Nano)plastics in the environment – Sources, fates and effects. *Science of the Total Environment*, 566–567, 15–26.

Cui, P., Wei, D., Ji, J., Huang, H., Jia, E., Dou, S., Wang, T., Wang, W., Li, M. (2019). Planar p–n homojunction perovskite solar cells with efficiency exceeding 21.3%. *Nature Energy*, 4, 150–159.

Derraik, J.G.B. (2002). The pollution of the marine environment by plastic debris: A review. *Marine Pollution Bulletin*, 44, 842–852.

Devriese, L.I., Van der Meulen, M.D., Maes, T., Bekaert, K., Paul-Pont, I., Frère, L., Robbens, J., Vethaak, A.D. (2015). Microplastic contamination in brown shrimp (*Crangon crangon*, Linnaeus 1758) from coastal waters of the southern North Sea and Channel area. *Marine Pollution Bulletin*, 98, 179–187.

Fa, W., Gong, C., Tian, L., Peng, T., Zan, L. (2011). Enhancement of photocatalytic degradation of poly (vinyl chloride) with perchlorinated iron (II) phthalocyanine modified nano-TiO_2. *Journal of Applied Polymer Science*, 122, 1823–1828.

Feng, X., Long, R., Liu, C., Liu, X. (2022). Visible-light-driven removal of tetracycline hydrochloride and microplastics (HDPE) by nanoflower hybrid heterojunction NH2-MIL-88B (Fe)/MoS_2 via enhanced electron-transfer. *Separation and Purification Technology*, 302, 122138.

Ferreira Soares, D.C.F., Domingues, S.C., Viana, D.B., Tebaldi, M.L. (2020). Polymer-hybrid nanoparticles: Current advances in biomedical applications. *Biomedicine and Pharmacotherapy*, 131, 110695.

Foley, C.J., Feiner, Z.S., Malinich, T.D., Höök, T.O. (2018). A meta-analysis of the effects of exposure to microplastics on fish and aquatic invertebrates. *Science of the Total Environment*, 631, 550–559.

Guo, D., Ren, S., Lu, S., Feng, Z., Tang, D., Liu, S., Zhou, L., Duan, N., Svyatkovskiy, A., Fu, S. (2020). Graphcodebert: Pretraining code representations with data flow. *arXiv*, arXiv:2009.08366.

Gupta, V. and Singh, S. (2023). Corona-poling enhanced photocatalytic degradation of methyl-violet and rhodamine B pollutants using ferroelectric nanoparticles. *Chemistry of Inorganic Materials*, 1, 100008. doi: 10.1016/j.cinorg.2023.100008.

Han, M., Zhu, S., Xia, C., Yang, B. (2022). Photocatalytic upcycling of poly (ethylene terephthalate) plastic to high-value chemicals. *Applied Catalysis B*, 316, 121662.

He, Y., Rehman, A.U., Xu, M., Not, C.A., Ng, A.M.C., Djurišić, A.B. (2023). Photocatalytic degradation of different types of microplastics by TiOx/ZnO tetrapod photocatalysts. *Heliyon*, 9.

Kershaw, P.J. and Rochman, C.M. (2015). Sources, fate and effects of microplastics in the marine environment: Part 2 of a global assessment. Report, GESAMP.

Kolvenbach, B.A., Helbling, D.E., Kohler, H.P.E., Corvini, P.F.X. (2014). Emerging chemicals and the evolution of biodegradation capacities and pathways in bacteria. *Current Opinion in Biotechnology*, 27, 8–14.

Kosuth, M., Mason, S.A., Wattenberg, E.V. (2018). Anthropogenic contamination of tap water, beer, and sea salt. *PLOS ONE*, 13, e0194970.

Li, S., Xu, S., He, L., Xu, F., Wang, Y., Zhang, L. (2010). Photocatalytic degradation of polyethylene plastic with polypyrrole/TiO_2 nanocomposite as photocatalyst. *Polymer-Plastics Technology and Engineering*, 49, 400–406.

Li, X., Bing, L., Zhang, W., Lam, W. (2019). Exploiting BERT for end-to-end aspect-based sentiment analysis. *arXiv*, arXiv:1910.00883.

Liang, W., Luo, Y., Song, S., Dong, X., Yu, X. (2013). High photocatalytic degradation activity of polyethylene containing polyacrylamide grafted TiO_2. *Polymer Degradation and Stability*, 98, 1754–1761.

Luo, H., Yao, D., Zeng, K., Li, J., Yan, S., Zhong, D., Hu, J., Yang, H., Chen, H. (2022). Solar pyrolysis of waste plastics with photothermal catalysts for high-value products. *Fuel Processing Technology*, 230, 107205.

Maurer-Jones, M.A., Gunsolus, I.L., Murphy, C.J., Haynes, C.L. (2013). Toxicity of engineered nanoparticles in the environment. *Analytical Chemistry*, 85, 3036–3049.

Mehta, A., Rather, R.A., Belec, B., Gardonio, S., Fang, M., Valant, M. (2022). Plastic waste precursor-derived fluorescent carbon and construction of ternary FCs@ CuO@ TiO_2 hybrid photocatalyst for hydrogen production and sensing application. *Energies*, 15, 1734.

Nasir, V. and Sassani, F. (2021). A review on deep learning in machining and tool monitoring: Methods, opportunities, and challenges. *International Journal of Advanced Manufacturing Technology*, 115, 2683–2709.

Nicole, L., Laberty-Robert, C., Rozes, L., Sanchez, C. (2014). Hybrid materials science: A promised land for the integrative design of multifunctional materials. *Nanoscale*, 6, 6267–6292.

Pedà, C., Caccamo, L., Fossi, M.C., Gai, F., Andaloro, F., Genovese, L., Perdichizzi, A., Romeo, T., Maricchiolo, G. (2016). Intestinal alterations in European sea bass *Dicentrarchus labrax* (Linnaeus, 1758) exposed to microplastics: Preliminary results. *Environmental Pollution*, 212, 251–256.

Schymanski, D., Goldbeck, C., Humpf, H.U., Fürst, P. (2018). Analysis of microplastics in water by micro-Raman spectroscopy: Release of plastic particles from different packaging into mineral water. *Water Research*, 129, 154–162.

Shah, A.A., Hasan, F., Hameed, A., Ahmed, S. (2008). Biological degradation of plastics: A comprehensive review. *Biotechnology Advances*, 26, 246–265.

Sun, J., Dai, X., Wang, Q., Van Loosdrecht, M.C.M., Ni, B.J. (2019). Microplastics in wastewater treatment plants: Detection, occurrence and removal. *Water Research*, 152, 21–37.

Sun, D.W., Chen, K.L., Huang, J.H. (2021). Benzenesulfonyl chloride-incorporated g-C3N4 for photocatalytic hydrogen generation by using the hydrolysate of poly (lactic acid) as sacrificial reagent. *Applied Catalysis A*, 628, 118397.

Ter Halle, A., Jeanneau, L., Martignac, M., Jardé, E., Pedrono, B., Brach, L., Gigault, J. (2017). Nanoplastic in the North Atlantic subtropical gyre. *Environmental Science and Technology*, 51, 13689–13697.

Tian, L., Chen, Q., Jiang, W., Wang, L., Xie, H., Kalogerakis, N., Ma, Y., Ji, R. (2019). A carbon-14 radiotracer-based study on the phototransformation of polystyrene nano-plastics in water versus in air. *Environmental Science: Nano*, 6, 2907–2917.

Toussaint, B., Raffael, B., Angers-Loustau, A., Gilliland, D., Kestens, V., Petrillo, M., Rio-Echevarria, I.M., Van den Eede, G. (2019). Review of micro- and nanoplastic contamination in the food chain. *Food Additives and Contaminants: Part A*, 36, 639–673.

Turchi, C.S. and Ollis, D.F. (1990). Photocatalytic degradation of organic water contaminants: Mechanisms involving hydroxyl radical attack. *Journal of Catalysis*, 122, 178–192.

Uekert, T., Kasap, H., Reisner, E. (2019). Photoreforming of nonrecyclable plastic waste over a carbon nitride/nickel phosphide catalyst. *Journal of the American Chemical Society*, 141, 15201–15210.

Van Cauwenberghe, L. and Janssen, C.R. (2014). Microplastics in bivalves cultured for human consumption. *Environmental Pollution*, 193, 65–70.

Vandermeersch, G., Van Cauwenberghe, L., Janssen, C.R., Marques, A., Granby, K., Fait, G., Kotterman, M.J.J., Diogène, J., Bekaert, K., Robbens, J. et al. (2015). A critical view on microplastic quantification in aquatic organisms. *Environmental Research*, 143, 46–55.

Vaschetto, M.E., Monkman, A.P., Springborg, M. (1999). First-principles studies of some conducting polymers: PPP, PPy, PPV, PPyV, and PANI. *Journal of Molecular Structure: THEOCHEM*, 468, 181–191.

Von Moos, N., Burkhardt-Holm, P., Köhler, A. (2012). Uptake and effects of microplastics on cells and tissue of the blue mussel *Mytilus edulis* L. after an experimental exposure. *Environmental Science and Technology*, 46, 11327–11335.

Wilkes, R.A. and Aristilde, L. (2017). Degradation and metabolism of synthetic plastics and associated products by Pseudomonas sp.: Capabilities and challenges. *Journal of Applied Microbiology*, 123, 582–593.

Wright, S.L., Thompson, R.C., Galloway, T.S. (2013). The physical impacts of microplastics on marine organisms: A review. *Environmental Pollution*, 178, 483–492.

Zhang, K., Hamidian, A.H., Tubić, A., Zhang, Y., Fang, J.K.H., Wu, C., Lam, P.K.S. (2021). Understanding plastic degradation and microplastic formation in the environment: A review. *Environmental Pollution*, 274, 116554.

Zhao, X., Li, Z., Chen, Y., Shi, L., Zhu, Y. (2008). Enhancement of photocatalytic degradation of polyethylene plastic with CuPc modified TiO_2 photocatalyst under solar light irradiation. *Applied Surface Science*, 254, 1825–1829.

Zhao, Y., Zhang, F., Zhang, J., Zou, K., Zhang, J., Chen, C., Long, M., Zhang, Q., Wang, J., Zheng, C. (2021). Preparation of composite photocatalyst with tunable and self-indicating delayed onset of performance and its application in polyethylene degradation. *Applied Catalysis B*, 286, 119918.

Zhao, X., Korey, M., Li, K., Copenhaver, K., Tekinalp, H., Celik, S., Kalaitzidou, K., Ruan, R., Ragauskas, A.J., Ozcan, S. (2022). Plastic waste upcycling toward a circular economy. *Chemical Engineering Journal*, 428, 131928.

15

Surface Modifications of $BiVO_4$ Semiconductor Materials for Photocatalytic Degradation of Micro- and Nano-Plastic

The proliferation of micro- and nano-plastics, owing to minuscule size and widespread occurrence in the environment, has raised significant concerns regarding their adverse effects, encompassing ecological disruption, health risks and the potential for contaminant spread, making their mitigation and removal vital for the protection of both environmental and human well-being. Thereby, this escalating problem of micro- and nano-plastic pollution necessitates innovative solutions for their efficient removal and remediation. Surface modifications of $BiVO_4$ semiconductors represent a promising avenue for developing smart materials aimed at the efficient photocatalytic degradation of these contaminants. $BiVO_4$, a well-known semiconductor, has exhibited remarkable photocatalytic properties owing to its suitable bandgap and electronic structure. However, to enhance its efficacy in degrading micro- and nano-plastics, surface modifications are crucial. These modifications can include doping with metals or nonmetals, deposition of co-catalysts like noble metals or carbon-based materials and the introduction of nanostructures like nanoparticles or nanowires. These tailored surface modifications can significantly enhance the disassociation of photoinduced electron–hole pairs, increase the surface area available for reactions and introduce additional active sites for the adsorption and subsequent degradation of micro- and nano-plastics.

This chapter explores the synergistic effects of pivotal surface-engineered $BiVO_4$ semiconductors, and the production of advanced photocatalytic materials capable of effectively addressing the critical issue of plastic pollution in both macro- and nano-forms.

Chapter written by Nikita YADAV, Vaishali GUPTA and Ojasvi SAINI.

15.1. Introduction to micro- and nano-plastic pollution

15.1.1. *Overview of micro- and nano-plastic pollution: a growing environmental concern*

Micro- and nano-plastic pollution represents an emergent and escalating environmental challenge that has gained substantial attention nowadays. These minute plastic particles, often measuring < 5 mm in size, have permeated terrestrial and aquatic ecosystems, leading to a pervasive risk to biodiversity and environmental integrity (Lai et al. 2022). The omnipresence of these particles in many environmental compartments, including air, soil and water, has triggered a rising concern due to their possible adverse effects both on ecosystems and public well-being (Yee et al. 2021). Micro-plastics originate from varied causes, such as from the collapse of larger plastic material due to weathering, mechanical abrasion and exposure to ultraviolet energy. Moreover, industrial procedures, such as the fragmentation of plastic materials throughout production, contribute significantly to their generation. Nano-plastics, on the other hand, can arise from the breakdown of micro-plastics, as well as from intentional manufacturing processes to produce nano-sized plastic particles. Micro- and nano-plastics exhibit remarkable environmental persistence, owing to their resistance to biodegradation. Once released into the environment, they infiltrate various ecosystems, including terrestrial, freshwater and marine environments. Their small size facilitates widespread dispersion, allowing them to infiltrate organisms across multiple trophic levels and permeate diverse habitats, from oceanic depths to the atmospheric realm (Vikas Madhav et al. 2020).

These minute plastic particles pose a myriad of ecological risks. Their ingestion by marine and terrestrial organisms can lead to physical harm, hindering digestive processes, causing internal injuries and serving as vectors for the transport of toxins (Usman et al. 2020). Moreover, the adsorption of persistent organic pollutants (POPs) and other harmful substances onto micro- and nano-plastics further amplifies their toxicity, potentially affecting entire food webs (Panel and Chain 2016).

The proliferation of micro- and nano-plastic pollution characterizes a multifaceted and interconnected environmental encounter with extensive consequences. The urgency to address this issue is underscored by its budding influence on ecosystems, biodiversity and public well-being (Yee et al. 2021; Lai et al. 2022). Mitigating the environmental footprint of micro- and nano-plastics requires a concerted interdisciplinary effort, involving scientists, policymakers and the public, to advance effective approaches for prevention, remediation and sustainable plastic management. The escalating prevalence of micro- and nano-plastic contamination signifies an urgent need for interdisciplinary research, policy interventions and innovative technologies. Addressing this environmental menace requires concerted efforts aimed at elucidating their impact, devising efficient monitoring techniques and implementing sustainable solutions to mitigate their proliferation in natural environments.

15.1.2. *Definition and classification*

Micro-plastics are particles with a diameter of < 5 mm, encompassing principal micro-plastics that have been produced at this scale as well as residual micro-plastics that accumulate over when bigger plastic objects disintegrate in the environment. These particles can be further categorized based on their source into primary micro-plastics, such as microbeads utilized in cosmetics, and secondary micro-plastics, which result from the disintegration of larger plastic items like bottles, bags and fishing gear (Ivleva 2021).

Nano-plastics are even smaller particles with dimensions typically < 100 nm. They can be the product of the degradation of larger plastics or can be deliberately produced at the nanoscale for definite applications. Nano-plastics are particularly concerning due to their potential to enter cells and tissues, raising questions about their impact on living organisms at the molecular level (Joksimovic et al. 2022).

15.1.2.1. *Classification based on the composition*

Micro- and nano-plastics exhibit a wide range of compositions, as they can be derived from various types of polymers used in the formation of plastic things. The composition of these particles plays a significant role in determining their physical and chemical properties, as well as their possible environmental and biological impacts.

15.1.2.2. *Polyethylene (PE)*

It is a common polymer used in plastic manufacturing. PE is composed of long chains of repeating ethylene monomers. The polymerization process involves breaking the double bond in ethylene and connecting the carbon atoms to form a long-chain structure. PE is known for its simplicity in structure, consisting of repeating ethylene units, which are two carbon atoms linked by a double bond (C_2H_4). This basic structure contributes to its remarkable properties and widespread applicability. PE is lightweight, making it easy to transport and handle. It is resistant to moisture, chemicals and UV radiation, contributing to its durability. The polymer can be transparent or translucent, depending on its density and processing conditions. It is versatile and can be found in numerous forms, such as low-density polyethylene (LDPE) and high-density polyethylene (HDPE). LDPE is characterized by a more branched structure, resulting in a lower density and greater flexibility. It is commonly used for packaging films, plastic bags and various containers. HDPE has a more linear structure, providing higher density and stiffness. It is employed in the production of bottles, pipes and containers for liquids (Yani et al. 2020).

PE is used in packaging materials, plastic bags, bottles and a variation of other consumer and industrial products. HDPE is commonly employed in the manufacturing of decanters for beverages, household substances, etc. It is widely used in the construction industry for water and gas pipes due to its high strength and resistance to corrosion (Ferreira et al. 2022). LDPE is used in the production of toys, containers and various household items.

However, PE is considered relatively inert and stable, which contributes to its durability but also leads to slow degradation in the environment (Ghatge et al. 2020). The disintegration of PE items can result in the release of micro-plastics and nano-plastics, contributing to environmental pollution in the form of fragments, fibers or particles, depending on their source and processing (Yao et al. 2022).

15.1.2.3. *Polypropylene*

Polypropylene (PP) is another widely used polymer, known for its durability and resistance to heat. PP is a linear hydrocarbon polymer, consisting of repeated propylene monomer units (Maddah 2016). The chemical formula for propylene is (C_3H_6). The polymerization process involves linking these propylene monomers together, creating long chains of PP. The majority of PP produced is isotactic, meaning that the methyl groups attached to the polymer chain are oriented on the same side. This arrangement contributes to the polymer's crystalline structure and high melting point, known as isotactic PP (iPP). An atactic PP form of PP has randomly oriented methyl groups, resulting in an amorphous structure with lower melting points.

PP is known for its high strength and stiffness, making it suitable for a variety of structural and load-bearing applications. It has excellent chemical resistance, resisting many acids, bases and solvents. PP is lightweight and has a relatively low density. It exhibits good fatigue resistance and is resistant to fatigue failure under repeated stress.

PP is widely used in packaging materials, including films, containers and woven bags, due to its excellent barrier properties and durability. It is mostly utilized in the textile industry for the production of carpets, upholstery and geotextiles. PP is employed in the manufacturing of various automotive components, including bumpers, interior trims and battery cases. It is used in medical devices such as syringes, vials and containers due to its biocompatibility and resistance to sterilization methods.

Micro- and nano-plastics made from PP can be present as fibers, films or particles. The disintegration of bigger PP items contributes to the formation of secondary micro-plastics. PP is considered a relatively environmentally friendly plastic compared to some other polymers. It has a lower environmental impact in terms of production energy and greenhouse gas emanations. Like other plastics, improper disposal of PP items can contribute to environmental pollution. Efforts to improve recycling rates and develop more sustainable practices are ongoing. PP is recyclable, and recycling codes #5 (PP) are commonly associated with PP products. PP recycling involves processes such as melting and extrusion to create new products. Challenges in PP recycling include the need for efficient sorting and separation, as well as addressing contamination issues to produce high-quality recycled materials.

15.1.2.4. *Polystyrene*

Polystyrene (PS) is a lightweight and rigid polymer. It is used in disposable foam products, packaging materials and insulation. PS is a polymer composed of styrene monomers. The chemical formula for styrene is C_8H_8. The polymerization process involves linking styrene monomers together to form a long-chain structure. PS is rigid, transparent (in the case of GPPS), and has a relatively low melting point. It is a good insulator of heat and electricity, making it suitable for insulation applications (Höppener et al. 2023). PS is lightweight and can be easily molded into various shapes, contributing to its versatility in manufacturing.

There are mostly two types of PS used. The first, general purpose polystyrene (GPPS), is clear, hard and brittle. It has good water resistance, high dielectric strength and excellent transparency. It is also inexpensive and has good physical and mechanical properties. This is the most common form of PS, known for its clarity and rigidity. GPPS is often used in applications where transparency is important, such as in disposable utensils and packaging. Second, high impact polystyrene (HIPS) is a modified form of PS with enhanced impact resistance. It is used in applications where toughness is a key requirement, such as in the production of toys and appliance parts.

PS is widely used in packaging materials, including foam trays, clamshell containers and disposable cups. Expanded polystyrene (EPS), a foam form of PS, is used for thermal insulation in construction and as packaging material for fragile items. PS is used in the production of an assortment of consumer belongings, consisting disposable cutlery, CD cases and electronic housings. HIPS is commonly used in the production of toys, model kits and other items requiring durability. PS is used in the healthcare industry for items like Petri dishes and laboratory consumables.

PS has faced environmental scrutiny due to its persistence in the environment and challenges in recycling. EPS foam, often used in packaging, is lightweight but not easily biodegradable and can contribute to litter and environmental pollution.

Micro- and nano-plastics originating from PS can be in the form of small beads, fragments or fibers. These particles are often released into the environment through the breakdown of larger PS items. PS is recyclable, but there are challenges associated with its recycling due to issues such as contamination and the need for specialized recycling facilities. Efforts to improve PS recycling technologies and increase recycling rates are ongoing.

15.1.2.5. *Polyvinyl chloride*

Polyvinyl chloride (PVC) is composed of repeating vinyl chloride monomer units. The chemical formula for vinyl chloride is C_2H_3Cl. The polymerization process involves linking these vinyl chloride monomers together to form long-chain PVC. PVC is a tough and rigid material in its rigid form, offering high mechanical strength. It meets the requirements for utilization in many different kinds of circumstances due to its exceptional endurance to chemicals. PVC can be formulated to be flame-retardant, contributing to its use in construction and electrical applications. It is an excellent electrical insulator. Generally, two types of PVC used in daily life activities. The first, rigid PVC (uPVC), is characterized by its rigidity and is commonly used in construction materials, such as pipes, fittings and window frames. The second, flexible PVC (PVC-P) is more plasticized, allowing it to be more pliable. It is used in applications such as inflatable structures, clothing and electrical cable insulation.

uPVC is widely used in the construction industry for pipes, fittings, window profiles and siding due to its durability and corrosion resistance. PVC is used for electrical cable insulation and in the production of electrical conduits. Flexible PVC is employed in the production of clothing, footwear and accessories. PVC is used in medical applications for items such as tubing, blood bags and medical devices along with the automotive field for interior parts, wiring insulation and gaskets.

Micro- and nano-plastics derived from PVC can exhibit diverse forms, including particles and fibers. The degradation of PVC-based products contributes to the occurrence of these particles in the environment. PVC has faced environmental concerns due to the release of chlorine gas during manufacturing and potential issues related to phthalate plasticizers used in flexible PVC formulations. The disposal of PVC in landfills can contribute to the release of chlorine gas, although modern waste management practices aim to mitigate these issues. PVC is recyclable, but challenges exist in its efficient recycling due to issues such as the need for specialized recycling facilities and the presence of additives. There is growing interest in the advancement of alternative materials and more sustainable practices to condense the environmental impact of PVC. Bio-based PVC and recycled PVC are areas of research and development to enhance the sustainability of this polymer. Efforts are ongoing to address the environmental impact of PVC through recycling initiatives and the development of alternative materials.

15.1.2.5.1. Classification based on the origin

Micro- and nano-plastics can be classified based on their origin, distinguishing between primary and secondary sources. This classification helps in understanding the routes via which the natural setting is exposed to these debris particles and designing effective mitigation strategies. The following section discusses the detailed description of these origins.

15.1.2.5.2. Primary micro-plastics

Particles formed deliberately minuscule for particular purposes are known as primary micro-plastics. They are designed to be small from the outset for specific functions. For example, microbeads (small plastic spheres often utilized in personal care products such as exfoliating scrubs and toothpaste), microfibers (small plastic fibers released during the washing of synthetic textiles like polyester and nylon) and nurdles (tiny polymer pellets that are utilized as a base material to make plastic goods).

Commonly used in consumer products, resulting in intentional release into the environment. Microbeads, microfibers and nurdles can enter aquatic bodies directly through wastewater discharge or indirectly through runoff. They can be devoured by marine organisms, leading to potential ecological and biological impacts.

15.1.2.5.3. Secondary micro-plastics

These particles originate from the disintegration of bigger plastic objects due to physical processes such as weathering, UV radiation and mechanical pressure in the environment. Secondary micro-plastics contribute significantly to the global micro-plastic load in the environment. The breakdown of plastics can release additives and chemicals into the environment, potentially causing harm.

15.1.2.5.4. Classification based on the size

Micro- and nano-plastics, minute plastic particles with diameters ranging from micrometers to nanometers, exhibit a diverse array of shapes, influencing their behavior, environmental distribution and potential biological impacts. Among the various shapes, fibers stand out with their elongated, thread-like structure, often originating from the degradation of textiles during washing or purposefully designed for specialized applications. These fibers, due to their form, pose unique challenges as they can be easily ingested by a broad spectrum of organisms, leading to potential entanglement and physical harm. Remains, characterized by asymmetrical figures resulting from the degradation of larger plastic items, contribute significantly to the overall micro-plastic load. Spherical micro-plastics, such as microbeads intentionally added to cosmetics or those formed through weathering, exhibit a rounded, ball-like shape, influencing their buoyancy and settling characteristics in aquatic environments. Film-shaped micro- and nano-plastics, often derived from the breakdown of plastic bags and packaging materials, have a thin, sheet-like structure that may affect their transport in air and water. Beyond these, micro- and nano-plastics can assume various other shapes, including rods and irregular forms, influencing their interactions with organisms and ecosystems. This classification based on shape is vital for understanding the complex dynamics of these plastics in the environment, informing strategies to mitigate their ecological impact and probable risks to living organisms.

15.1.3. *Occurrence and distribution of micro- and nano-plastic in environmental matrices*

15.1.3.1. *Presence in aquatic environments*

The primary reservoirs of these plastics are aquatic settings, where they result from the breakdown of bigger plastic debris or are introduced directly through sources such as industrial effluents, stormwater runoff and inappropriate removal of plastic waste (Joksimovic et al. 2022). The buoyancy and durability of these particles facilitate their widespread dispersion, leading to contamination in oceans, rivers, lakes and even groundwater. The aquatic ecosystem's vulnerability to micro- and nano-plastic pollution is particularly worrisome, as these particles can exist for chronic periods and accumulate within the tissues of aquatic organisms, ultimately entering the food chain (Panel and Chain 2016).

15.1.3.2. *Atmospheric transport and terrestrial impact*

Beyond aquatic ecosystems, micro- and nano-plastics have been detected in the atmosphere, representing their probable for long-range transport and deposition in terrestrial environments (Vikas Madhav et al. 2020). Atmospheric transport may occur via procedures such as aero-dispersal, facilitating the global circulation of these particles. Once deposited on land, micro- and nano-plastics can interact with soil, impacting soil health, nutrient cycling and potentially entering the terrestrial food web (Du et al. 2021). The adverse effects of this pervasive pollution on terrestrial ecosystems and agricultural systems are not yet fully understood but warrant careful investigation.

15.1.3.3. *Biological and ecological implications*

The biological and ecological implications of plastic pollution are multifaceted. In aquatic ecosystems, these particles can be ingested by a diverse range of organisms, from planktonic organisms to larger marine species. The ingestion of micro- and nano-plastics can cause physical impairment, nutritional deficiencies and the likely transfer of noxious substances related with the plastics (Campanale et al. 2020).

Moreover, the persistence of these particles increases concerns about their long-term environmental impact and the potential for bioaccumulation within food webs (Bhuyan 2022).

15.2. Semiconductor photocatalysis in environmental remediation: fundamentals and principles

Semiconductors play a pivotal role in environmental clean-up processes, particularly in addressing pollution challenges. Their unique electronic properties enable them to serve as effective photocatalysts in advanced oxidation processes (AOPs). For instance, semiconductor materials like titanium dioxide (TiO_2) can harness solar or artificial light to generate electron–hole pairs, initiating powerful oxidative reactions that break down contaminants into less detrimental substances (Ibhadon and Fitzpatrick 2013). This capability is increasingly utilized for the disintegration of organic contaminants, such as industrial dyes, pesticides and even micro- and nano-plastics. The semiconductor-based photocatalysis provides a sustainable and energy-efficient approach for environmental remediation, offering a promising avenue for mitigating the impact of pollution on environment and public health. Semiconductor photocatalysis is a promising field in environmental remediation, especially in the context of addressing pollution challenges such as the degradation of these plastics. This technology relies on the use of semiconductor materials to catalyze chemical reactions driven by light energy. The principles and fundamentals of semiconductor photocatalysis play a vital role in understanding its application in the photocatalytic degradation of contaminants like plastics (Biancullo et al. 2019; Pandey et al. 2023).

Semiconductor photocatalysis involves the use of semiconductor materials, commonly oxides such as titanium dioxide (TiO_2) and zinc oxide (ZnO), to catalyze chemical reactions due to incidence of light (Sagadevan et al. 2022). The key component of this process is the generation of electron–hole pairs when the semiconductor material is irradiated with light of appropriate energy. These electron–hole pairs participate in redox reactions, leading to the breakdown of organic pollutants. Numerous factors influence the effectiveness of semiconductor photocatalysis in the degradation of micro- and nano-plastics, including semiconductor material, light source (the energy and wavelength of the light source determine the activation of the photocatalyst), surface area (higher surface area enhances the adsorption capacity and photocatalytic activity), and catalyst loading (the amount of photocatalyst used can influence the effectiveness of the procedure) (Bouazza et al. 2009; Zhou et al. 2022).

When a semiconductor material absorbs photons, holes are left in its valence band and electrons are promoted to the conduction band, starting the photocatalytic procedure. Next, with species adsorbed on the outermost layer of semiconductors the photogenerated electrons and holes execute the redox processes. In the case of environmental remediation, the target is often the degradation of pollutants. Semiconductor photocatalysis offers a sustainable and effective solution for the degradation of these plastic particles. During the process of photocatalysis, plastics may adsorb onto the surface of the semiconductor material due to the occurrence of functional groups on their surface (Goodarzi et al. 2023). This adsorption facilitates contact between the plastic and the photocatalyst. Upon contact to light, the photocatalyst generates electron–hole pairs. These generated charge carriers then drift to the semiconductor surface. Thereafter, the electrons and holes partake in redox reactions with adsorbed species, including the plastics. This results in the disintegration of the plastics into smaller, less harmful compounds. The ultimate goal of the process is the comprehensive mineralization of plastics into harmless end products, such as water, carbon dioxide and inorganic ions.

15.2.1. *Mechanisms of photocatalytic degradation*

The photocatalytic degradation of micro- and nano-plastics involves complex processes driven by the interaction between semiconductor photocatalysts and the plastic pollutants under light irradiation. Several mechanisms contribute to the disintegration of plastics into smaller, less harmful compounds. The functional groups existing on the surface of nanomaterials can interact with the surface of the semiconductor photocatalyst through physical adsorption forces, facilitating the attachment of plastic particles to the photocatalyst. Photocatalytic degradation can proceed through multiple pathways, depending on the specific properties of the plastic, the semiconductor photocatalyst and the environmental circumstances (Zhang et al. 2019). Some pathways may involve a direct attack on the plastic surface, while others may involve the generation of intermediates that undergo further reactions. The detailed process of photocatalysis is shown in Figure 15.1:

1) Photogeneration of electron–hole pairs: when exposed to light, photons with an intensity comparable to or exceeding the semiconductor photocatalyst's bandgap are absorbed by TiO_2. This results in the transfer of electrons from the valence band to the conduction band as a result due to absorption, resulting in the formation of electron–hole pairs. The photogenerated electrons and holes are crucial for initiating redox reactions on the semiconductor surface. Photogenerated electrons in the conduction band and holes in the valence band drift to the exterior of the semiconductor. The electrons participate in reduction reactions, while the holes engage in oxidation reactions. In the context of micro- and nano-plastic degradation, the redox reactions involve the attack on the plastic surface by reactive species formed during these reactions.

2) Reactive oxygen species (ROS) generation: redox reactions on the semiconductor surface led to the formation of highly reactive species, such as superoxide radicals (O_2^-), hydroxyl radicals (OH^-) and other oxygen species. These ROS play an important role in breaking down the polymer chains of plastics by attacking the carbon–carbon and carbon–oxygen bonds.

3) Cleavage of polymer chains: the generated reactive species attack the plastic surface, leading to the cleavage of polymer chains. This process results in the formation of smaller fragments and intermediates, which are more susceptible to further degradation.

4) Mineralization: the ultimate goal of photocatalytic degradation is the mineralization of plastics into harmless inorganic compounds. Mineralization comprises the comprehensive transition of plastic fragments into simpler, environmentally benign end products such as water, carbon dioxide and inorganic ions.

5) Catalyst recycling: semiconductor photocatalysts are typically not consumed during the degradation process and can be recovered and reused. Catalyst recycling is an environmentally friendly aspect of photocatalysis, contributing to the sustainability of the degradation process.

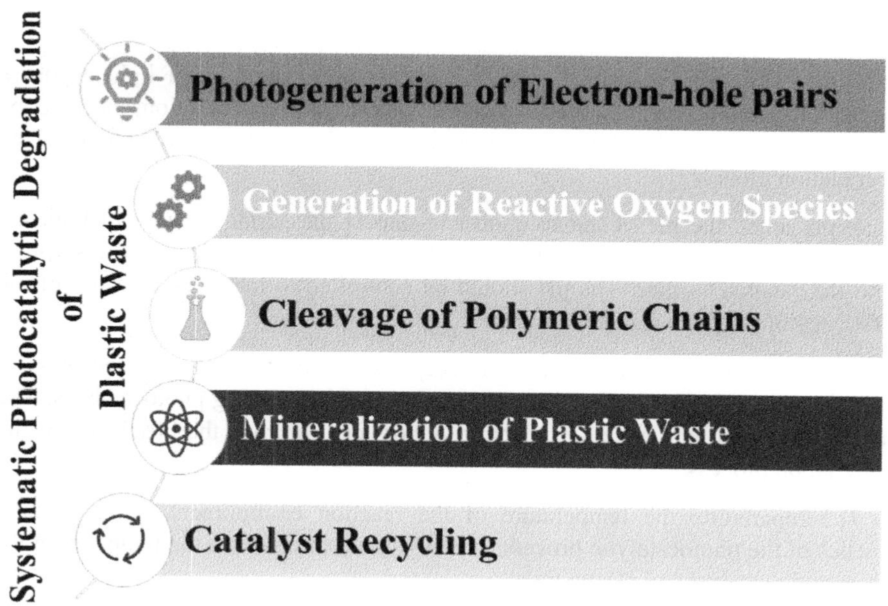

Figure 15.1. *Systematic demonstration of photocatalytic degradation of plastic waste. For a color version of this figure, see www.iste.co.uk/singh/photodegradation.zip*

15.2.2. *Factors influencing photocatalytic efficiency*

Photocatalysis is a process that encompasses the usage of a catalyst to hasten a chemical response by absorbing light. In the context of micro- and nano-plastic degradation, photocatalysis has been explored as a potential method to break down these pollutants into smaller, less harmful components. Several factors influence the photocatalytic efficiency of micro- and nano-plastic degradation as shown in Figure 15.2, and understanding these factors is vital for augmenting the efficacy of the process, such as:

1) Type of catalyst: the choice of photocatalyst is a critical factor. Titanium dioxide (TiO_2) is a commonly used photocatalyst for micro- and nano-plastic degradation due to its high stability and photo-activity. Other catalysts, such as ZnO and various metal oxides, have also been investigated.

2) Catalyst surface area: the surface area of the photocatalyst is unswervingly associated to its efficiency. High surface area offers more binding sites for the adsorption of micro- and nano-plastic particles, enhancing the degradation process.

3) Wavelength of light: the wavelength of light used for photocatalysis affects the efficiency of the process. The photocatalyst should absorb light in the UV or visible range, as micro- and nano-plastics typically do not absorb light in these regions.

4) Intensity and duration of light: the intensity and duration of light exposure significantly impact the photocatalytic efficiency. Optimal conditions need to be established to ensure that the catalyst is activated and remains active throughout the degradation process.

5) pH level: the pH of the solution can impact the surface charge of both the photocatalyst and micro-/nano-plastics, affecting their interaction and adsorption onto the catalyst surface. The pH should be controlled to make the best use of the efficiency of the photocatalytic process.

6) Presence of electron donors and acceptors: electron donors and acceptors in the solution can influence the electron transfer processes during photocatalysis. The presence of sacrificial agents or electron donors can improve the efficiency of the degradation process.

7) Temperature: the temperature of the reaction environment can affect the kinetics of the photocatalytic process. It is important to optimize the temperature for the specific catalyst and plastic type.

8) Concentration of micro-/nano-plastics: the concentration of micro- and nano-plastics in the solution can influence the competence of the photocatalytic process. High concentrations may lead to inundation of active sites on the catalyst exterior, reducing effectiveness.

9) Matrix effects: the occurrence of other materials in the water, such as organic matter or ions, can influence the photocatalytic degradation. Understanding and mitigating these matrix effects is essential for real-world applications.

10) Catalyst immobilization: the method used to restrain the photocatalyst (e.g. on a sustenance material) can affect its stability and accessibility to micro- and nano-plastics.

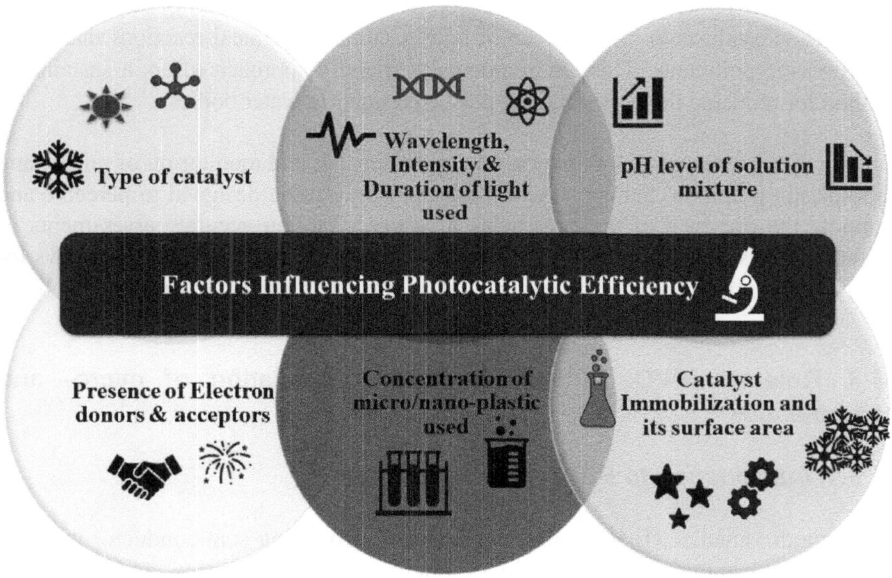

Figure 15.2. *Factors affecting the process of photocatalytic efficiency for the degradation of plastic waste. For a color version of this figure, see www.iste.co.uk/singh/photodegradation.zip*

15.2.3. *Role of semiconductors in environmental clean-up*

Semiconductors play a vital role in addressing the pressing environmental issue of micro- and nano-plastic pollution through innovative technologies designed for their removal. One of the key applications involves the development of advanced filtration and separation systems based on semiconductor materials. These materials, when engineered at the micro- or nanoscale, exhibit unique properties that can be exploited for efficient elimination of micro- and nano-plastics from water sources.

Semiconductor membranes with tailored nanostructures can act as selective filters, allowing for the precise separation of micro- and nano-plastic particles from water. These membranes can be integrated into water treatment processes, providing an effective barrier to prevent the entry of plastic pollutants into ecosystems. The semiconductor-based filtration systems offer advantages such as high efficiency, durability and scalability, making them promising tools for large-scale environmental clean-up efforts (Bakhtiarnia et al. 2022).

In addition to physical filtration, semiconductors contribute to the development of innovative oxidation processes for the disintegration of micro- and nano-plastics. Semiconductor photocatalysis, for instance, involves the utilization of semiconductor materials as catalysts in the presence of light to initiate chemical reactions that break down plastic pollutants. This environmentally friendly approach offers a sustainable method for reducing the persistence of plastic particles in water bodies.

Moreover, semiconductors play a role in the sensing and monitoring of micro- and nano-plastic pollution. Semiconductor-based sensors can be designed to perceive and compute the presence of these particles in water, enabling accurate assessments of pollution levels. Real-time monitoring facilitated by semiconductor sensors allows for timely interventions and the implementation of effective clean-up strategies.

15.3. Role of $BiVO_4$ in photocatalytic degradation of micro- and nano-plastics

15.3.1. *Introduction to $BiVO_4$ semiconductors*

Bismuth vanadate ($BiVO_4$) has emerged as a promising semiconductor material for the photocatalytic disintegration of micro- and nano-plastics, addressing the pressing environmental concerns associated with plastic pollution. $BiVO_4$ possesses distinct structural and electronic properties that make it well-suited for effective photocatalysis.

15.3.1.1. *Structural properties*

$BiVO_4$, bismuth vanadate, crystallizes in the monoclinic scheelite structure. $BiVO_4$ adopts a tetragonal scheelite structure in which the bismuth (Bi) and vanadium (V) ions are arranged in a well-defined lattice. This lattice structure provides a stable framework for the material. The monoclinic symmetry of $BiVO_4$ indicates that the crystal structure has one unique axis of unequal length and two axes of equal length (but different from the first). This asymmetry is important for the material's catalytic properties. The unit cell of $BiVO_4$ consists of repeated building blocks that make up the entire crystal lattice. Understanding the unit cell is crucial for studying the material's properties and predicting its behavior in different conditions. Bismuth ions are typically eightfold coordinated, forming a distorted cube/octahedron. Vanadium ions, on the other hand, are fivefold coordinated, forming a square pyramid. The coordination environment influences the reactivity of the ions and their ability to participate in catalytic reactions. The crystal structure provides accessible catalytic sites for the adsorption of micro- and nano-plastics, facilitating their subsequent degradation. The availability of active sites is crucial for the initiation of photocatalytic reactions on the material's surface.

15.3.1.2. *Electronic properties*

$BiVO_4$'s electrical characteristics are essential to its photocatalytic properties. The band gap of $BiVO_4$, which controls the amount of power needed to transfer an electron from the valence band to the conduction band, is one of its significant electrical characteristics. With a band gap ranging from 2.4 to 2.8 eV, classified it as visible light. Because of this, $BiVO_4$ has an excellent absorption of visible light, which makes it appropriate for photocatalytic uses. $BiVO_4$'s band gap aligns with the visible light area of the electromagnetic spectrum, enabling it to absorb light in this range. This property is essential for applications such as photocatalysis, where the absorption of visible light drives electron excitation and subsequent redox reactions. When $BiVO_4$ absorbs photons with energy greater than its band gap, electron–hole pairs are generated. These charge carriers play a crucial role in photocatalytic processes, participating in redox reactions that lead to the degradation of pollutants or the generation of useful products, such as hydrogen in the case of water splitting. The electronic structure of $BiVO_4$ facilitates redox reactions, making it particularly suitable for applications where oxidation-reduction processes are involved. This property is exploited in photocatalytic water splitting and the degradation of pollutants. The efficiency of a semiconductor in electronic applications is influenced by the mobility of charge carriers and the rate of recombination. Engineering the electronic properties of $BiVO_4$, including the reduction of carrier recombination, is a focus of research to enhance its overall performance. The photodegradation of micro- and nano-plastics using $BiVO_4$ as a photocatalyst involves complex mechanisms driven by the absorption of light, formation of electron–hole pairs and subsequent redox reactions.

The photocatalytic efficacy of $BiVO_4$ is also impacted by its surface area and morphology. Nanoscale engineering of $BiVO_4$ particles can enhance their surface area, providing additional binding sites for the interaction with micro- and nano-plastics. Additionally, surface defects and doping strategies are explored to further improve the material's charge carrier separation and overall photocatalytic performance.

15.3.2. *Significance of $BiVO_4$ in photocatalysis*

$BiVO_4$ is a semiconductor material that has grown substantial importance in the arena of photocatalysis owing its unique properties and potential applications. The significance of $BiVO_4$ in photocatalysis lies in its versatility, efficiency in visible light absorption and applicability to a wide range of environmental and energy-related processes (Trinh et al. 2021). The key significance of $BiVO_4$ in photocatalysis lies in the following aspects:

1) Visible light absorption: $BiVO_4$ exhibits excellent light absorption in the visible spectral range of the electromagnetic spectrum. This is crucial for photocatalysis as it allows the material to harness sunlight, the most abundant and environmentally friendly light source, for initiating catalytic reactions.

2) Band gap structure: the band gap of $BiVO_4$ is within the visible light range, typically around 2.4–2.8 eV. This enables the creation of electron–hole pairs under visible light irradiation, facilitating efficient photoinduced redox reactions.

3) Water splitting for hydrogen production: one of the notable applications of $BiVO_4$ is in the photocatalytic water splitting for hydrogen formation. The material can absorb visible light and catalyze the oxidation of water to generate oxygen and protons, contributing to the development of sustainable and clean energy sources.

4) Environmental remediation: $BiVO_4$ is employed in the photocatalytic degradation of organic pollutants. Its photocatalytic activity can initiate the breakdown of various organic compounds, contributing to the treatment of water and air pollutants. This is mainly relevant for identifying environmental issues linked with industrial pollutants.

5) Antibacterial and antifungal properties: $BiVO_4$ has shown promise in the photocatalytic inactivation of bacteria and fungi. This makes it a potential candidate for applications in water purification and the development of antibacterial surfaces, thereby addressing issues related to waterborne diseases and microbial contamination.

6) Photocatalytic solar cells: $BiVO_4$ is explored for its potential use in the development of photocatalytic solar cells. Its ability to generate charge carriers under sunlight can contribute to the efficient transition of solar energy into electrical energy, providing an avenue for sustainable and renewable power generation.

7) Visible light-driven organic synthesis: $BiVO_4$ can be employed in visible light-driven organic synthesis reactions. This has implications for the green and sustainable synthesis of various organic compounds, reducing the dependence on traditional, energy-intensive methods.

15.3.3. *Advantages and limitations of $BiVO_4$ for this application*

Advantages of $BiVO_4$ for photocatalytic degradation of micro- and nano-plastics are as follows:

– Visible light absorption: $BiVO_4$ possesses a bandgap within the visible light spectrum, allowing it to efficiently absorb sunlight. This property is advantageous for practical applications, as it enables the utilization of abundant and environmentally friendly light sources.

– Environmental compatibility: $BiVO_4$ is environmentally benign and does not introduce toxic elements into the environment during the photocatalytic process. Its use aligns with the goal of developing sustainable and eco-friendly technologies for plastic degradation.

– Redox activity: the electronic structure of $BiVO_4$ facilitates redox reactions, making it highly effective in initiating oxidation and reduction processes during photocatalysis. This is crucial for the breakdown of complex polymer structures in micro- and nano-plastics.

– Catalytic stability: $BiVO_4$ exhibits good stability and can withstand multiple cycles of photocatalytic activity without significant degradation. This characteristic is important for practical applications where sustained activity over time is required.

– Water compatibility: given its band structure, $BiVO_4$ is particularly suitable for applications in aqueous environments, making it relevant for the degradation of micro- and nano-plastics in water bodies.

– Potential for scalability: $BiVO_4$ is relatively easy to synthesize and has the potential for scalability in production. This makes it attractive for large-scale applications in water treatment and environmental remediation.

– Multifunctional properties: in addition to its role in micro- and nano-plastic degradation, $BiVO_4$ can be employed for other photocatalytic usage, such as water splitting for hydrogen production and the disintegration of organic pollutants. This multifunctionality enhances its versatility in addressing various environmental challenges.

Limitations of $BiVO_4$ for photocatalytic degradation of micro- and nano-plastic are as follows:

– Limited photocatalytic efficiency: while $BiVO_4$ is effective, its photocatalytic efficiency may be lower compared to some other semiconductor materials. Ongoing research focuses on improving efficiency through material modifications and engineering.

– Band gap limitation: the bandgap of $BiVO_4$, though suitable for visible light absorption, may limit its performance in certain environmental conditions or under specific light spectra. Strategies to widen or tailor the bandgap are explored to enhance its applicability.

– Charge carrier recombination: the reformation of photogenerated electron–hole pairs can decrease the overall efficacy of the process. Efforts are made to minimize charge carrier recombination through strategies such as surface modification and doping.

– Dependency on environmental factors: the efficiency of $BiVO_4$ can be impacted by environmental conditions, including pH, temperature and the existence of certain ions. Optimization for specific conditions may be necessary for optimal performance.

– Need for catalyst immobilization: in order to maximize its effectiveness, $BiVO_4$ is often immobilized on a support material. The choice and preparation of the support material are critical and can affect overall performance.

– Synthesis challenges: achieving uniform and well-defined $BiVO_4$ nanostructures can be challenging, and the synthesis process may require careful control to obtain desired properties. Advances in synthetic techniques are addressing these challenges.

15.4. Surface modifications of $BiVO_4$ for enhanced catalytic activity

15.4.1. *Overview of surface modification techniques*

Surface modifications of $BiVO_4$ play a pivotal role in enhancing its catalytic activity, especially in the realm of photocatalysis. $BiVO_4$, a semiconductor material, has garnered attention for applications in solar energy conversion, water splitting and environmental pollutant degradation (Bano et al. 2021). However, in order to better its performance, investigators have discovered innumerable surface modification techniques. One approach involves doping, wherein foreign elements like metals (e.g. Fe, Cr and Mn) or nonmetals (e.g. N) are introduced into the $BiVO_4$ crystal lattice to modify its electronic structure, enhancing charge carrier mobility and reducing recombination rates. Co-catalyst deposition is another strategy, wherein materials such as noble metals (Pt and Au), metal oxides (CoOx and NiOx) or graphene-based compounds are deposited on the $BiVO_4$ surface to facilitate specific catalytic reactions and improve charge separation. Surface sensitization involves introducing dyes or quantum dots to extend the photocatalytic activity into the visible and near-infrared regions. Additionally, surface passivation techniques, such as coating with silicon oxide or aluminum oxide, aim to reduce defects and enhance stability. Ion exchange involves replacing ions within the lattice to modify the band structure, while texturing or morphology control techniques, including hydrothermal synthesis and template-assisted methods, aim to increase surface area and improve light harvesting. Chemical functionalization introduces specific functional groups to alter the surface reactivity, and in situ growth of nanostructures, such as nanorods or nanowires, can enhance charge transport. The choice of surface modification procedure rest upon the desired application, and often, a combination of methods is employed to synergistically optimize the catalytic performance of $BiVO_4$.

Several surface modification techniques have been explored to tailor the properties of $BiVO_4$, and they can be generally written off as into physical, chemical and hybrid approaches. Choosing the appropriate surface modification technique depends on the specific application and desired properties. Researchers often combine multiple techniques to achieve synergistic effects and optimize the catalytic performance of $BiVO_4$ for a given application. The following sections discussed the different modification techniques, helpful in the degradation of micro- and nano-plastic pollution from environmental matrices.

15.4.2. *Chemical modifications: metal and nonmetal doping and co-catalyst deposition*

Chemical modifications of $BiVO_4$ through metal and nonmetal doping, as well as co-catalyst deposition, represent effective strategies to enhance its catalytic activity, particularly in photocatalytic applications. Doping involves introducing foreign elements into the crystal lattice of $BiVO_4$ to alter its electronic structure and improve its catalytic performance. Metal doping, such as with transition metals such as Fe, Cr and Mn, introduces additional charge carriers, enhances charge mobility and reduces recombination rates, thereby improving the overall efficiency of photocatalysis. Nonmetal doping, commonly involving nitrogen (N), can modify the band construction, extending the absorption edge into the visible light range and enhancing the material's ability to harness solar energy.

Co-catalyst deposition is another powerful approach for surface modification. Noble metals such as platinum (Pt) and gold (Au) or metal oxides such as cobalt oxide (CoOx) and nickel oxide (NiOx) are often deposited onto the surface of $BiVO_4$. These co-catalysts serve multiple purposes, including facilitating charge transfer, promoting specific catalytic reactions and enhancing overall photocatalytic activity. For instance, platinum acts as a co-catalyst that facilitates the reduction of oxygen during water-splitting reactions, while metal oxide co-catalysts can act as electron sinks, reducing recombination of charge carriers.

The combination of metal and nonmetal doping with co-catalyst deposition can lead to synergistic effects, optimizing the catalytic performance of $BiVO_4$. These modifications aim to address challenges such as limited light absorption, rapid charge recombination and inefficient surface reactions. By tailoring the electronic structure and surface characteristics of $BiVO_4$, these chemical modifications contribute to improved efficiency in applications such as solar water splitting and photocatalytic breakdown of pollutants. Researchers continue to explore and refine these strategies to advance the formation of $BiVO_4$-based nanomaterials for sustainable energy and environmental applications.

15.4.3. *Physical modifications*

Physical modifications of $BiVO_4$, specifically through surface functionalization, constitute a significant approach to tailor its properties and enhance its performance in several usages, predominantly in the arena of photocatalysis. Surface functionalization involves the introduction of explicit functional groups or compounds onto the surface of $BiVO_4$ to influence its chemical reactivity, electronic structure and interaction with target molecules. This method aims to improve the material's surface properties without significantly altering its bulk structure.

One common form of surface functionalization involves chemical groups such as hydroxyl (-OH), amino ($-NH_2$) or carboxyl (-COOH) groups. These functional groups can be introduced through various techniques, including wet chemistry methods, sol-gel processes or chemical vapor deposition (CVD). The presence of these functional groups can alter the surface charge, enhance adsorption properties and facilitate the interaction of $BiVO_4$ with reactants. The benefits of surface functionalization include increased surface area, improved charge separation and transfer and enhanced catalytic activity. By introducing functional groups, surface defects may also be passivated, reducing reformation of charge carriers and improving the stability of $BiVO_4$ under photocatalytic conditions.

In addition to organic functional groups, inorganic compounds such as metal oxides or nanoparticles can be physically deposited onto the $BiVO_4$ surface. This approach aims to create a composite material with synergistic properties, combining the advantages of both the $BiVO_4$ semiconductor and the deposited material. For instance, the deposition of TiO_2 nanoparticles on $BiVO_4$ can enhance photocatalytic action by encouraging charge transfer and improving light absorption.

The choice of surface functionalization method and the type of functional groups or compounds introduced depend on the specific application and the desired improvements in $BiVO_4$'s catalytic performance. Surface functionalization provides a versatile means to tailor the surface properties of $BiVO_4$, making it a promising avenue for optimizing its efficiency in photocatalytic usages, including water splitting and environmental pollutant degradation.

15.4.4. *Hybrid and composite materials*

Hybrid and composite materials involving $BiVO_4$ refer to the combination of $BiVO_4$ with other materials to create structures that harness the synergistic properties of each component. This approach is often employed to enhance the inclusive performance of $BiVO_4$ in applications such as photocatalysis, electrocatalysis and energy storage. Hybridization with other materials agrees for the incorporation of varied functionalities, addressing specific challenges associated with $BiVO_4$, such as limited light absorption, rapid charge recombination or inefficient catalytic activity.

One common strategy involves the hybridization of $BiVO_4$ with other semiconductors or nanostructures. For instance, coupling $BiVO_4$ with TiO_2, ZnO or other metal oxides can create a composite material with improved charge carrier dynamics and enhanced photocatalytic activity. The combination of different semiconductors can extend the absorption spectrum, enhance light-harvesting capabilities and reduce charge recombination rates, ultimately leading to increased efficiency in solar-driven catalytic processes.

In addition to semiconductor hybrids, $BiVO_4$ can be combined with carbon-based materials, such as graphene or carbon nanotubes, to form hybrid composites. Carbon-based materials offer excellent electron conductivity and structural support, promoting efficient charge transfer and improving the overall stability of the composite. The 2D nature of graphene, for example, can enhance the surface area of the composite, facilitating more active spots for catalysis.

Metal and metal-oxide nanoparticles are also commonly integrated into $BiVO_4$ structures. Noble metal nanoparticles, such as Pt or Au, can act as co-catalysts to facilitate specific reactions and reduce overpotential in electrocatalytic applications. Metal oxide nanoparticles, like CoOx or NiOx, can serve similar roles and contribute to improved charge separation.

The synthesis of these hybrid and composite materials often involves various methods, including sol-gel processes, hydrothermal methods, CVD and physical mixing. The desired specific characteristics and the compatibility of the materials being combined determines the synthesis procedure. The advantages of hybrid and composite materials of $BiVO_4$ lie in the synergistic effects achieved by combining materials with complementary properties. These materials can exhibit enhanced photoactivity, improved charge transport and increased stability compared to their individual components. In future developments, the advancements of novel hybrid and composite materials will likely to play a crucial role in augmenting the capacity of $BiVO_4$ for various technological applications.

15.4.5. *Advances in surface modification technologies*

The developments in surface modification technologies have significantly contributed to tailoring material properties for specific applications, ranging from improved catalysis to enhanced biomedical devices. These technologies aim to alter the surface characteristics of materials without fundamentally changing their bulk properties, enabling fine-tuning of surface-related features such as reactivity, wettability and biocompatibility, as shown in Figure 15.3. Several key advances have emerged in recent years:

1) Nano-structuring and nanocoating: nano-structuring involves the fabrication of nano-sized features on the material surface, while nanocoating refers to the application of thin layers of nanomaterials. Enhanced surface area, improved mechanical properties and controlled release of substances, making these technologies valuable in catalysis, drug delivery and sensor applications.

2) Plasma surface treatment: plasma treatments involve exposing materials to ionized gases, leading to surface modifications through chemical reactions and physical interactions. Improved adhesion, wettability and surface energy, making it useful in applications such as polymer surface activation for better adhesion of coatings or biomedical device integration.

3) CVD: this encompasses the deposition of thin sheets on surfaces by chemical reactions in the vapor phase. Precise control over film thickness, uniformity and composition, enabling the creation of functional coatings for corrosion resistance, wear protection and electronic applications.

4) Atomic layer deposition (ALD): this is a type of CVD that involves the sequential deposition of individual atomic layers. Precise control over thickness and conformal coatings, leading to applications in semiconductor manufacturing, energy storage and catalysis.

5) Surface functionalization with biomolecules: this includes introduction of biomolecules onto surfaces for applications in biomedicine, biosensing and tissue engineering, which resulted in enhanced biocompatibility, specific molecular recognition and controlled interactions with biological systems.

6) Photocatalytic surface modifications: this includes integration of photocatalytic materials, such as titanium dioxide or ZnO, onto surfaces for self-cleaning and antimicrobial properties. Enhanced degradation of organic pollutants reduced microbial adhesion and improved surface cleanliness.

7) Smart coatings and responsive surfaces: these are the coatings that retort to external inducements such as temperature, pH or light.

8) Graphene and carbon nanotube incorporation: this includes the integration of graphene or carbon nanotubes onto surfaces for improved electrical conductivity and mechanical strength, which resulted in enhanced mechanical and electrical properties, making them appreciated in applications such as sensors, composites and energy-storage devices.

9) Electrochemical surface modifications: these include electrochemical methods, including electroplating, anodization and electropolishing, to modify surfaces, which resulted in controlled surface morphology, improved corrosion resistance and tailored surface topography for various applications.

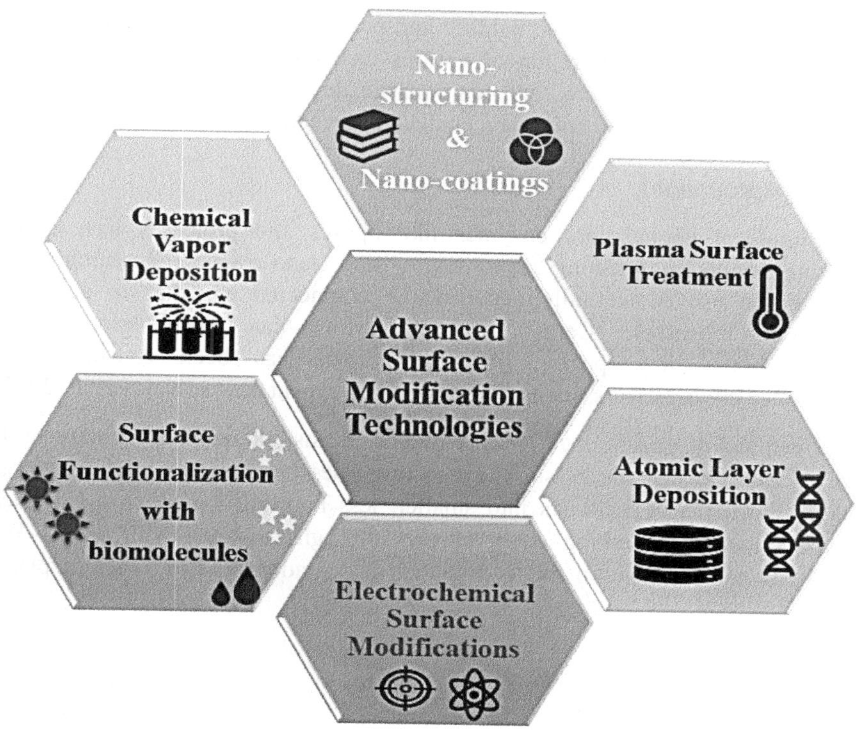

Figure 15.3. *Some of the most commonly used methods for surface modifications of $BiVO_4$ semiconductor-based materials for the effective plastic degradation. For a color version of this figure, see www.iste.co.uk/singh/photodegradation.zip*

These advances in surface modification technologies underscore the importance of tailoring materials at the nanoscale and molecular levels to achieve desired properties. As these technologies continue to evolve, they hold promise for further advancements in fields such as electronics, medicine, energy and environmental remediation.

15.5. Applications and challenges in real-world scenarios

15.5.1. *Practical applications in micro- and nano-plastic degradation*

Micro- and nano-plastics, characterized by their small size and widespread distribution, have become a major environmental concern globally. These minute plastic particles, resulting from the disintegration of bigger plastic materials, pose substantial threats to ecosystems, wildlife and public health. Efforts to address this issue have led to the exploration of various technologies and strategies for the practical application of micro- and nano-plastic degradation. In this context, the following outlines the applications and challenges faced in real-world scenarios.

15.5.1.1. *Applications*

1) Biological degradation: enzymes like lipases and proteases have shown promising results in breaking down micro- and nano-plastics. Researchers are exploring the use of naturally occurring or engineered enzymes to accelerate degradation processes. Certain microorganisms can metabolize plastics. Bioremediation using specific microbial strains has been considered for contaminated environments, such as wastewater treatment plants (Nakei et al. 2022).

2) Chemical degradation: AOPs like photocatalysis and ozonation comprise the usage of ROS to degrade plastics. These processes have shown effectiveness in breaking down plastic particles into smaller, less harmful components. Chemical hydrolysis involves breaking down plastics through reactions with water. This method has gained attention for its potential in degrading a wide range of plastic polymers.

3) Physical methods: *ultrasonication* involves the use of high-frequency ultrasound waves that can break down micro- and nano-plastics into smaller fragments, facilitating subsequent degradation processes. This noninvasive technique has been explored for applications in water treatment. Plasma-based techniques involve exposing plastic surfaces to ionized gases, leading to chemical modifications and degradation. This approach holds promise for surface-level degradation of plastics.

4) Combination approaches: combining biological, chemical and physical methods can enhance the overall efficiency of plastic degradation. Hybrid approaches aim to leverage the strengths of different techniques to address the complex nature of plastic pollution.

15.5.1.2. *Challenges*

Scaling up the practical applications of micro- and nano-plastic degradation from laboratory settings to real-world scenarios poses significant challenges. While various promising technologies have shown efficacy in controlled environments, transitioning to large-scale implementation introduces complexities that must be addressed. The following outlines key challenges associated with scaling up and implementing micro- and nano-plastic degradation technologies:

1) Selectivity and specificity: these include ensuring that degradation processes target only plastics while sparing essential organic and inorganic components is a significant challenge. Achieving high selectivity and specificity is crucial to avoid unintended environmental consequences.

2) Economic viability: many degradation technologies are resource-intensive and may have high upfront costs. Scaling up these technologies while maintaining economic viability is challenging, particularly when considering the need for large-scale infrastructure and operational expenses.

3) Energy requirements: some degradation processes, such as AOPs and certain physical methods, can be energy intensive. Scaling up may require substantial energy inputs, raising concerns about the overall environmental sustainability of the process.

4) Infrastructure development: constructing facilities capable of handling large volumes of plastic waste and implementing the necessary equipment for degradation processes involve significant infrastructure development. Designing and building such facilities require careful planning and substantial investment.

5) Logistical challenges: efficiently collecting and transporting micro- and nano-plastics from various sources to centralized degradation facilities present logistical challenges. Developing a streamlined system for waste collection is essential for effective scaling up.

6) Technological optimization: achieving consistent and efficient degradation at larger scales may require ongoing technological optimization. Fine-tuning the parameters of degradation processes, such as reaction times and concentrations, becomes more complex when dealing with larger volumes of waste.

7) Environmental impact assessment: as the scale of degradation processes increases, understanding the environmental impact of by-products becomes more critical. Comprehensive assessments of potential by-products and their ecological implications are necessary to ensure that scaled-up operations do not inadvertently introduce new environmental challenges.

8) Regulatory compliance: compliance with environmental regulations and standards is crucial for the responsible implementation of micro- and nano-plastic degradation technologies. Navigating regulatory frameworks at larger scales may involve complex permitting processes and approvals.

9) Monitoring and control: ensuring consistent and reliable performance across a large-scale operation requires robust monitoring and control systems. Implementing measures to detect and address variations in the degradation process is essential for maintaining effectiveness.

10) Collaboration and knowledge transfer: successful scaling up requires collaboration among scientists, engineers, policymakers and industry stakeholders. Facilitating knowledge transfer and interdisciplinary cooperation is essential for addressing the multifaceted challenges associated with large-scale implementation.

Addressing these challenges will require a coordinated effort involving research institutions, industry partners and regulatory bodies. Continuous innovation, technological advancements and a commitment to sustainability are key factors in overcoming the obstacles associated with scaling up micro- and nano-plastic degradation technologies for practical applications.

15.6. Conclusion

The utilization of semiconductor materials, specifically bismuth vanadate, in photocatalytic degradation processes emerges as a precise and promising solution for the treatment of micro- and nano-plastics from water bodies. $BiVO_4$ demonstrates exceptional photocatalytic action under visible light, allowing efficient removal of plastic particles. The unique electronic structure of $BiVO_4$ facilitates the creation of ROS, leading to the breakdown of plastic pollutants into environmentally benign by-products. Moreover, the precision of semiconductor photocatalysis allows for selective targeting of plastic particles, minimizing the impact on non-targeted substances in water. This selectivity enhances the environmental compatibility of the technology and underscores its potential for integration into water remediation processes on a larger scale.

As we navigate the complexities of environmental stewardship, the precision offered by $BiVO_4$ in photocatalytic degradation presents a promising avenue for mitigating the detrimental effects of micro- and nano-plastic contamination in water ecosystems. Continued research and innovation in semiconductor photocatalysis, with a focus on materials like $BiVO_4$, hold the key to achieve and pave the way for a targeted and impactful response to the global challenge of micro- and nano-plastic pollution in our water resources.

15.7. References

Bakhtiarnia, S., Sheibani, S., Billard, A., Aubry, E., Arab Pour Yazdi, M. (2022). Deposition of nanoporous $BiVO_4$ thin-film photocatalyst by reactive magnetron sputtering: Effect of total pressure and substrate. *Transactions of Nonferrous Metals Society of China*, 32(3), 957–971. doi: 10.1016/S1003-6326(22)65846-1.

Bano, K., Mittal, S.K., Singh, P.P., Kaushal, S. (2021). Sunlight driven photocatalytic degradation of organic pollutants using a $MnV_2O_6/BiVO_4$ heterojunction: Mechanistic perception and degradation pathways. *Nanoscale Advances*, 3(22), 6446–6458. doi: 10.1039/d1na00499a.

Bhuyan, M.S. (2022). Effects of microplastics on fish and in human health. *Frontiers in Environmental Science*, 10(March), 1–17. doi: 10.3389/fenvs.2022.827289.

Biancullo, F., Moreira, N.F.F., Ribeiro, A.R., Manaia, C.M., Faria, J.L., Nunes, O.C., Castro-Silva, S.M., Silva, A.M.T. (2019). Heterogeneous photocatalysis using UVA-LEDs for the removal of antibiotics and antibiotic resistant bacteria from urban wastewater treatment plant effluents. *Chemical Engineering Journal*, 367, 304–313. doi: 10.1016/j.cej.2019.02.012.

Bouazza, N., Ouzzine, M., Lillo-Ródenas, M.A., Eder, D., Linares-Solano, A. (2009). TiO_2 nanotubes and CNT-TiO_2 hybrid materials for the photocatalytic oxidation of propene at low concentration. *Applied Catalysis B: Environmental*, 92(3–4), 377–383. doi: 10.1016/j.apcatb.2009.08.017.

Campanale, C., Massarelli, C., Savino, I., Locaputo, V., Uricchio, V.F. (2020). A detailed review study on potential effects of microplastics and additives of concern on human health. *International Journal of Environmental Research and Public Health*, 17(4), 1212. doi: 10.3390/ijerph17041212.

Du, S., Zhu, R., Cai, Y., Xu, N., Yap, P.S., Zhang, Y., He, Y., Zhang, Y. (2021). Environmental fate and impacts of microplastics in aquatic ecosystems: A review. *RSC Advances*, 11(26), 15762–15784. doi: 10.1039/d1ra00880c.

Ferreira, T., Mendes, G.A., de Oliveira, A.M., Dias, C.G.B.T. (2022). Manufacture and characterization of polypropylene (PP) and high-density polyethylene (HDPE) blocks for potential use as masonry component in civil construction. *Polymers*, 14(12), 2463. doi: 10.3390/polym14122463.

Ghatge, S., Yang, Y., Ahn, J.H., Hur, H.G. (2020). Biodegradation of polyethylene: A brief review. *Applied Biological Chemistry*, 63(1), 27. doi: 10.1186/s13765-020-00511-3.

Goodarzi, N., Ashrafi-Peyman, Z., Khani, E., Moshfegh, A.Z. (2023). Recent progress on semiconductor heterogeneous photocatalysts in clean energy production and environmental remediation. *Catalysts*, 13(7), 1102. doi: 10.3390/catal13071102.

Höppener, E.M., Shahmohammadi, M.S., Parker, L.A., Henke, S., Urbanus, J.H. (2023). Classification of (micro)plastics using cathodoluminescence and machine learning. *Talanta*, 253, 123985. doi: 10.1016/j.talanta.2022.123985.

Ibhadon, A.O. and Fitzpatrick, P. (2013). Heterogeneous photocatalysis: Recent advances and applications. *Catalysts*, 3(1), 189–218. doi: 10.3390/catal3010189.

Ivleva, N.P. (2021). Chemical analysis of microplastics and nanoplastics: Challenges, advanced methods, and perspectives. *Chemical Reviews*, 121(19), 11886–11936. doi: 10.1021/acs.chemrev.1c00178.

Joksimovic, N., Selakovic, D., Jovicic, N., Jankovic, N., Pradeepkumar, P., Eftekhari, A., Rosic, G. (2022). Nanoplastics as an invisible threat to humans and the environment. *Journal of Nanomaterials*, 2022(12). doi: 10.1155/2022/6707819.

Lai, H., Liu, X., Qu, M. (2022). Nanoplastics and human health: Hazard identification and biointerface. *Nanomaterials*, 12(8), 1298. doi: 10.3390/nano12081298.

Maddah, H.A. (2016). Polypropylene as a promising plastic: A review. *American Journal of Polymer Science*, 6(1), 1–11. doi: 10.5923/j.ajps.20160601.01.

Nakei, M.D., Misinzo, G., Tindwa, H., Semu, E. (2022). Degradation of polyethylene plastic bags and bottles using microorganisms isolated from soils of Morogoro, Tanzania. *Frontiers in Microbiology*, 13(December), 1–15. doi: 10.3389/fmicb.2022.1077588.

Pandey, P., Dhiman, M., Kansal, A., Subudhi, S.P. (2023). Plastic waste management for sustainable environment: Techniques and approaches. *Waste Disposal and Sustainable Energy*, 5(2), 205–222. doi: 10.1007/s42768-023-00134-6.

Panel, E., and Chain, F. (2016). Presence of microplastics and nanoplastics in food, with particular focus on seafood. *EFSA Journal*, 14(May), e04501. doi: 10.2903/j.efsa.2016.4501.

Sagadevan, S., Fatimah, I., Egbosiub, T.C., Alshahateet, S.F., Lett, J.A., Weldegebrieal, G.K., Le, M.V., Johan, M.R. (2022). Photocatalytic efficiency of titanium dioxide for dyes and heavy metals removal from wastewater. *Bulletin of Chemical Reaction Engineering & Catalysis*, 17(2), 430–450. doi: 10.9767/BCREC.17.2.13948.430-450.

Trinh, D.T.T., Channei, D., Nakaruk, A., Khanitchaidecha, W. (2021). New insight into the photocatalytic degradation of organic pollutant over $BiVO_4/SiO_2/GO$ nanocomposite. *Scientific Reports*, 11(1), 1–11. doi: 10.1038/s41598-021-84323-5.

Usman, S., Razis, A.F.A., Shaari, K., Amal, M.N.A., Saad, M.Z., Isa, N.M., Nazarudin, M.F., Zulkifli, S.Z., Sutra, J., Ibrahim, M.A. (2020). Microplastics pollution as an invisible potential threat to food safety and security, policy challenges and the way forward. *International Journal of Environmental Research and Public Health*, 17(24), 1–24. doi: 10.3390/ijerph17249591.

Vikas Madhav, N., Gopinath, K.P., Krishnan, A., Rajendran, N., Krishnan, A. (2020). A critical review on various trophic transfer routes of microplastics in the context of the Indian coastal ecosystem. *Watershed Ecology and the Environment*, 2, 25–41. doi: 10.1016/j.wsee.2020.08.001.

Yani, I., Rosiliani, D., Khona'ah, B., Almahdini, F.A. (2020). Identification and plastic type and classification of PET, HDPE, and PP using RGB method. *IOP Conference Series: Materials Science and Engineering*, 857(1), 012015. doi: 10.1088/1757-899X/857/1/012015.

Yao, Z., Seong, H.J., Jang, Y.S. (2022). Environmental toxicity and decomposition of polyethylene. *Ecotoxicology and Environmental Safety*, 242(August), 113933. doi: 10.1016/j.ecoenv.2022.113933.

Yee, M.S.L., Hii, L.W., Looi, C.K., Lim, W.M., Wong, S.F., Kok, Y.Y., Tan, B.K., Wong, C.Y., Leong, C.O. (2021). Impact of microplastics and nanoplastics on human health. *Nanomaterials*, 11(2), 1–23. doi: 10.3390/nano11020496.

Zhang, F., Wang, X., Liu, H., Liu, C., Wan, Y., Long, Y., Cai, Z. (2019). Recent advances and applications of semiconductor photocatalytic technology. *Applied Sciences (Switzerland)*, 9(12). doi: 10.3390/app9122489.

Zhou, X., Shen, B., Lyubartsev, A., Zhai, J., Hedin, N. (2022). Semiconducting piezoelectric heterostructures for piezo- and piezophotocatalysis. *Nano Energy*, 96(December), 107141. doi: 10.1016/j.nanoen.2022.107141.

List of Authors

Devjani ADHIKARI
Department of Chemistry
S.S.M.M.U.S.S.S. Govt. P.G. College
Dwarahat
India

Akil AHMAD
Department of Chemistry
College of Science and Humanities
Prince Sattam Bin
Abdulaziz University
Al-Kharj
Saudi Arabia

Kalpana AWASTHI
Department of Physics
Kashi Naresh Government
Post Graduate College
Gyanpur
India

Roopam GAUR
Special Centre for Nanoscience
Jawaharlal Nehru University
New Delhi
India

Preeti GUPTA
Department of Chemistry
Deen Dayal Upadhyay
Gorakhpur University
India

Vaishali GUPTA
Special Centre for Nanoscience
Jawaharlal Nehru University
New Delhi
India

Abhijeet KUMAR
Department of Chemistry
School of Physical Sciences
Mahatma Gandhi Central University
Bihar
India

Manish KUMAR
Department of Chemistry
L.N.T. College
B.R.A. Bihar University
Muzaffarpur
India

Sunil KUMAR
Department of Chemistry
L.N.T. College
B.R.A. Bihar University
Muzaffarpur
India

Sujata KUNDAN
Department of Chemistry and Chemical Sciences
Central University of Jammu
India

Youngil LEE
University of Ulsan
South Korea

Nisha MANDLOI
Faculty of Pharmacy
Medi-Caps University
Indore
India

Aakanksha MEWAL
Faculty of Pharmacy
Medi-Caps University
Indore
India

Sudhakar PANDEY
Department of Physics
Narottam Singh Padam Singh Government Degree College
Mangaraha
India

Niranjan PATRA
Department of Chemistry
Koneru Lakshmaiah Education Foundation
Vaddeswaram
India

Gudiguntla RAVI
Department of Chemistry
Koneru Lakshmaiah Education Foundation
Vaddeswaram
India

Rachana SAIN
School of Materials Sciences and Technology
Indian Institute of Technology
Banaras Hindu University
India

Ojasvi SAINI
University of Petroleum and Energy Studies
Dehradun
India

Sudarshan SARKAR
School of Materials Sciences and Technology
Indian Institute of Technology
Banaras Hindu University
India

Nisha SHANKHWAR
Special Centre for Nanoscience
Jawaharlal Nehru University
New Delhi
India

Poonam SHARMA
Faculty of Pharmacy
Medi-Caps University
Indore
India

Chandrakala SINGH
Department of Chemistry
Radhe Hari Government
P.G. College
Kashipur
India

Geeta SINGH
Department of Chemistry
Deen Dayal Upadhyay
Gorakhpur University
India

Laxman SINGH
Department of Chemistry
Siddhartha University
Sidharthnagar
India

Pinki SINGH
Special Centre for Nanoscience
Jawaharlal Nehru University
New Delhi
India

Satyendra SINGH
Special Centre for Nanoscience
Jawaharlal Nehru University
New Delhi
India

Sweta SINGH
Department of Physics
School of Physical Sciences
Mahatma Gandhi Central University
Bihar
India

Muddada Jaya SURYA
Department of Chemistry
Koneru Lakshmaiah Education
Foundation
Vijayawada
India

Jewel THOMAS
School of Nanoscience and
Nanotechnology
Mahatma Gandhi University
Kerala
India

Madhu TIWARI
Department of Chemistry
Narottam Singh Padam Singh
Government Degree College
Mangaraha
India

Ajit Kumar VARMA
Faculty of Pharmacy
Dr. A.P.J. Abdul Kalam University
Indore
India

Arun Pratap VERMA
Department of Chemistry
Deen Dayal Upadhyay
Gorakhpur University
India

Monika VERMA
Department of Chemistry and
Chemical Sciences
Central University of Jammu
India

Vinita
Department of Chemistry
Deen Dayal Upadhyay
Gorakhpur University
India

Nikita YADAV
Amity School of Earth and
Environmental Sciences
Amity University Haryana
Gurugram
India

Pravesh Kumar YADAV
School of Materials Sciences and
Technology
Indian Institute of Technology
Banaras Hindu University
India

Thakur Prasad YADAV
Department of Physics
University of Allahabad
Prayagraj
India

Yashaswini
Department of Applied Chemistry
Delhi Technology University
India

Index

G, H

I, L

M, N

O, P

R, S

T, W, Z

Other titles from

ISTE

in

Materials Science

2024

JANNOT Yves, DEGIOVANNI Alain
Thermal Properties Measurement of Materials (2nd Edition)

VIGNES Alain
Silicon, From Sand to Chips 1: Microelectronic Components
Silicon, From Sand to Chips 2: Microelectronic Chips, Solar Cells, MEMS

2022

GADAUD Pascal
Crystal Elasticity

2019

SALENÇON Jean
Viscoelastic Modeling for Structural Analysis

2018

ABABOU Rachid
Capillary Flows in Heterogeneous and Random Porous Media 1

GODIN Nathalie, REYNAUD Pascal, FANTOZZI Gilbert
Acoustic Emission and Durability of Composite Materials
(Durability and Ageing of Organic Composite Materials Set – Volume 3)

JANNOT Yves, DEGIOVANNI Alain
Thermal Properties Measurement of Materials

2017

CAILLETAUD Georges, SAÏ Kacem, TALEB Lakhdar
Multi-mechanism Modeling of Inelastic Material Behavior

CARVELLI Valter, JAIN Atul, LOMOV Stepan
Fatigue of Textile and Short Fiber Reinforced Composites
(Durability and Ageing of Organic Composite Materials Set – Volume 1)

DELHAES Pierre
Materials and Thermodynamics: Living and Economic Systems

FRESSENGEAS Claude
Mechanics of Dislocation Fields

GIGLIOTTI Marco, LAFARIE-FRENOT Marie-Christine,
GRANDIDIER Jean-Claude, MINERVINO Matteo
Mechanical Behavior of Organic Matrix Composites
(Durability and Ageing of Organic Composite Materials Set – Volume 2)

2016

TORRENTI Jean-Michel, LA TORRE Francesca
Materials and Infrastructures 1
(Research for Innovative Transports Set – Volume 5A)
Materials and Infrastructures 2
(Research for Innovative Transports Set – Volume 5B)

RÉMOND Yves, AHZI Saïd, BANIASSADI Majid, GARMESTANI Hamid
Applied RVE Reconstruction and Homogenization of Heterogeneous Materials

2015

JOUD Jean-Charles, BARTHES-LABROUSSE Marie-Geneviève
Physical Chemistry and Acid-Base Properties of Surfaces

LALEVEE Jacques, FOUASSIER Jean-Pierre
Dyes and Chromophores in Polymer Science

2014

DAVIM J. Paulo
Machinability of Advanced Materials

HAMAIDE Thierry, DETERRE Rémi, FELLER Jean-François
Environmental Impact of Polymers

LAFLEUR Pierre G., VERGNES Bruno
Polymer Extrusion

2013

BATHIAS Claude
Fatigue Limit in Metals

LEXCELLENT Christian
Shape-memory Alloys Handbook

2012

BONELLI Stéphane
Erosion of Geomaterials

BRACCINI Muriel, DUPEUX Michel
Mechanics of Solid Interfaces

DAVIM J. Paulo
Wear of Advanced Materials

DELHAES Pierre
Carbon Science and Technology: From Energy to Materials

GOUDEAU Philippe, GUINEBRETIERE René
X-Rays and Materials

OLLIVIER Jean-Pierre, TORRENTI Jean-Michel, CARCASSES Myriam
Physical Properties of Concrete and Concrete Constituents

RECHO Naman
Fracture Mechanics and Crack Growth

SAANOUNI Khemais
Damage Mechanics in Metal Forming: Advanced Modeling and Numerical Simulation

VANBÉSIEN Olivier
Artificial Materials

VERDU Jacques
Oxidative Ageing of Polymers

2011

BATHIAS Claude, PINEAU André
Fatigue of Materials and Structures: Application to Damage and Design
Fatigue of Materials and Structures: Application to Design and Damage

HICHER Pierre-Yves
Multiscale Geomechanics: From Soil to Engineering Projects

IONESCU Ioan R., BOUVIER Salima, CAZACU Oana, FRANCIOSI Patrick
Plasticity of Crystalline Materials: From Dislocations to Continuum

PIJAUDIER-CABOT Gilles, DUFOUR Frédéric
Damage Mechanics of Cementitious Materials and Structures

PRIESTER Louisette
Grain Boundaries and Crystalline Plasticity

VIGNES Alain
Extractive Metallurgy 1: Basic Thermodynamics and Kinetics
Extractive Metallurgy 2: Metallurgical Reaction Processes
Extractive Metallurgy 3: Processing Operations and Routes

2010

BATHIAS Claude, PINEAU André
Fatigue of Materials and Structures: Fundamentals

CHATEIGNER Daniel
Combined Analysis

CHEVALIER Yvon, VINH TUONG Jean
Mechanical Characterization of Materials and Wave Dispersion
Mechanical Characterization of Materials and Wave Dispersion: Instrumentation and Experiment Interpretation

DELHAES Pierre
Carbon-based Solids and Materials

2009

ALVAREZ-ARMAS Iris, DEGALLAIX-MOREUIL Suzanne
Duplex Stainless Steels

DAVIM J. Paulo
Machining Composite Materials

GALERIE Alain
Vapor Surface Treatments

2008

BATTAGLIA Jean-Luc
Heat Transfer in Materials Forming Processes

BLONDEAU Régis
Metallurgy and Mechanics of Welding

FRANÇOIS Dominique
Structural Components

REYNE Maurice
Plastic Forming Processes

TAKADOUM Jamal
Materials and Surface Engineering in Tribology

2007

BOCH Philippe †, NIEPCE Jean-Claude
Ceramic Materials

CRISTESCU Constantin
Materials with Rheological Properties

2006

LASSEN Tom, RECHO Naman
Fatigue Life Analyses of Welded Structures